1st Nommensen International Conference on Technology and Engineering 2017

Advancements in Technology and Engineering

IOP Conference Series: Materials Science and Engineering
Volume 237

Medan, Indonesia
11 - 12 July 2017

Editors:

Petr Valasek
Abraham Kabutey
Gurkan Gundil
Guo-En Chang
Poki Chen
Dr. Hugeng
Mula Sigiro
Himsar Ambarita
Turnad Lenggo Ginta

ISBN: 978-1-5108-4869-6
ISSN: 1757-8981

Printed from e-media with permission by:

Curran Associates, Inc.
57 Morehouse Lane
Red Hook, NY 12571

Some format issues inherent in the e-media version may also appear in this print version.

Printed by Curran Associates, Inc. (2017)

For permission requests, please contact the Institute of Physics at the address below.

Institute of Physics
Dirac House, Temple Back
Bristol BS1 6BE UK

Phone: 44 1 17 929 7481
Fax: 44 1 17 920 0979

techtracking@iop.org

Additional copies of this publication are available from:

Curran Associates, Inc.
57 Morehouse Lane
Red Hook, NY 12571 USA
Phone: 845-758-0400
Fax: 845-758-2633
Email: curran@proceedings.com
Web: www.proceedings.com

Table of contents

Volume 237

1st Nommensen International Conference on Technology and Engineering
11–12 July 2017, Medan, Indonesia

Accepted papers received: 4 September 2017
Published online: 22 September 2017

Electrical and Electronic Engineering

Civil and Enviromental Engineering

Food and Agriculture Technology

Material Sciences and Engineering

IOP Conf. Series: Materials Science and Engineering **237** (2017) 011001 doi:10.1088/1757-899X/237/1/011001

PREFACE

As Chairman of the 1st Nommensen International Conference on Technology and Engineering, I would like to welcome you all speakers and participants to our campus in Medan. This city is fascinating with its culinary tourism offering tropical fruit like durian and various food and cake that spoil our tongue. Beside its culinary tourism, the city of Medan is a dynamic business city with relevant past in agriculture and plantation. Hope you will have the opportunity to enjoy the food and your time while staying in this city.

This conference is the first international conference conducted by our university. The theme at this first attempt is focusing on "Advancements in Technology and Engineering." The theme is selected with the objective to bring more advancement in technology to the current development in this city and the whole country of Indonesia. The great effort dedicated by our government to expedite the construction of massive infrastructures requires more technology advancements. Our contribution to this conference however small is also of valuable input to the current government effort in developing this country.

I would like to take this opportunity to thank all the committee, speakers, authors, reviewers and participants who dedicated their effort for the successful execution of this conference. Without your contribution, we simply could not have had this conference.

We received 48 articles submission in this beginning. They come from various countries like Czech, Taiwan, Malaysia, Australia, and Turkey in addition to those from Indonesia. We categorized the papers under five groups, namely: Mechanical Engineering and Technology, Electrical Engineering, Civil and Environmental Engineering, Material Sciences and Engineering, and lastly Food and Agricultural Technology. Some paper can be categorized conveniently into one of these groups. Others bring its own difficulties because they might be put under more than one group. Still, the committee has done a great job to sent your paper to the right reviewer. All papers regardless of their standing or initial classification, were available for general discussion at the task force meeting.

We are fortunate to have four excellent keynote speakers at the moment. They are David Herak from CULS, Yupiter HP Manurung from UTM Malaysia, Cheng Yuan Chang from CYCU Taiwan and Himsar Ambarita from USU Indonesia. David Herak is currently doing extensive work in biofuels and renewable energy. Yupiter Manurung interests are among others manufacturing technology specializing in welding and laser as well as production engineering. Cheng Yuan Chang is leading research work on noise and vibration control technology. Himsar Ambarita currently leads a research centre focusing on sustainable energy and biomaterial. I would like to give thank to the four of you for the interesting keynote speech to this conference.

Finally I hope that all participants enjoy a successful conference, make a lot of new contacts, engage in fruitful discussions and have a pleasant stay in Medan.

Dr. Richard AM. Napitupulu
CHAIRMAN

IOP Conf. Series: Materials Science and Engineering **237** (2017) 011001 doi:10.1088/1757-899X/237/1/011001

1st NICTE

Conference Organization

INITIATOR INSTITUTION

Faculty of Engineering
Nommensen HKBP University

ORGANIZING INSTITUTION

Nommensen HKBP University
Czech University of Life Sciences Prague
Chung Yuang Christian University

Supporting Institution

NT CORP

PBN

Honorary Chair :

Ir. Nurdin Tampubolon, MM
Dr. Ir. Sabam Malau

International Advisory Board :

Prof. Menghui Li, CYCU – Taiwan
Prof. Frantisek Kumhala, CULS – Czech Republic
Prof. Agustinus P Irawan, UNTAR – Indonesia
Assoc. Prof. Jiri Masek, CULS – Czech Republic

Editorial Board :

Assoc. Prof. Petr Valasek, PhD., CULS – Czech
Ing. Abraham Kabutey, PhD., CULS – Czech
Assoc. Prof. Dr. Gurkan. Gundil, OMU – Turkey
Prof. Guo-En Chang, PhD., NCCU – Taiwan
Prof. Poki Chen, PhD., NTUST – Taiwan
Assoc. Prof. Dr. Hugeng, UMN – Indonesia
Dr. Mula Sigiro, PhD., NHU – Indonesia
Himsar Ambarita, Dr.Eng., USU – Indonesia
Dr. Turnad Lenggo Ginta, UTP – Malaysia

Chairman :

Dr. Richard AM. Napitupulu, ST.MT.

Co-Chairman :

Dr. Ir. Sindak Hutauruk, MSEE

Secretary :

Ir. Partahi Lumbangaol,MEngSc.

Panel & Scientific Session :

Dr. Mula Sigiro, PhD
Dr. Mukhtar Panjaitan

Treasurer :

Yetty R. Saragih, ST.MT.

Committee Member :

Parulian Siagian, ST.MT.
Libianko Sianturi, ST. MT.
Ir. Timbang Pangaribuan, MT.
Charles Manurung, ST.MT.

Proof Reader :

Fenti Napitupulu, SPd,MPd.

Reviewer :

Dr. Tumiur Gultom, UNIMED – Indonesia
Dr. Janter Simanjuntak, UNIMED – Indonesia
Dr. Rondang Tambun, USU – Indonesia
Assoc. Prof. Dr. Hugeng, UMN – Indonesia
Dr. Mukhtar Panjaitan, NHU – Indonesia
Dr. Samse Pandiangan, NHU – Indonesia
Dr. Sindak Hutauruk, NHU – Indonesia
Dr. Richard AM. Napitupulu, NHU – Indonesia
Himsar Ambarita, Dr.Eng., USU – Indonesia
Dr. Sabam Malau, NHU – Indonesia
Prof. David Herak, CULS – Czech
Dr. Turnad Lenggo Ginta, UTP – Malaysia
Prof. Yupiter Manurung, UTM – Malaysia
Prof. Cheng Yuan Chan, CYCU – Taiwan
Prof. Shyh-Leh Chen, CCU – Taiwan
Dr. Irwan Purnana, LIPI – Indonesia
Dr. Suganda Girsang, Binus – Indonesia
Assoc. Prof. Petr Valasek, CULS – Czech
Ing. Abraham Kabutey, PhD., CULS – Czech
Assoc. Prof. Jiri Masek,– Czech Republic
Prof. Poki Chen, PhD., NTUST – Taiwan
Timbang P. MT, NHU – Indonesia
Tulus B. Sitorus Meng, USU – Indonesia
Partahi L. MEngSc, NHU – Indonesia

Administrative and Supporting Staff :

Roslin Pasaribu
Parulian Sirait, SKom.
Tamara Hutagalung, SPd.
Poltak Siahaan
Sungguh Rahmat Bohalima
Andre Gultom
Benget Siringo-ringo
Alberto Hutauruk
Swardy Sibarani
Benhart Marpaung

IOP Conf. Series: Materials Science and Engineering **237** (2017) 011002 doi:10.1088/1757-899X/237/1/011002

Peer review statement

All papers published in this volume of *IOP Conference Series: Materials Science and Engineering* have been peer reviewed through processes administered by the proceedings Editors. Reviews were conducted by expert referees to the professional and scientific standards expected of a proceedings journal published by IOP Publishing.

IOP Conf. Series: Materials Science and Engineering **237** (2017) 012001 doi:10.1088/1757-899X/237/1/012001

Suitability of aquatic biomass from Lake Toba (North Sumatra, Indonesia) for energy generation by combustion process

A Brunerová[1,*], H Roubík[2], D Herák[3]
[1]Department of Material Science and Manufacturing Technology, Faculty of Engineering, Czech University of Life Sciences Prague, Kamýcká 129, CZ 165 00 Prague, Czech Republic
[2]Department of Sustainable Technologies, Faculty of Tropical Agri Sciences, Czech University of Life Sciences Prague, Czech Republic
[3]Department of Mechanical Engineering, Faculty of Engineering, Czech University of Life Sciences Prague, Czech Republic

* brunerova@tf.czu.cz

Abstract. Several aquatic plant species were identified as aquatic pollution of Lake Toba, North Sumatra (Indonesia); specifically, water hyacinth *Eichhornia crassipes* and aquatic weeds *Hydrilla verticillata* and *Myriophyllum spicatum* due to their high biomass yield which causes impenetrable mats at the bottom and surface of the lake. That complicates other vegetation growth and utilization of water areas for fishing or recreation. In attempt to clean the lake and prevent plants expansion, great amount of plants populations are removed from water but subsequent efficient utilization of such aquatic biomass is not ensured. Present research investigated energy potential of aquatic biomass originated from mentioned aquatic plants from Lake Toba and its possible utilization for energy production by direct combustion. Performed chemical analysis contained from determination of moisture, ash and volatile matter contents and calorific values. Evaluation of results proved highest suitability and energy potential of *Eichhornia crassipes* with gross calorific value (GCV) 16.31 $MJ\cdot kg^{-1}$, followed by *Hydrilla verticillata* with GCV 15.24 $MJ\cdot kg^{-1}$. Samples of *Myriophyllum spicatum* exhibited unsatisfactory results due to its low GCV (11.27 $MJ\cdot kg^{-1}$) in combination with high ash content (36.99%) which indicates complications during combustion, thus, low energy production efficiency and overall unsuitability for combustion purposes.

1. Introduction

Agricultural sector in Republic of Indonesia is characterized by production of abundant amount of various agricultural residues [1]; it corresponds to approximately 146.7 million tons of such residues produced every year [2]. Such residues represent waste biomass with great potential for energy generation by combustion methods and can be used as a source of clean and renewable energy [1; 3]. Nevertheless, not only large–scale manufactures producing agricultural residues are source of suitable biomass for energy production; various kinds of waste biomass with no other purposes can be also found in Indonesia, namely, herbaceous, wood, fruit or aquatic biomass [3].

Aquatic biomass includes biological wastes originated from aquatic plant species; most of them provide benefits for environment during their lives, however, some specific species represent serious issues due to their profuse vegetation and invasive behaviour [4, 5]. Several aquatic weed species wildly growing in water areas in Indonesia were considered as undesirable members of local ecosystem by previous studies [6, 7, 8] that identified as local aquatic pollution mainly water hyacinth *Eichhornia crassipes* and aquatic weeds *Hydrilla verticillata* and *Myriophyllum spicatum.*

There was monitored and evaluated presence of mentioned plants populations, which are considered as an aquatic pollution, in the water areas of Lake Toba (see in Figure 1), the largest lake (surface area 800 km^2) in tropical Asia located in the middle of North Sumatra [5, 8, 9]. Lake Toba, represents frequently visited tourist destination and is considered as a natural wonder, therefore there is an effort to eliminate the population of undesirable aquatic weed species [10]. Mentioned aquatic plant species are characterized by extremely high biomass yield which forms desirable thick impenetrable mats at the bottom and surface (large floating islands of plants) of the lake [10, 11] which exclude other vegetation growing and complicate utilization of water areas (activities such as fishing, irrigation, energy generation or recreation) [12]. To prevent all mentioned negative impacts of identified aquatic plant species, the existing plants populations are removed from water by manual or mechanical harvesting (heavy machinery) in attempt to clean the lake and prevent further plants reproduction [8, 13]. Identical harvesting techniques were also documented in case of aquatic pollution of Lake Rawa Pening, Central Java [14] and Lake Tempe, South Sulawesi [15].

Figure 1. Aquatic pollution of Lake Toba: a) *Eichhornia crassipes*, b) *Hydrilla verticillata*, c) *Myriophyllum spicatum* (manual harvesting) [Authors data]

Such a harvested aquatic biomass is occasionally utilized as cheap fertilizer or livestock feed by small scale farmers but majority is left without subsequent effective utilization [14, 16, 17]. Nowadays, investigations focused on effective utilization of aquatic biomass were performed, mainly focused on biofuels production. Specifically, a transformation of present aquatic plans species into the liquid biofuel (bioethanol) and gaseous biofuel (biogas) was investigated directly in Indonesia [18, 19]. Investigations focused on production of solid biofuel (in form of briquettes) were performed with satisfactory results in Nigeria [20], Kenya [21] and Zimbabwe [22] where the aquatic pollution of water areas is also widespread. Mentioned studies proved potential of investigated aquatic plant for biofuel production within testing of mechanical and chemical quality of produced briquette samples.

Main aim of present study was to monitor and evaluate the energy potential of aquatic waste biomass originated from Lake Toba, Indonesia. Especially wildly growing aquatic plant species *Eichhornia crassipes*, *Hydrilla verticillata* and *Myriophyllum spicatum.* Experimental investigation was focused on potential of mentioned aquatic plant species for direct combustion and subsequent energy generation.

IOP Conf. Series: Materials Science and Engineering **237** (2017) 012001 doi:10.1088/1757-899X/237/1/012001

2. Materials and Methods

Whole experimental investigation was performed within mechanical and chemical analysis of selected investigated materials describing their suitability for energy generation by direct combustion, as well as evaluation of observed results was conducted to technical mandatory standard which describes all processes of specific measurements and tests in detail and states required levels of specific biofuels quality indicators. Namely, following standards were used in the present research: EN ISO 17225–1 (2015), EN 15234–1 (2011), ISO 16559 (2014), ISO 18134–2 (2017), EN ISO 18122 (2015), EN ISO 18123 (2015), EN 14918 (2010) and ISO 1928 (2010).

2.1. Materials and samples

Wildly growing aquatic plant species originating from water areas of Lake Toba, North Sumatra, (Indonesia)were used as investigated biological materialsfor experimental measurements of present study; namely, *Eichhornia crassipes*, *Hydrilla verticillata* and *Myriophyllum spicatum*(expressed in Figure 2). Whole bodies of mentioned plants were harvested manually from water areas and collected during summer season from August to September 2016 in district of Laguboti, Regency of Toba Samosir and in surroundings of Samosir Island, Regency of Samosir.

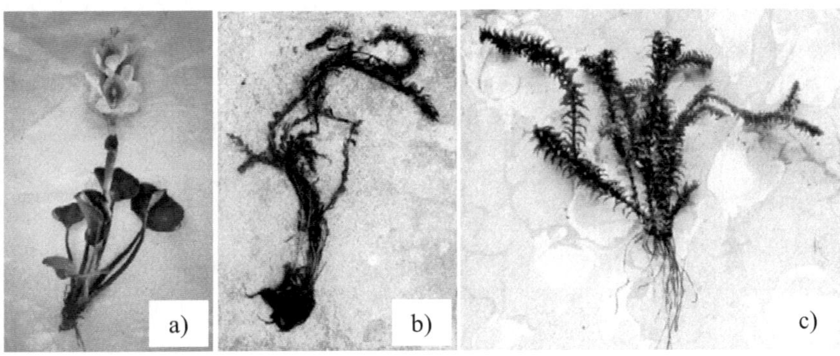

Figure 2. Investigated aquatic plants:a) *E.crassipes*, b) *H. verticillata*, c) *M. spicatum*[Authors data]

2.2. Experimental methodology

Primarily, raw samples were subjected to determination of initial moisture content directly after harvesting by using of laboratory oven Daihan Labtech, type LDO–080F (New Delhi, India). Secondary, previously dried materials were grinded into fine powder (<0.1 mm) and used for determination of ash and volatile matter contents by using of thermo gravimetric analyser LECO, type TGA 701 (Saint Joseph, United States). Subsequently, a gross calorific value determination was performed by using of isoperibol calorimeter LECO, type AC 600 (Saint Joseph, United States) and net calorific value was calculated using appropriate formulas according to standard ISO 1928 (2010).

3. Results and discussion

Presence of investigated aquatic plants species and aquatic pollution was proved in case of Lake Toba by initial visual evaluation of target area. A great amount of vegetation growing in Lake Toba in the form of floating macrophytes (*E. crassipes*) and submerged macrophytes (*M. spicatum* and *H. verticillata*) was observed.

Chemical analysis of selected aquatic plants performed by experimental testing exhibited result values of main chemical quality indicators which are resulting in the issue of plants suitability for energy production by direct combustion. Detail values of specific tested materials parameters are noted in Table 1 and statistical analysis of obtained result values is expressed in Table 2 (see below).

Initial moisture content of all investigated samples exhibited extremely high values. Such results were expected if considered the taxonomic classification of plants. However, low amount of dry matter content (12.2% in average) could be considered as a limitation due to higher demand on biomass quantity collection. Recommended moisture content of feedstock material for solid biofuel production is <10% [23], therefore, proper preparation (drying) of raw material before utilization for combustion purposes would be necessary. However, body structure of selected aquatic plants is easily dryable, thus, solar power can be used for that purpose.

Table 1. Indicators of investigated aquatic plants chemical quality

	GCV ($MJ \cdot kg^{-1}$)	SD ($MJ \cdot kg^{-1}$)	NCV ($MJ \cdot kg^{-1}$)	SD ($MJ \cdot kg^{-1}$)	AC (%)	SD (%)	MC (%)	SD (%)	VMC (%)	SD (%)
E.crassipes	**16.31**	0.10	**15.09**	0.09	11.60	0.29	85.3	0.02	87.41	1.20
H. verticillata	15.24	0.42	13.93	0.38	15.53	1.11	88.5	0.02	80.41	4.30
M. spicatum	11.27	0.15	10.58	0.14	**36.99**	0.15	89.6	0.02	58.36	0.79

(GCV – gross calorific value, NCV – net calorific value, SD – standard deviation, AC – ash content, MC – moisture content, VMC – volatile matter content)

Table 2. Statistical analyses of measured data

	GCV		NCV		AC	
	t_{rat} (–)	t_{crit} (–)	t_{rat} (–)	t_{crit} (–)	t_{rat} (–)	t_{crit} (–)
E. crassipes *H. verticillata*	4.29	0.96	5.08	0.96	5.94	0.96
E. crassipes *M. spicatum*	48.73	0.96	46.65	0.96	5.47	0.96
H. verticillata *M. spicatum*	15.40	0.96	14.18	0.96	7.66	0.96

(GCV – gross calorific value, NCV – net calorific value, AC – ash content)

Statistical analysis using standard two tailed paired t–test (Table 2) between determined data for investigated aquatic plants showed that compared result values of GCV, NCV and ash content are significantly different from each other. The significance of the analysis results was based on values of t_{crit} being lower than t_{rat} values with level of significance 0.05.

Evaluation of ash content result values indicated low level of present parameter; such results are undesirable because they indicate complication during combustion of materials, thus, related lower combustion efficiency and lower energy potential. Ash content is primarily related to kind of biomass; wood biomass always exhibited lower ash content in compare with other kinds [24]. Considering herbaceous biomass, a rice husks representing one of major agriculture residues in Asia, also exhibited high result values of ash content equal approximately 20% [25]. Ash content can be also influenced by contamination of biomass by external particles. In case of investigated aquatic plants, it could have been caused by sand or soil impurities contamination.

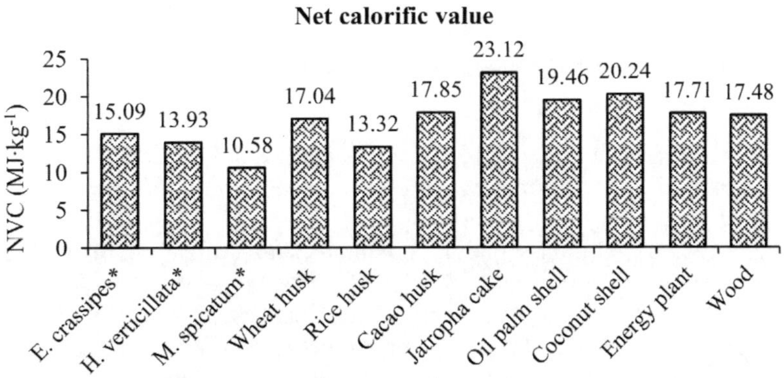

Figure 3. Energy potential of various kinds agriculture residues [*Authors data; 25–30]

Mentioned relation between high ash content and low combustion efficiency was directly proved by result values of calorific values. Focused on the energy potential, *E. crassipes* sample sex hibited highest level of net calorific value, followed by *H. Verticillata* samples and lowest level was achieved by samples of *M. spicatum* (extremely high ash content). Comparison of obtained result values with other kinds of waste biomass commonly used for energy production(expressed in Figure 3)describes that investigated materials exhibited lower level of energy potential, nevertheless, *E. crassipes* and *H. verticillata* exhibited satisfactory level, but level of net calorific value of *M. Spicatum* was very low and represented unsatisfactory result. Overall evaluation of obtained result values of chemical quality indicators proved suitability of *E. crassipes* and *H. verticillata* for combustion purposes with subsequent energy production. Samples of *M. spicatum* exhibited unsatisfactory results due to its extremely high ash content in combination with low net calorific value which in dequate combustion properties of material, thus, it would represent low quality biofuel.

4. Conclusion

Summarizing of all observed findings related to chemical composition and energy potential of investigated aquatic plants, the utilization of *Eichhornia crassipes* and *Hydrilla verticillata* for energy production by combustion can be recommended. In general, energy potential (calorific values) of investigated aquatic plants occurred at lower level in comparison with other biomass types. Such a limitation can be improved for example by transformation of biomass into solid biofuel which increases calorific values of final products. Although, several limitations were identified within selected plants energy potential, it is also important to consider that utilization of such waste biomass is an important step of proper waste management within the actual effort to clean water areas of Lake Toba.This factmust be considered as positive aspect of such materials utilization for energy production, as well as, low financial demands for the acquisition of such materials. Extensive aquatic plants populations already exist and after harvesting becoming aquatic waste biomass; potential renewable source of energy with no financial demands for its cultivation.

Acknowledgment

Appreciation belongs to research team members Surya Ningsi Hutauruk and Herti Novalia Hutapea from Institut Teknologi Del for assistance due to data collection and participation at the conference NICTE 2017. The research was supported by Internal Grant Agency of the FE, CULS Prague, grant number 2017:31140/1312/3112 and by the Internal Grant Agency of the CULS Prague, grant number 20173005 (31140/1313/3108).

References

[1] Singh R and Setiawan A D, 2013, Biomass energy policies and strategies: Harvesting potential in India and Indonesia. *Renewable and Sustainable Energy Reviews* 22 pp 332–345

[2] Abdullah K, 2002, Biomass energy potentials and utilization in Indonesia. *IPB and Indonesian Renewable Energy Society* [IRES], Bogor

[3] Kuvarakul T, Pratidina A, Anggraeni D and Saraswati H, 2015, Renewable Energy Guideline on Biomass and Biogas Power Project Development in Indonesia, p 204.

[4] Scheffer M, 1998, Ecology of shallow lakes *Population and community biology* series 22

[5] Smart R M and Doyle R D, 1995, Ecological theory and management of submersed aquatic plant communities *Aquatic Plant Control Research Program* (Vicksburg: U.S. Army Engineer Waterways Experiment Station)

[6] Pancho J V and Soerjani M, 1978, Aquatic Weeds of Southeast Asia. *BIOTROP, Indonesia: Bogor* p 130.

[7] Nasution Z, 2003, Land and Forest Management in the Lake Toba Catchment Area. *Universiti Sains Malaysia.*

[8] Nasution M I T, Awal S M S and Permana D M, 2016, The Methods of Preventing Water Hyacinth as Aquatic Pollution in Lake Toba Caused by Agricultural Waste International *Journal of Environmental Science and Development* vol. 7(8) pp 630–633

[9] Gopal B and Ghosh D, 2009, Lakes and Reservoirs of Asia. *Encyclopedia of Inland Waters* pp 501–512 (Boston: Elsevier)

[10] Langeland K A, 1990, Hydrilla: A continuing problem in Florida waters *University of Florida Circular* No. 884 (Gainesville: University of Florida)

[11] Rahman I A, Saad B, Shaidan S and Sya Rizal E S, 2005, Adsorption characteristics of malachite green on activated carbon derived from rice husks produced by chemical–thermal process., *Bioresource Technology* 96 pp 1578–1583.

[12] Kannan R R, Rajasimman M, Rajamohan N and Sivaprakash A B, 2010, Equilibrium and kinetic studies on sorption of malachite green using Hydrilla Verticillata biomass. *International Journal of Environmental Research* vol 4(4) pp 817–824.

[13] Tjitrosemito S, 1994, Integrated management of paddy and aquatic weeds in Asia. In: Proceedings of the International Seminar "*Biological Control and Integrated Management of Paddy and Aquatic Weeds in Asia*" J., Asian and Pacific Council., Tsukuba, Japan

[14] Little E C S, 1968, Handbook of utilization of aquatic plants. *Plant Production and Protection Division* Rome, FAO, p 123.

[15] Ali M S S, Majika A and Salman D, 2017, Food Consumption and Production in Tempe Lake, South Sulawesi, Indonesia. *Journal of Asian Rural Studies* vol. 1(1) pp 43–52.

[16] Ling S W, 1960, Control of aquatic vegetation. In: *Third International Inland Fisheries Training Centre* Rome, FAO p 12.

[17] Bagnall L O, 1971, Processed aquatic plants for animal nutrition. *Annu.Res.Rep.Inst.Food Agric.Sci.Univ.Fla.*vol 49

[18] Aswathy U S. Rajeev K, Sukumaran G, Devi L, Rajasree K P, Singhania R R and

Pandey A, 2010, Bio-ethanol from water hyacinth biomass: an evaluation of enzymatic saccharification strategy *Bioresource Technology* vol. 101, pp 925–930.

[19] Annachhatre A P and Khanna P, 1987, Methane recovery from water hyacinth through whole-cell immobilization technology *Biotechnology and Bioengineering* vol. 29 pp 805-818.

[20] Ighodalo O A, Zoukumor K, Egbon C, Okoh S and Odu K, 2011, Processing Water Hyacinth into Biomass Briquettes for Cooking Purposes. *Journal of Emerging Trends in Engineering and Applied Sciences* 2(2) pp 305–307

[21] Rodrigues A J, Omondi M, Hayombe P O, Akuno W, Kerich D and Maobe I, 2014, Converting Water Hyacinth to Briquettes: A Beach Community Based Approach. vol 4531 pp 358–378.

[22] Munjeri K, Ziuku S, Maganga H, Siachingoma B and Ndlovu S, 2016, On the potential of water hyacinth as a biomass briquette for heating applications. *International Journal of Energy and Environmental Engineering* vol 7(1) pp 37–43.

[23] Kaliyan N and Morey R V, 2009, Factors affecting strength and durability of densified biomass products. *Biomass and Bioenergy* vol. 33(3) pp. 337–359.

[24] Johansson L S, Leckner B, Gustavsson L, Cooper D, Tullin C and Potter A, 2004, Emission characteristics of modern and old-type residential boilers fired with wood logs and wood pellets. *Atmospheric Environment* 38(25) pp 4183–4195.

[25] Brunerová A, Malaťák J, Müller M, Valášek P and Roubík H, 2017, Tropical waste biomass potential for solid biofuels production. *Agronomy Research* 15(2) pp 359–368.

[26] Brunerová A, Brožek M and Müller M, 2017, Utilization of waste biomass from post–harvest lines in the form of briquettes for energy production. *Agronomy Research* vol 15 pp 344–358.

[27] Mahlia T M I, Abdulmuin M Z, Alamsyah T M I and Mukhlishien D, 2001, An alternative energy source from palm wastes industry for Malaysia and Indonesia. *Energy Conversion and Management* vol 42(18) pp 2109–2118.

[28] Özyuguran A and Yaman S, 2017, Prediction of Calorific Value of Biomass from Proximate Analysis. *Energy Procedia* vol 107 pp 130–136.

[29] Brunerová A, Müller M and Brožek M, 2017, Potential of wild growing Japanese Knotweed (*Reynoutria Japonica*) for briquette production. In: *Engineering For Rural Development*. Jelgava: Latvia, pp 561–568.

[30] Malaťák J, Bradna J and Velebil J, 2016, Combustion of briquettes from oversize fraction of compost from wood waste and other biomass residues. *Agronomy Research* 14 pp 525–532.

IOP Conf. Series: Materials Science and Engineering **237** (2017) 012002 doi:10.1088/1757-899X/237/1/012002

Analysis of production flow process with lean manufacturing approach

***Ikhsan* Siregar**[1,*], ***Abdillah* Arif Nasution**[2], ***Aji* Prasetio**[1] **and *Kharis* Fadillah**[1]
[1]Department of Industrial Engineering, Universitas Sumatera Utara, Jalan Almamater, Kampus USU, Medan 20155, Indonesia.
[2]Department of Accounting, Faculty of Economy and Bussiness, Universitas Sumatera Utara, Jalan Almamater, Kampus USU, Medan 20155, Indonesia.

* ikhsan.siregar@usu.ac.id

Abstract. This research was conducted on the company engaged in the production of Fast Moving Consumer Goods (FMCG). The production process in the company are still exists several activities that cause waste. Non value added activity (NVA) in the implementation is still widely found, so the cycle time generated to make the product will be longer. A form of improvement on the production line is by applying lean manufacturing method to identify waste along the value stream to find non value added activities. Non value added activity can be eliminated and reduced by utilizing value stream mapping and identifying it with activity mapping process. According to the results obtained that there are 26% of value-added activities and 74% non value added activity. The results obtained through the current state map of the production process of process lead time value of 678.11 minutes and processing time of 173.94 minutes. While the results obtained from the research proposal is the percentage of value added time of 41% of production process activities while non value added time of the production process of 59%. While the results obtained through the future state map of the production process of process lead time value of 426.69 minutes and processing time of 173.89 minutes.

1. Introduction

Product quality and short lead time are the keys to success in winning the competition. Companies should always try to focus on waste minimization that occurs in the entire production process. Forms of improvement on the production line with the application of lean manufacturing method are used to identify waste along the value stream and to find non value added activities. Lean manufacturing itself, has its weaknesses and strengths, as stated in previous research, the study stated that small and medium enterprises (SMEs) have different characteristics for lean manufacturing [1]. Through previous research also mentioned that with lean manufacturing can simulate lean index model for manufacturing case [2]. Lean manufacturing can also to increase productivity that has been done in companies engaged in producing sheet metal [3]. Some even use lean manufacturing to analyze production runs that has been done in Malaysian companies [4].

In addition, other studies in technological developments have conducted research to explore and evaluate previous work that focuses on the relationship between Lean and manufacturing production [5]. Prior research also related to lean manufacturing discusses detailed implementation of management care and lean manufacturing techniques in maintenance workshops to eliminate losses from damage. And to improve employee productivity and motivation in dairy-producing companies, it is clear from this research that lean manufacturing is not only for increased productivity but also for maintenance management [6]. Some even say lean manufacturing is a miracle methods, this is because it can reduce waste that occurs in the company [7]. In addition, lean manufacturing can also be collaborated with ergonomics, this has been done in previous research, in the study explained that the improvement can be done continuously both from production and ergonomics that involve operators in work [8].

This research was conducted on the company engaged in the production of Fast Moving Consumer Goods (FMCG). Problems faced by the company are often there are activities that cause waste and non value added activity. From the observation result, which becomes non value added activity one of them is the activity of moving the dough from the mixing station to the stamping station. This removal activity is more than 10 meters away resulting in a lot of time needed to produce the product. The waste that occurs in the production process can be seen in Table 1 below.

Table 1. Waste occurred in the production process

No.	Work Center	Cause of Waste	Impact
1.	Mixing	Waiting for materials and mixing machine for repair.	Resulting in longer production process time.
2.	Stamping	Fixed the less dough on the crusher machine and the number of unnecessary activities.	Adds inputs and extends production leadtime.
3.	Drying (Oven)	There is a coil that undergoes work in process.	Company inventory increases.
4.	Wrapping	There is a damaged coil and plastic film.	The company output is not maximal.
5.	Packaging	Transporting the product to the finished material warehouse.	The production lead time becomes long.

Based on the above facts, it is necessary to research that can reduce non value added activity to increase the value added (product added) product, and shorten the lead time, so that impact on the increase of company productivity, and with lean manufacturing approach is expected to find the right solution to know the type And the root cause of non-value-added activities on the company's production floor, so that waste can be reduced.

2. Research Methods

2.1 Research Location

This research was conducted at PT. ABC is engaged in the production of Fast Moving Consumer Goods (FMCG). The company is located Medan Star Industrial Estate, Tj. Morawa, Deli Serdang, North Sumatra.

2.2 Types and Research Objects

Type of research conducted is action research (action research) is a study conducted to obtain practical findings for operational purposes (Sinulingga, 2013). This study aims to obtain a solution that will be used on the company as a form of improvement from the original system. The object of research observed is the production process that occurs in the production.

2.3 Research Design

Sources of data used in the study are primary data and secondary data. Primary data obtained by observation or direct measurement. Data belonging to this category is:

a) Production time data.
b) Rating factor.
c) Allowance.

Secondary data is obtained based on company documentation data. Data belonging to this category is:

a) Data of production amount per month.
b) The order of production process.
c) Number of production operators.
d) Clock and work shift data.
e) The number of machines from each work station.

Observation and measurement of production process time is done by using stopwatch time study method. In addition, the method of data collection is done by conducting interviews or question and answer with the responsible of the production department on matters relating to research and collect secondary data taken from company documents related to research.

2.4 Formation of Current State Mapping

The steps to form the current state mapping are as follow:

a) Determination of Value Stream Manager.
Value Stream Manager is someone who is responsible for the condition of the factory and understanding the system and production process in the factory. Through Value Stream Manager could be known all the information and secondary data company. Value stream managers are selected based on the following aspects: have aspects of individual personality; Having knowledge and experience or expert in lean engineering for all departments in the company; have responsibility for the flow of materials and the flow of information that takes place in the production process.
b) Formation of SIPOC Diagram
SIPOC diagram illustrates the information about Supplier, Input, Process, Output and Customer involved in the production process.
c) Standard time calculation
d) Mapmaking for each process category
e) Making a whole map flow
f) Identification of waste (non value added activity) on Current State Map with PAM (Process Activity Mapping)

2.5 Current State Map Analysis

An analysis of the improvement of the current state map is done by identifying any waste that exists along the value stream of current state. Then will be searched the root of the problem and how to overcome it. Some steps are done ie:

a) Breakdown of value added and non value added activities.
b) Time cycle analysis
c) PAM (Process Activity Mapping) analysis with 5W1H.

2.6. Future State Map Establishment

Future state map is a picture of the situation to be achieved by the company going forward. Some steps that need to be done are:

a) Preparation of corrective action based on analysis
b) Making PAM (Process Activity Mapping) proposal.

IOP Conf. Series: Materials Science and Engineering **237** (2017) 012002 doi:10.1088/1757-899X/237/1/012002

c) Overall map flow map of proposal

3. Results and Discussion

3.1. Current State Mapping Results

At this stage, each process along the value stream is combined with material flow and information flow so that it becomes a single flow in the factory. The results obtained through the current state map of the production process of process lead time value of 678.11 minutes and processing time of 173.94 minutes. Comparison between value added and non value added activity can be seen in Figure 1.

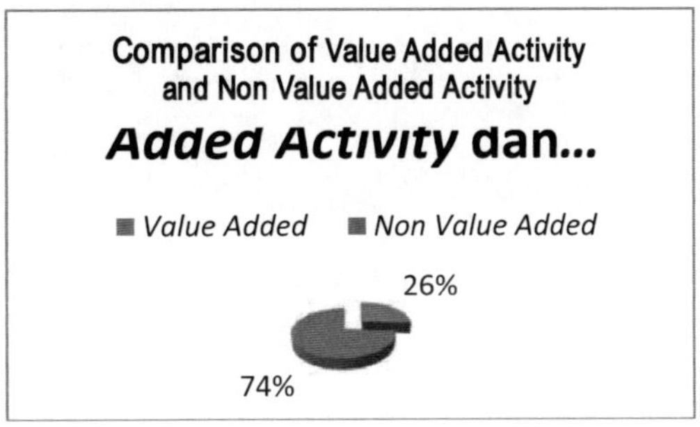

Figure 1. Value Added Activity and Non Value Added Activity Pie Chart

Based on pie chart above can be seen that the percentage of value added time equal to 26% from activity of production process while non value added time from production process equal to 74%. The Current State Mapping result is shown in Figure 2.

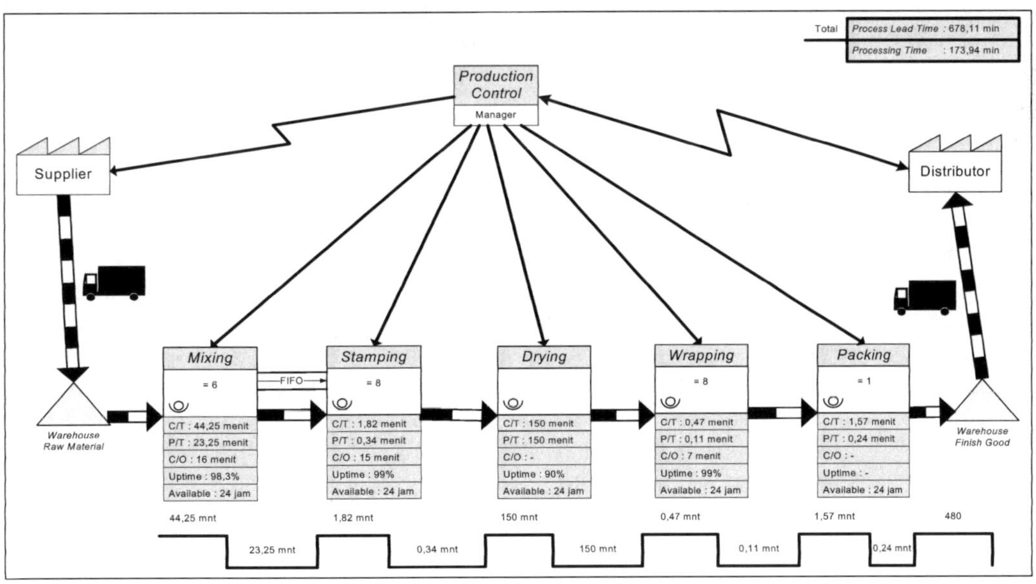

Figure 2. Production Process Current State Map

IOP Conf. Series: Materials Science and Engineering **237** (2017) 012002 doi:10.1088/1757-899X/237/1/012002

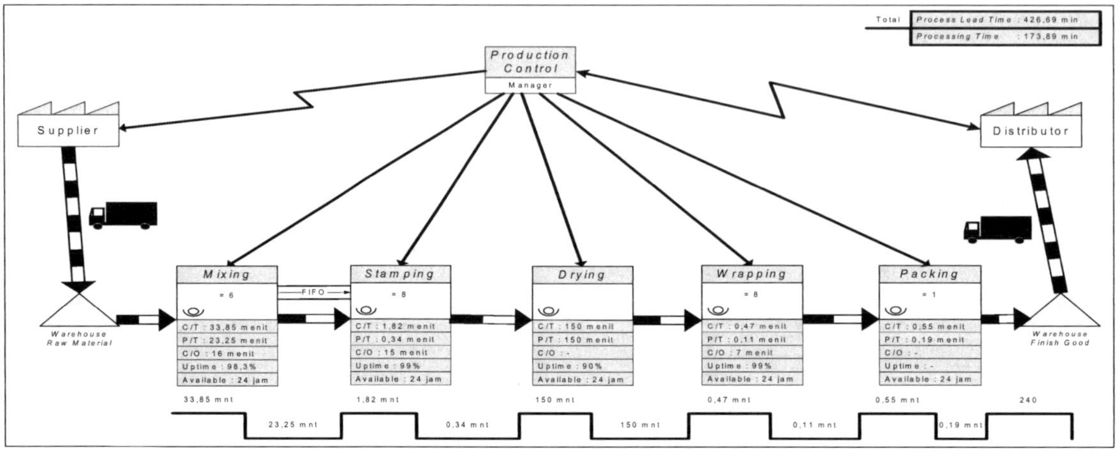

Figure 3. Production Process Future State Map

Based on Figure 2 above can be seen that there is a high process lead time that is equal to 78.11 minutes and *pprocessing* time of 173.94 minutes. Process lead time is the time generated from the entire production process starting from raw material until the product is produced. Figured 2 described the current production process on a map. Based on that, shown that the manufacturing have 5 workstations; consist of mixing, stamping, drying, wrapping, and packing.

3.2. Future State Mapping Result

The portfolio of future state maps based on the results of the proposed improvements made earlier. The result obtained through the future state map of the production process is the process lead time value of 173.89 minutes. Future State Map can be seen in Figure 3 above. Figure 3 is based on the simulation of the future mapping based on the current state mapping shown in figure 2. Based on Figure 3 above can be seen that the results obtained through the future state map of the production process of process lead time value of 426.69 minutes and processing time of 173.89 minutes. Process lead time can be reduced by 251.42 minutes. From this situation conclude that the future map shown the better progress compare on before.

4. Conclusion

The problem faced by the company is the amount of non value added activity or waste in the production process so as to extend the production time. Based on activity mapping process and current state map, non value added activity is much higher than value added activity. The proposed improvement plan on the future state map is expected to decrease the lead time of the company's production process. Proposed improvements to existing future state maps have not covered all the waste that may exist in the company. Therefore the company needs to make continuous improvement by creating a new current state map and making new proposals.

5. References

[1] Moeuf, A., Tamayo, S., Lamouri, S., Pellerin, R., & Lelievre, A. (2016). Strengths and weaknesses of small and medium sized enterprises regarding the implementation of lean manufacturing. *IFAC-Papers On Line*, *49*(12), 71-76.

[2] Oleghe, O., & Salonitis, K. (2016). Variation modeling of lean manufacturing performance using fuzzy logic based quantitative lean index. *Procedia CIRP*, *41*, 608-613.

[3] Choomlucksana, J., Ongsaranakorn, M., & Suksabai, P. (2015). Improving the productivity of sheet metal stamping subassembly area using the application of lean manufacturing principles. *Procedia Manufacturing*, *2*, 102-107.

[4] Rohani, J. M., & Zahraee, S. M. (2015). Production line analysis via value stream mapping: A lean manufacturing process of color industry. *Procedia Manufacturing*, *2*, 6-10.

[5] Hartini, S., & Ciptomulyono, U. (2015). The relationship between lean and sustainable manufacturing on performance: literature review. *Procedia Manufacturing*, *4*, 38-45.

[6] Arslankaya, S., & Atay, H. (2015). Maintenance management and lean manufacturing practices in a firm which produces dairy products. *Procedia-Social and Behavioral Sciences*, *207*, 214-224.

[7] Wyrwicka, M. K., & Mrugalska, B. (2017). Mirages of Lean Manufacturing in Practice. *Procedia Engineering*, *182*, 780-785.

[8] dos Santos, Z. G., Vieira, L., & Balbinotti, G. (2015). Lean Manufacturing and Ergonomic Working Conditions in the Automotive Industry. *Procedia Manufacturing*, *3*, 5947-5954.

IOP Conf. Series: Materials Science and Engineering **237** (2017) 012003 doi:10.1088/1757-899X/237/1/012003

Improvement of production layout based on optimum production balancing scale results by using Moodie Young and Comsoal method

Ikhsan Siregar, Tri Ulina Anastasia *Sipangkar* and Aji *Prasetio*
Department of Industrial Engineering, Faculty of Engineering, Universitas Sumatera Utara, Jalan Almamater Kampus USU , Medan 20155, Indonesia.

Abstract. This research was conducted at a make to order production system company which is engaged in the car body of the vehicle. One of the products produced is dump truck which is one kind of transportation for the transport of goods equipped with hydraulics to facilitate goods' loading and unloading process. The company has 7 work stations with different cycle times. Companies often experience delays in order delivery. The production process on the production floor has not been done optimally where there is a build up of work in process in some work centres. The build up of work in process (WIP) products is seen in the welding and painting stations. Stacking that occurs on the production line may cause the company to be liable for damages due to delays in product completion. The WIP occurs due to unbalanced paths can be seen from the variance of cycle time of each station is very diverse. The time difference of each work element is due to the allocation of work elements to each work centre unevenly. On the basis of the allocation of uneven work elements, the dump truck assembly line is made. The analysis is done by using Moodie Young and Comsoal method to do the balancing of production line. The result of layout improvement by using systematic layout planning (SLP) method is change the composition of the work centre from 7 into 4 work centre which enables the movement of material to be more effective and efficient so that it can get an efficient and effective production trajectory and can solve existing problems. The result of the track balancing is then used as a guide in constructing a new layout based on the balancing result with the most optimum method.

1. Introduction

Production activities are closely related to the balance of the assembly line. The numbers of element of assembly are grouped into several work stations. In completing the activity of the working element, the available time is determined by the speed of the assembly trajectory. Each work station that is as much as possible should have the same cycle time to avoid idle time between stations. The purpose of this research is to minimize idle time between workstations so that the efficiency of each works station.

Previous research that discussed about line balancing used mathematical model [1] [2]. The purpose of line balancing according to previous research is to design the production line to the maximum, and to seek profit under uncertainty [3]. The problem of the balance of the production path is always there in assembly companies, many researchers conduct research based on case studies, such as in electric factories [4]. In addition, the purpose of line balancing is to minimize, in addition to the number of workstations, the number of high-skilled workers required to complete operations and the number of assembly equipment correctly [5]. Line balancing is structured to achieve efficiency and effectiveness [6] [7]. Even the use of line balancing is used in the aircraft industry [8]. In addition line balancing is also used for balancing validation by simulating labour flexibility [9]. The reason why this research is done, it is to look at the comparison between the two methods in the line balancing used and use it in applying layouts that others have not previously seen researchers do before. Moodie and Young method is applied to all heuristics, then the results are compared based on the evaluation criteria and Comsoal is the software to calculate it. This research using this method cause the characteristic problem in the company can solve by using this method. This research contribution was reduced work centre, and to get more efficiency on time.

2. Research Methods

The type of research used is descriptive research in the form of job and activity analysis. The form of descriptive research in the form of job and activity analysis is research that aims to investigate in detail the activities and work of a person or group of people to get recommendations for various purposes, such as workload balance and efficiency in the use of time.

The research instrument is a tool used in data collection. Instruments used in this research are stationery and motion time study to calculate the time of each working element. Theoretical framework is a conceptual model that shows the logical relationship between the factors / variables that have been identified important for analyzing the research problem.
This research identifies each element of work. After that, calculate the standard time that will be used in the repair. Instruments used in the research include stopwatch and pencil and data collection form. The data collected to conduct this research is in the form of primary data and secondary data. Primary data consists of data elements - elements of work in the assembly, the operating process of each element of work and the initial work path of the company. Secondary data consists of operator number data, machine number data, and effective working hour data.
The operating time of each working element is calculated from the average yield of 10 times the measurement. Then calculated normally with the following formula:

Normal Time = Cycle Time x Rating Factor .... (1)

Next is calculated the default time of each working elements by using the following formula:

Standard Time = Normal Time x 100/100 - allowance .... (2)

This standard time will be used for the next calculation process is to determine the balance of the trajectory on the transformer assembly. Then calculate the value of balance delay, path efficiency, and smoothness index of the actual trajectory of the company which is a parameter to see how far the track is balanced. After that done track balancing by using Moodie Young and COMSOAL method and then calculated value of balance delay, track efficiency, and smoothness index of each method. Both methods are compared, than selected methods have lower balance delay value, higher line efficiency value and lower smoothness index value.

3. Results and Discussion

The imbalance of assembly lines in production activities on the factory floor can be seen from the unemployment of several work stations, while in other work stations still work in full. This is due to the time required by a station to complete the work faster than the speed of a predetermined trajectory.

This company is a company with make to order production system which is engaged in the car body. The company produces various types of car body. One of the products produced is a dump truck which is a means of transportation for the transport of goods equipped with hydraulics to facilitate the process of loading and unloading of goods.

The company has 7 work stations where each work station has different cycle times. The work stations at the company can be seen in Table 1.

Table 1. Cycle Time of Each Work Station

Station	Description	Time (Minutes)
I	Sheet Cutting	32
II	Sheet Forming	333
III	Subframe Assy	456
IV	Welding	1253
V	Painting	1108
VI	Hydraulic	824
VII	Finishing	34

Table 1 shows that there is a time difference on each work station. As a result of the difference in working time, there is accumulation of work in process product (WIP) at the next station which has fewer working time. The biggest time difference occurs at Station IV (Welding Station) and Station V (Painting Station). At work station VI does not occur accumulation because time used at work station V bigger than time work station VI so no WIP stacking on production floor.
With the problem of stacking on the production floor, it should be noted that the production flow becomes one of the causes of accumulation in the layout of the production flow. Here's the layout of the company's production floor.

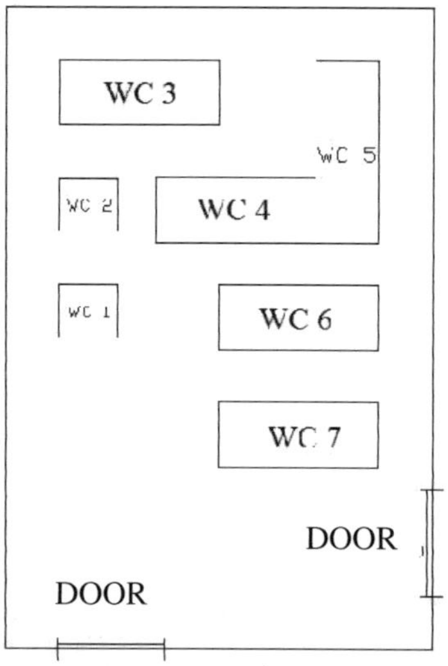

Figure 1. Company's Layout

Figure 1 shows the production layout of a company that is still not regular and has a distance between stations that make the process of closing the product move becomes longer.

The working element data and the operation time of each element are:

IOP Conf. Series: Materials Science and Engineering **237** (2017) 012003 doi:10.1088/1757-899X/237/1/012003

Table 2. Data of Work Elements in Dump Truck Production Process

EK	Time (minutes)	EK	Time (minutes)
I	7	XVI	1253
II	25	XVII	149
III	68	XVIII	273
V	79	XX	81
VI	28	XXI	523
VII	31	XXII	40
VIII	29	XXIII	552
IX	29	XXIV	74
X	78	XXV	74
XI	56	XXVI	39
XII	55	XXVII	28
XIII	80	XXVIII	17
XIV	63	XXIX	19
XV	124	XXX	15

In table 2. There are seen 30 work elements that have different working element time. Actual trajectory at company, all working elements are allocated in 7 works centre.

Initial Trajectory

The work elements in the initial work centre as well as the time of each work centre can be seen in table 3

Table 3. Initial Trajectory

Work centre	Working Element	Element Working Time (minutes)	Work Centre Working Time (Minutes)
I	1	245	245
II	2	69	340
	3	70	
	4	81	
	5	29	
	6	32	
	7	29	
	8	30	
III	9	80	469
	10	57	
	11	57	
	12	82	
	13	65	
	14	128	
IV	15	1271	1271
V	16	153	1137
	17	280	
	18	84	
	19	83	
	20	537	
VI	21	41	604

	22		563	
	23	75		
	24	76		
	25	40		
VII	26	28		270
	27	17		
	28	19		
	29	15		

In Table 3. There are 7 trajectories with 30 work elements at the beginning of the work station. Then the calculation of balance delay, line efficiency and smoothness index to see the value of the initial trajectory efficiency.

3.1 Calculating Balance Delay, Line Efficiency and Smoothness Index

a. Calculation of Balance Delay

Calculation of balance delay of a path, using the formula

$$D = \frac{n.C - \sum S}{n.C} x100\%$$

Where on the actual track is known

n = 7
C = 1271 minutes
∑Sti = 4336 minutes

Then, balance delay:

$$D = \frac{7x1271 - 4336}{7x1271} x100\%$$

$$D = 51{,}27\ \%$$

b. Line Efficiency

Line efficiency calculation using formula:

$$E = \frac{\sum S}{n.C} x100\%$$

In the actual trajectory, it is known:

n = 7
∑Sti = 4336 minutes
CT = 1271 minutes

Then, line efficiency is :

$$E = \frac{4336}{7x1271} x100\% = 48{,}74\ \%$$

c. *Smoothness Index*

Smoothness Index calculation using the formula

$$S = \sqrt{\sum (S\ \ m\ \ - S\ \ )^2}$$

Where

SI = *smoothing index*
Sti*max* = largest working time
Sti = i station working time
Then *smoothing index*;

$$S = \sqrt{(1271 - 245)^2 + \cdots + (1271 - 270)^2}$$

$$S = \sqrt{4027487} = 2006.85$$

Balancing trajectory using Moodie Young Method.

On balancing trajectory by using Moodie Young method obtained balancing result of 4 work centres on production floor. Calculation of Balance Delay, Line Efficiency and Smoothness Index.

a. Balance Delay Calculation

Calculation of balance delay of a path, using the formula

$$D = \frac{n.C - \sum S}{n.C} x100\%$$

Where on the actual track is known
n = 4
C = 1271 minutes
∑Sti = 4336 minutes
Then, balance delay

$$D = \frac{4x1271 - 4336}{4x1271} x100\%$$

$$D = 14{,}72\ \%$$

b. Line Efficiency

Line efficiency calculation using formula:

$$E = \frac{\sum S}{n.C} x100\%$$

In the actual trajectory, it is known:
n = 4
∑Sti = 4336 minutes
CT = 1271 minutes
Then, line efficiency

$$E = \frac{4336}{4x1271} x100\% = 85{,}29\ \%$$

c. Smoothness Index

Smoothness Index calculation using the formula

$$S = \sqrt{\sum (S \quad m \quad - S \quad)^2}$$

Where;
SI = smoothing index
Sti*max* = largest working time
Sti = i station working time
Then smoothing index;

$$S = \sqrt{(1271-973)^2 + \cdots + (1271-955)^2}$$

$$S = \sqrt{139884} = 374.02$$

Based on result of balancing of trajectory balancing result by using method Moodie Young to repair layout at production floor. The results of the improved floor layout of production are as follows:

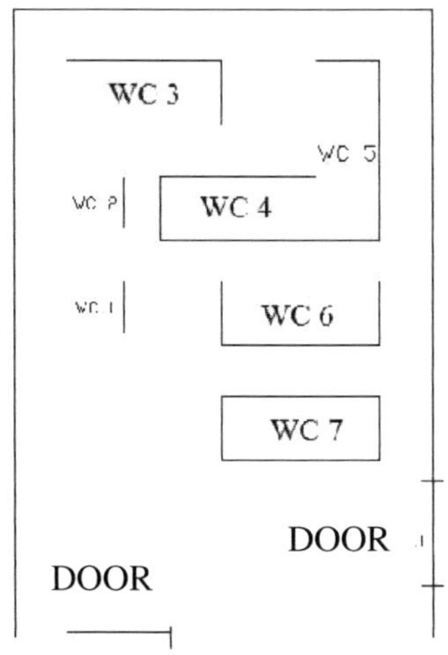

Comparison between existing and proposed are on proposed layout there is 4 work centre, the existing is 7 work centre.

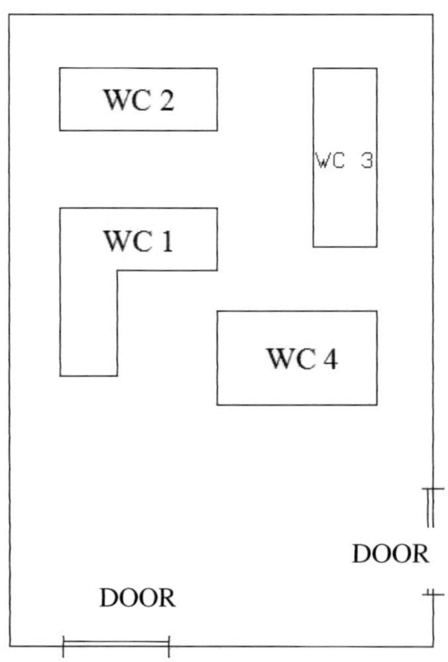

Figure 2. Proposed Layout

Figure 2 shows the proposed layout based on track balancing result using optimum method that is Moodie Young method. The coordinate point of each department in the proposed layout can be seen in Table 4.

IOP Conf. Series: Materials Science and Engineering **237** (2017) 012003 doi:10.1088/1757-899X/237/1/012003

Tabel 4. Production Area Location

No	Work Centre Area	Coordinate Point	
		X (m)	Y (m)
1	Work Centre1	12,5	31
.2	Work Centre 2	12,5	52
3	Work Centre 3	28,5	42
4	Work Centre 4	27,5	22,5

4. Conclusion

Based on the results of data processing and analysis that has been done can be taken several conclusions as follows:

A. The actual number of work centres is 7 work stations with balance delay 51,27%, line efficiency 48,74% and smoothing index 2006.85.
B. Proposals using Moodie Young obtained work centre reduction into 4 work centres with a balance delay of 14.72%, line efficiency 85.29% and smoothing index 374.02.
C. Selected improvements using the Moodie Young method due to providing better results in efficiency and balance delay.
D. The result of layout improvement by using systematic layout planning (SLP) method is change the composition of the work centre from 7 into 4 work centre which enables the movement of material to be more effective and efficient.

References

[1] Pape, T., 2015, Heuristics and lower bounds for the simple assembly line balancing problem type 1: Overview, computational tests and improvements, European Journal of Operational Research 240, pp. 32–42.

[2] Jonnalagedda, V., 2014, Application of Simple Genetic Algorithm to U-Shaped Assembly Line Balancing Problem of Type II, Proceedings of the 19th World Congress, The International Federation of Automatic Control,ape Town, South Africa. pp. 24-29.

[3] Bentaha,M., L., 2014, Disassembly Line Balancing and Sequencing under Uncertainty, 21st CIRP Conference on Life Cycle Engineering., Procedia CIRP 15, pp. 239 – 244.

[4] Lam, N,. T., 2016, Lean line balancing for an electronics assembly line, 13th Global Conference on Sustainable Manufacturing - Decoupling Growth from Resource Use, Procedia CIRP 40, pp. 437 – 442.

[5] Mura, M. D., 2016, Worker skills and equipment optimization in assembly line balancing by a genetic approach, 6th CIRP Conference on Assembly Technologies and Systems (CATS), Procedia CIRP 44, pp.102 – 107.

[6] Altekin, F., T., A Piecewise Linear Model for Stochastic Disassembly Line Balancing, IFAC-Papers On Line 49-12, pp. 932–937.

[7] Oesterle, J., 2016, Hybrid multi-objective optimization method for solving simultaneously the line balancing, equipment and buffer sizing problems for hybrid assembly systems, 49th CIRP Conference on Manufacturing System.

[8] Borreguero, 2015, T., Enhanced Assembly Line Balancing and Scheduling Methodology for the Aeronautical Industry, The Manufacturing Engineering Society International Conference, MESIC 2015, Procedia Engineering 132, pp. 990 – 997.

[9] Pröpstera, M., 2015, Validation of Line Balancing by Simulation of Workforce Flexibility, 9th CIRP Conference on Intelligent Computation in Manufacturing Engineering, Procedia CIRP 33, pp. 93 – 98.

IOP Conf. Series: Materials Science and Engineering **237** (2017) 012004 doi:10.1088/1757-899X/237/1/012004

Numerical estimation of deformation energy of selected bulk oilseeds in compression loading

C Demirel[1,*], A Kabutey[2], D Herak[2], G A K Gurdil[1]

[1] Ondokuz Mayis University, Department of Agricultural Machinery, Faculty of Agriculture, Samsun, Turkey
[2] Czech University of Life Sciences Prague, Department of Mechanical Engineering, Kamycka 129, 165 21 Prague, Czech Republic

* kabutey@tf.czu.cz

Abstract. This paper aimed at the determination of the deformation energy of some bulk oilseeds or kernels namely oil palm, sunflower, rape and flax in linear pressing applying the trapezoidal rule which is characterized by the area under the force and deformation curve.The bulk samples were measured at the initial pressing height of 60 mm with the vessel diameter of 60 mm where they were compressed under the universal compression machine at a maximum force of 200 kN and speed of 5 mm/min.Based on the compression test, the optimal deformation energy for recovering the oil was observed at a force of 163 kN where there was no seed/kernel cake ejection in comparison to the initial maximum force used particularly for rape and flax bulk oilseeds. This information is needed for analyzing the energy efficiency of the non-linear compression process involving a mechanical screw press or expeller.

1. Introduction

For agricultural activities such as land preparation, planting, weeding, irrigation, harvesting and among others energy is mainly utilized [1]. Similarly, energy is used for oil separation from vegetable oilseeds or kernels such as oil palm, sunflower, rape, flax and others by either mechanical expression or solvent method [2].

To improve the oil recovery efficiency especially the mechanical screw presses or expellers, it is important to understand the energy utilization of the bulk oilseeds under compression loading where the dependency between the force and deformation curve of the bulk material is described using the universal compression testing machine and pressing vessel with a plunger [3]. The area under the force and deformation curve is denoted as deformation energy which can be numerically determined using the trapezoidal rule [4, 5].

The knowledge of the linear compression loading is required for developing appropriate mathematical models that can be transferred to the non-linear environment involving mechanical screw presses or expellers [5].

The objective of the present study was to evaluate numerically the deformation energy of some bulk oilseeds/kernels under compression loading. The percentage oil of the bulk samples was also calculated.

2. Materials and Method

The bulk oilseeds or kernels of Oil palm (*Elaesisguineensis* Jacq.), Sunflower (*Helianthus annus*L.), Rape (*Brassica napus*L.) and Flax (*Linumprivatissimum* L.) were used for the compression test.

The moisture content of the bulk samples in the above order wasdetermined to be (8.57, 4.79, 4.62 and 7.32 %) on a wet basis were determined using the standard oven method with a temperature setting of 105 °C and drying time of 17 h [6]. The initial and final weights of the bulk samples before and after oven drying weredetermined using an electronic balance (Kern 440–35, Kern & Sohn GmbH, Balingen, Germany). The data was evaluated according to [7].

$$MC_{w.b.} = \left[\left(\frac{m_a - m_b}{m_a}\right)100\right] \quad (1)$$

where $MC_{w.b.}$ is the moisture content on a wet basis (%), m_a is the mass of the bulk samples in the initial state and m_b is the mass of the bulk samples after drying or heat treatment (g).

2.1 Compression loading test

The bulk samples were compressed under the universal compression testing machine (ZDM 50, Czech Republic) and pressing vessel of diameter 60 mm at a maximum force of 200 kN and speed of 5 mm/min to establish the dependency between the force and deformation curve. Each bulk sample initial pressing height was measured at 60 mm and repeated three times.

2.2 Deformation energy

The deformation energy was calculated according to [8 -10].

$$E = \int_0^x F(x)dx \approx \sum_{n=0}^{n=i-1}\left[\left(\frac{F_{n+1} + F_n}{2}\right)(x_{n+1} - x_n)\right] \quad (2)$$

Where E is the deformation energy (J), $F(x)$ represents polynomial and exponential functions in relation to the force and deformations (N.mm), $F_{n+1} + F_n$ and $x_{n+1} - x_n$ are values of the force (N) and deformation (mm), n is the number of observed values and $i$is 1,2,3..........i max observations.

2.3 Percentage oil yield

The percentage oil yield was determined according to[2].

$$OY = \frac{O_w}{O_m}100 \quad (3)$$

where OY is the oil yield (%), O_w is the mass of oil (g) and O_m is the mass of bulk oilseeds/kernels (g)

3. Results and Discussion

The relationship between the force and deformation curves of the different bulk oilseeds/kernels is described in Figure 1. At a maximum force of 200 kN and speed of 5 mm/min was the display of the serration pattern among the bulk oilseeds of rape and flax compared to that of bulk oil palm kernels and sunflower bulk oilseeds which showed a smooth curve. The serration effect thus influences the energy efficiency of the oil output [11].

The optimal energy without the serration behaviour was determined at a maximum force of 163 kN as shown in Figure 1. For the deformation energy calculation, the trapezoidal rule which is characterized by the area under the force and deformation curve was also fitted by the polynomial and exponential functions as described in Figure 2, where the integral of the function represents the deformation energy as indicated by equation 2. The estimated amounts of deformation energy of the various bulk oilseeds/kernels are presented in Table 1 and Figure 3 respectively. The bulk kernels of oil palm showed higher deformation energy followed by sunflower, rape and flax bulk oilseeds. The percentage oil yield of each bulk sample in that order was 27.32±0.79%, 21.08±0.59%, 23.04±0.15% and 14.19±0.61% respectively.

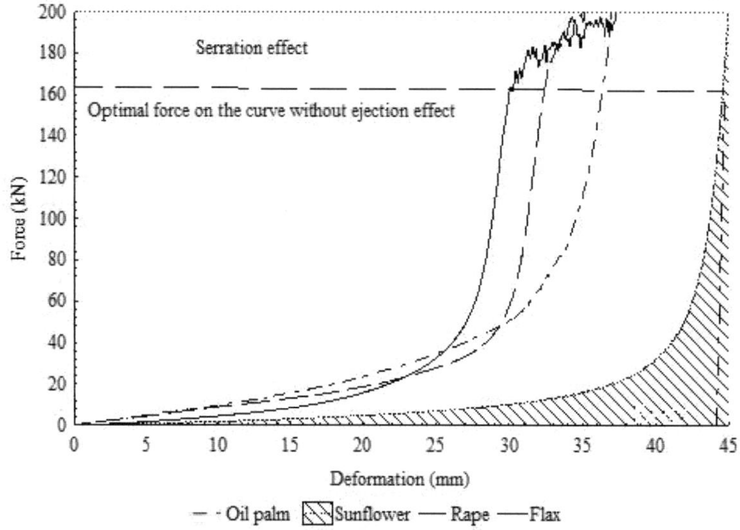

Figure 1. Force and deformation curves of bulk oilseed or kernel of oil palm, sunflower, rape and flax

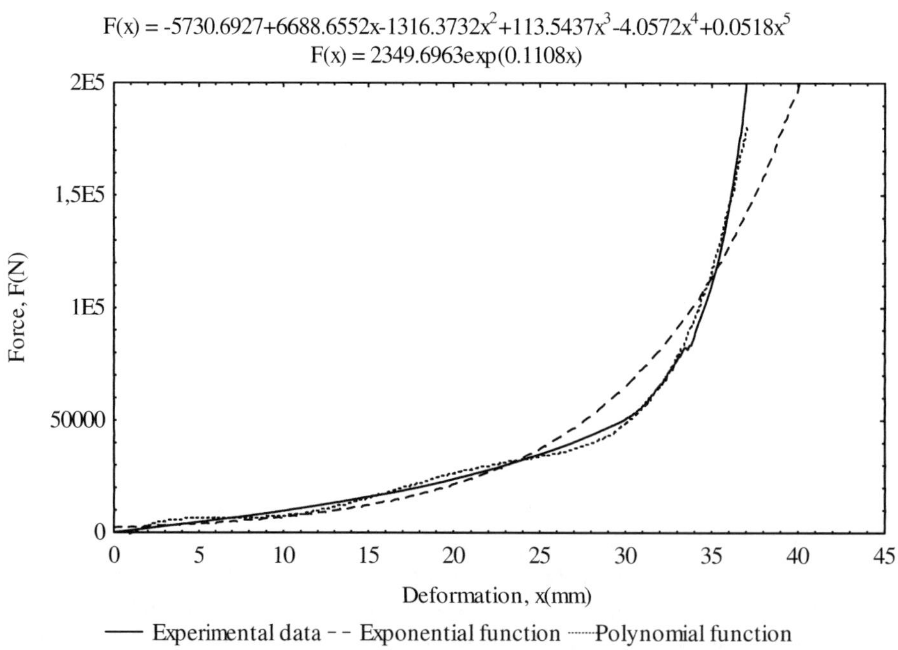

Figure 2. Experimental data of bulk oil palm kernels fitted by exponential and polynomial functions at force 200 kN

Table 1. Deformation energy of selected bulk oilseeds/kernels (Mean±Standard Deviation)

Bulk oilseeds/kernels	Force (kN)	Maximum deformation (mm)	Deformation energy (J) Trapezoidal rule	Polynomial function	Exponential function
[a]Oil palm	200	39.53±2.21	1245.55±20.41	1583.67±160.21	1284.67±32.35
[a]Sunflower		45.36±0.37	675.94±5.55	627.30±11.19	616.37±8.27
[b]Rape		35.64±0.59	1202.58±44.95	1154.33±17.21	1083±43.92
[b]Flax		37.96±0.60	2137.56±218.51	2128.33±244.27	2338±309.82
[a]Oil palm	163[c]	38.91±0.20	1133.72±15.44	1372.33±121.74	1185.33±27.06
[a]Sunflower		44.97±0.37	604.77±8.57	601.59±46.09	575.10±11
[b]Rape		33.03±0.56	718.27±29.41	721.53±24.93	699.27±17.19
[b]Flax		29.68±0.52	600.96±4.62	601.08±7.85	612.11±9.24

[a]No serration effect.
[b]Serration effect on the force-deformation curve.
[c]Optimal force for rape and flax bulk oilseeds.

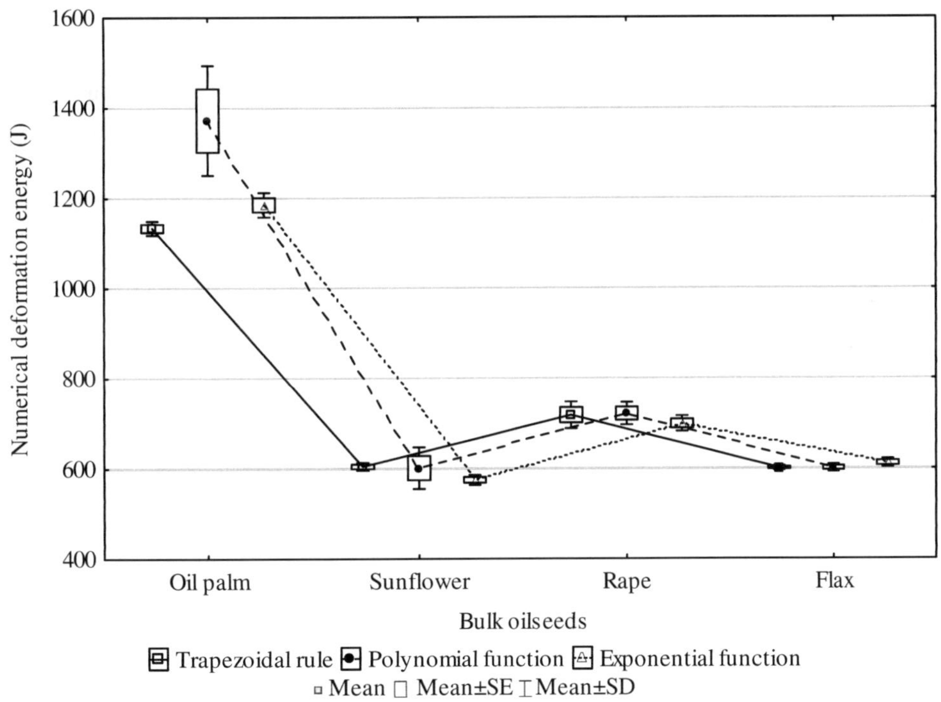

Figure 3. Box plot of numerical deformation energy (J) of bulk oilseeds/kernels at optimal force of 163 kN

4. Conclusion

Rape and flax bulk oilseeds experienced a serration effect on the force and deformation curves at an initial force of 200 kN compared to oil palm bulk kernels and sunflower bulk oilseeds which indicated smooth curve behaviour.

The optimal force of 163 kN without the serration effect was observed for rape and flax bulk oilseeds. The deformation energy for recovering the oil of oil palm bulk kernels, sunflower, rape and flax bulk oilseeds was 1245.55±20.41(J), 675.94±5.55 (J), 1202.58±44.95 (J) and 2137.56±218.51 (J) respectively. In future study, the present results will be analyzed theoretically using the tangent curve mathematical model applicable in MathCAD 14 software. The study among other studies is a step forward towards optimizing the vegetable oil processing technology.

References

[1] Jakayinfa S O and Bamgboye A I 2007 Development of equations for estimating energy requirements in palm-kernel oil processing operations *J. of Food Eng.*79(1):322–329

[2] Deli S, Farah Masturah M, Tajul Aris Y and Wan Nadiah W A 2011 The effects of physical parameters of the screw press oil expeller on oil yield from nigella sativa l. seeds *Int. Food Res.***18**:1367–1373

[3] Munson-Mcgee S H 2014 D-optimal experimental designs for uniaxial expression *J. of Food Pro. Eng.* **37**:248–256

[4] Grzegorz L 2007 Fracture toughness of pea: Weibull analysis *J. of Food Eng.***83**:436 – 443

[5] Herak D, Kabutey A, Divisova M and Simanjuntak S 2013 Mathematical model of mechanical behaviour of Jatropha curcas L. seeds under compression loading *Bio. Eng.***114**(3):279–288

[6] ISI 1996 Indian standard methods for analysis of oilseeds IS:3579 New Delhi Indian Stand. Ins.

[7] Blahovec J 2008 Agromaterials Study Guide. Czech University of Life Sciences Prague Prague

[8] Divisova M, Herak D, Kabutey A, Sleger V, Sigalingging R and Svatonova T 2014 Deformation curve characteristics of rapeseeds and sunflower seeds under compression loading. *Sci. Agric. Bohemica***45**(3):180–186

[9] Herak D, Kabutey A, Sedlacek A and Gurdil G 2012 Mechanical behaviour of several layers of selectedplant seeds under compression loading*Res. in Agric. Eng.***58**(1):24-29

[10] Kabutey A, Herak D, Dajbych O,Boatri W E and Sigalingging R 2014 Deformation energy of Jatropha curcas L. seeds under compression loading*Res. in Agric. Eng.***60**(2): 68-74

[11] Kabutey A, Herak D and Sedlacek A 2011 Behaviour of different moisture contents of Jatropha curcas L. seeds under compression loading *Res. in Agric. Eng.***57**(2):72–77

IOP Conf. Series: Materials Science and Engineering **237** (2017) 012005 doi:10.1088/1757-899X/237/1/012005

The influences of load mass changing on inverted pendulum stability based on simulation study

Timbang Pangaribuan[1,2], Nasruddin M N[1], Eddy Marlianto[1], Mula Sigiro[3]
[1]Department of Physics, Faculty of Mathematics and Natural Science, University of North Sumatera, Jl. Bioteknologi No.1, Medan 20155, Indonesia.
[2]Department of Electrical Engineering, Faculty of Engineering, Nommensen HKBP University, Jl. Sutomo No. 4A, Medan 20234, Indonesia.
[3]Department of Physics Education, Faculty of Teacher Training and Education, Nommensen HKBP University, Jl. Sutomo No. 4A, Medan 20234, Indonesia.

* timbang.pbn@gmail.com

Abstract. An inverted pendulum has nonlinear dynamic, so it is not easy to do in analysis to see its behavior. From many observations which have been made, there are two things that need to be added on the perfection of inverted pendulum. Firstly, when the pendulum has a large mass, and the second when the pendulum is given a load mass much larger than mass of the inverted pendulum. There are some question, first, how big the load mass can be given so that the movement of the inverted pendulum stay stable is. Second, how weight the changes and moves of load mass which can be given. For all the changes, it hopes the inverted pendulum is stay stable. Finally, the final result is still expected to be as stable, it must need conclude what kind of controller is capable of carrying such a mass burden, and how large the mass load limit can be given.

1. Introduction

A behaviour of nonlinear plant on an inverted pendulum can allow the system to become unstable, if its load mass changes suddenly occur on inverted pendulum. The load mass can change with constant values, or it will change depend on time, there is also the burden of mass change all at once moving. The load on cart mass can be made by giving additional load mass after the system is stable. For load mass changes and moves, the movement of that load mass can be done by giving disturbance to the mass of the cart, and it changes depend along with time changes.

If the mass of the cart has changed we can observe the changing of occurring on the position of pendulum, how is that possible to affect the stability of the inverted pendulum. The magnitude of the load mass changes can be made gradually to see the stages of the process, so at last we will find the limit of the mass imposition to the inverted pendulum.

There're so many researchers have designed the stability of an inverted pendulum by using different controllers. One of the techniques was standard pole-placement which proposed by Hari Vasudevan [1]. The controller design was done with the approach of linear system, it has realized to a nonlinear inverted pendulum. The other one has designed the stability of an inverted pendulum by using fuzzy controller which proposed by Ahmad M. El-Nagar [2]. Ronzhina M N [3], have designed the stability of an inverted pendulum by using full feedback from state space techniques, and Sorokin V S [4] has designed stability an inverted pendulum by using swing-up controller techniques. In all these cases, the behavior of the system stability has observed with well, but none of them observe the effect of changing load mass on system stability.

In the other case, the behavior of pendulum's response was observed through giving disturbance and changing the load mass, but the effect of movement of load mass still yet to be observed by them. All those cases have proposed by Prasad at all [5] and Arda and Kuscu [6].

Finally, it needs to observe the effect of mass moving on the stability of the pendulum and analyse the limit of mass on inverted pendulum with a large mass. This paper present a discussion to observe by the influence of changing the load mass on inverted pedulum stability based on simulation study, and expected will give a good contribution to the fields of control systems.

1.1 System Details

In addition, there are two purposes of the paper presented. First, the inverted pendulum's dynamic has been relegated separately by author, it has been also controlled and gave the result with a stable response. With reference to Figure 1, pendulum is assumed has a limit of movement from $-\theta_L$ to $+\theta_L$, and the action of the cart x is free in movement to stabilise the pendulum with apply the control force F. The action of control force generated by the optimal control techniques, and all the works described here based on simulation. In the beginning of the simulation, the cart mass, m_c is **8** kg, the pendulum mass m_p is **2** kg, the pendulum length, L is **1** m. The numerical work was performed by using a special program which developed by the author on using matlab software.

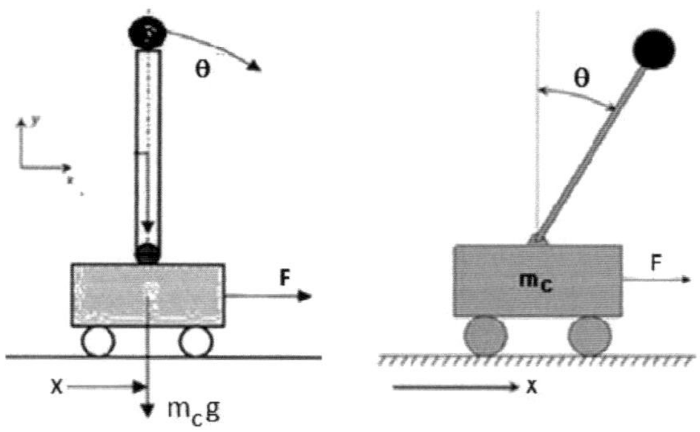

Figure 1. General Arrangement of An Inverted Pendulum

Second, based on the topic above, the load mass of the cart can be changed with the constant mass from 8 kg become 18, 28, 38 or 48 kg. The variable load mass can also be varied by using time, as shown the Figure 2 below. Mass of the cart changed and it changed according to a half sinus with magnitude started from 8 to 18, 20, 38 or 38 48 kg.

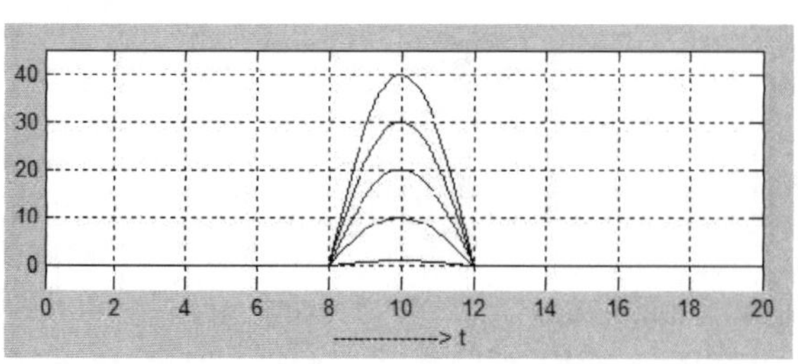

Figure 2. Varies of Pendulum's Mass

1.2 Modelling Inverted Pendulum

The dynamics of an inverted pendulum can be derived from its arrangement as shown in Figure 1. It consists of a cart, a pendulum and a driving force unit. The cart uses four wheels can move left or right on the floor freely. The pendulum is placed on the centre of the top surface of the cart. The cart able to rotate to the left or right as like as the cart movement. If all frictions are neglected on the system, the dynamic equation of the inverted pendulum have two equations as expressed below:

$$\ddot{\theta}(t) = \frac{m_p\,\dot{\theta}^2\,\mathrm{c}\quad\theta\,\mathrm{s}\quad\theta - (m_p + m_c)\frac{g}{L}\,\mathrm{s}\quad\theta - \frac{\mathrm{c}\quad\theta}{L}F}{(-1 + \mathrm{c}\quad{}^2\theta)\,m_p - m_c} \tag{1}$$

$$\ddot{x}(t) = \frac{L\,\dot{\theta}^2\,\mathrm{s}\quad\theta - (m_p + m_c)\,g\,\mathrm{t}\quad\theta - F}{(-1 + \mathrm{c}\quad{}^2\theta)\,m_p - m_c} \tag{2}$$

θ is the the position of pendulum, $\boldsymbol{x}$ is the position of the cart. The initial position of pendulum was $\theta = \theta_0$ in the unit (rad), the initial position of the cart with $\boldsymbol{x} = \boldsymbol{x}_0 = 0$ in the unit (m). The parameters $\boldsymbol{m_c}$ and $\boldsymbol{m_p}$ are respectively by the mass of the cart and pendulum in the unit (kg), and $\boldsymbol{g} = 9{,}81$ (m/s^2) is the grafity acceleration.

1.3 Optimal Control Design

From the dynamics of inverted pendulum equation (1) and (2), it needs an approach to generate the linear form with asumtions: for small θ, the value of sin $\theta = \theta$, and the value of cos $\theta = 1$. For those asumtions, the dynamics of inverted pendulum have changed into the equation (3) and (4) below:

$$\ddot{\theta}(t) = \frac{(m_p + m_c)\frac{g}{L}\theta - \frac{1}{L}F}{m_c} \tag{3}$$

$$\ddot{x}(t) = \frac{m_c\,g\,\theta - F}{m_c} \tag{4}$$

$$\begin{bmatrix}\dot{x}_1\\ \dot{x}_2\\ \dot{x}_3\\ \dot{x}_4\end{bmatrix} = \begin{bmatrix}0 & 1 & 0 & 0\\ \frac{(m_p+m_c)\,g}{L\,m_c} & 0 & 0 & 0\\ 0 & 0 & 0 & 1\\ \frac{m_p\,g}{m_c} & 0 & 0 & 0\end{bmatrix}\begin{bmatrix}x_1\\ x_2\\ x_3\\ x_4\end{bmatrix} + \begin{bmatrix}0\\ \frac{1}{L\,m_c}\\ 0\\ \frac{1}{m_c}\end{bmatrix}F \tag{5}$$

Equation (5) has the general form with $\dot{x} = A\,x + B\,u$. By using the Riccati's equation [9] to design the optimal feedback, we can choose the weight $\boldsymbol{Q}$ and $\boldsymbol{R}$ as below:

$$Q = [20\ 0\ 0\ 0\ ;\quad 0\ 1\ 0\ 0\ ;\quad 0\ 0\ 20\ 0\ ;\quad 0\ 0\ 0\ 1]^T \text{ and } R = [1]$$

We found the optimal feedback gain by using the matlab's command,

$$K = [k_1\quad k_2\quad k_3\quad k_4] = \mathrm{lqr}\,(A, B, Q, R) \tag{6}$$

and the result was $K = [32.6220\ \ 6.8964\ \ -1.4142\ \ -2.4721]$.

On insertion of those gain feedback in the form of closed loop system with $\boldsymbol{r}$(t) is the reference intput, the control force is,

$$F(t) = r(t) - K\,x(t) \tag{7}$$

Finally, the equation (8) and (9) are obtained for the closed loop system dynamics:

$$\ddot{\theta}(t) = \frac{m_p\,\dot{\theta}^2 c\quad \theta s\quad \theta - (m_p + m_c)\frac{g}{L} s\quad \theta - \frac{c\quad \theta}{L}\left[k_1\,\theta + k_1\dot{\theta} + k_3\,x + k_4\,\dot{x}\right]}{(-1 + c\quad {}^2\theta)\,m_p - m_c} \tag{8}$$

(1)

$$\ddot{x}(t) = \frac{L\,\dot{\theta}^2 s\quad \theta - (m_p + m_c)\,g\ t\quad \theta - \left[k_1\,\theta + k_1\dot{\theta} + k_3\,x + k_4\,\dot{x}\right]}{(-1 + c\quad {}^2\theta)\,m_p - m_c} \tag{9}$$

The examination of equations (8) and (9) have been done with $r = 0$ and the initial conditions $x(0) = \left[\frac{\pi}{8}\quad 0\quad 0\quad 0\right]^T$. It shows the state variables of inverted pendulum are coupled.

1.4 The Comparison Between Nonlinear and Linear Dynamics

If we applied the optimal feedback gain from the equation (7) for the linear system as given in equation (3) and (4), we can express both equations as below:

$$\ddot{\theta}(t) = \frac{(m_p + m_c)\frac{g}{L}\theta - \frac{1}{L}\left[k_1\,\theta + k_1\dot{\theta} + k_3\,x + k_4\,\dot{x}\right]}{m_c} \tag{10}$$

$$\ddot{x}(t) = \frac{m_c\,g\,\theta - \left[k_1\,\theta + k_1\dot{\theta} + k_3\,x + k_4\,\dot{x}\right]}{m_c} \tag{11}$$

The responses of the linier system $\theta(t)$ and $x(t)$ from equation (10) and (11) must resemble to the nonlinear system from equation (8) and (9), then the error between the responses nonlinear and linear system must be as small as possible.

2. Simulation

In the first step of the simulation, the initial position responses of pendulum and cart showed by the Figure 3. The simulations carried out by using Runge-Kutta integration.

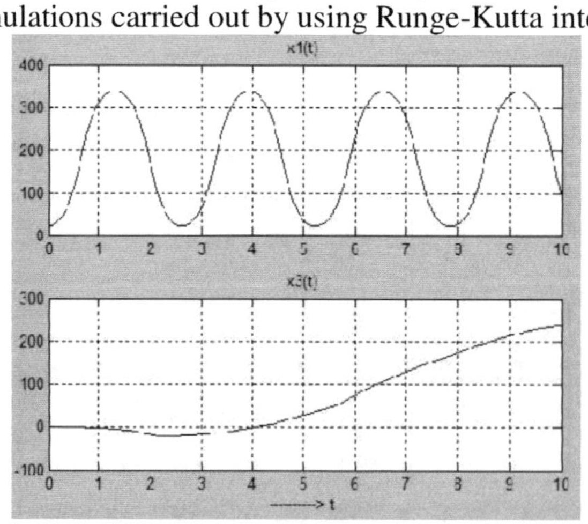

Figure 3. Initial Responses of Inverted Pendulum's Positions

$x_1(t)$ is $\theta(t)$ in degree (°), $x_3(t)$ is $x(t)$ in (m). From the results, it can be seen that the pendulum will continue oscillating from 22.5° to 337.5°. The movement of the cart starts from zero to infinity, so the pendulum and cart position become unstable.

In the second step of the simulation, it shows the optimal respond the linier and nonlinear system according to the Figure 4.

IOP Conf. Series: Materials Science and Engineering **237** (2017) 012005 doi:10.1088/1757-899X/237/1/012005

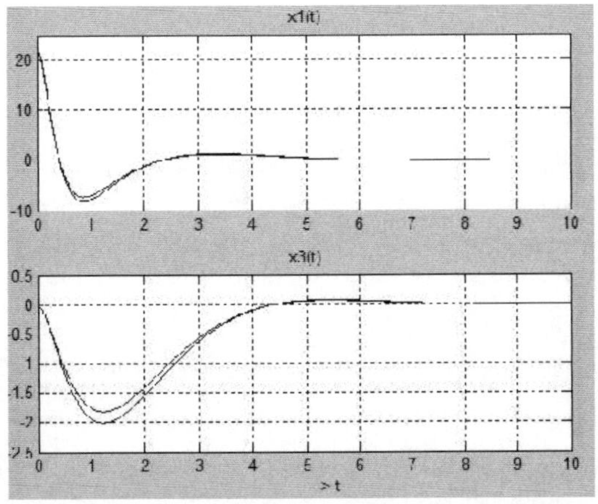

Figure 4. Comparison of Responses of Linear and Nonlinear System

From the results, we saw the response of the pendulum and the cart will be continued toward zero. These all responses mean, the system has stable, the both responses toward equilibrium pont. The position error between nonlinear and linear response are calculated by sum of square error, and those equal to 0.4865 % and 0.9815 % for each for the pendulum error and the cart error. Both the error are small enough, and all the responses are eligible.

3. The Influence of Load Mass Changes

The change of the load mass from 8 kg to 18, 28, 38 and 48 kg was appiled as given in Figure 5. It seems the response of the pendulum and cart for the different mass was stay stable.

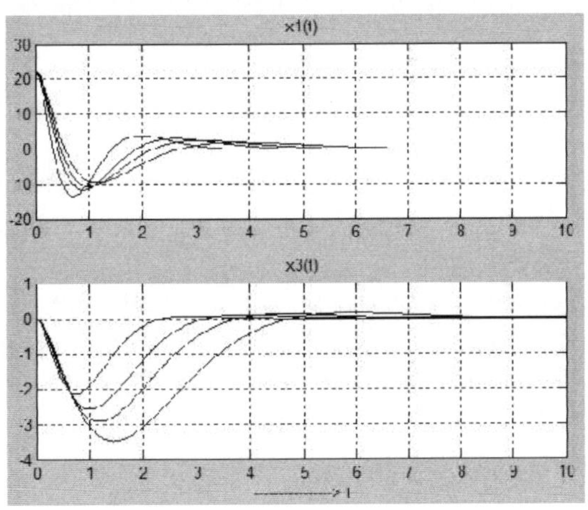

Figure 5. The effect of cart's mass change on position response

When the change of a different load mass appiled to the cart of inverted pendulum, the respon of position was found as shown in Figure 6. All responses show that all load mass changes can be overcome by controller, so the the inverted pendulum always stay stable, and the inverted pendulum has already designed to a strong controller.

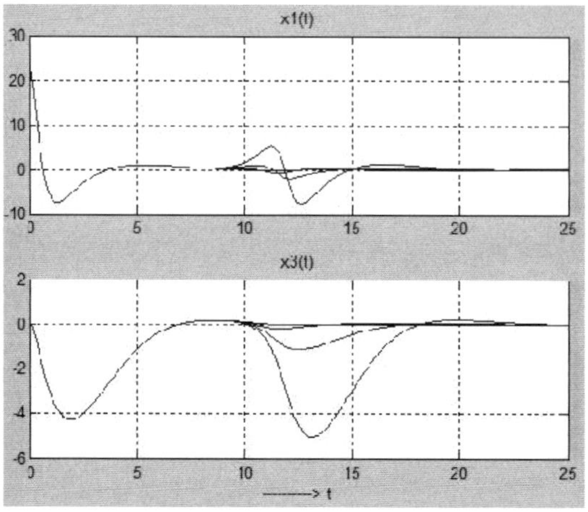

Figure 6. The effect of mass load disturbance changes on pendulum's position

4. Conclusion

As we have described in the introduction, we need to observe the influences of load mass moving on the stability of pendulum and analyse the limit of mass on inverted pendulum with a large mass, according to the given result on Figure 5 and Figure 6 we can say that we have producted a strong controller with big power and automatic. The limit of load mass which given by the cart has reached 50 kg.

Lastly, the controller has been designed to the strong controller by using the optimal techniques. The weight for optimal conditions to be selected with trial and error several times, so we found the best feedback gain to eliminate all the disturbances.

References

[1] Hari Vasudevan 2015 Design For Control Of Wheeled Inverted Pendulum Platforms (Journal of Mechanisms and Robotics Volume 7) (ASME)

[2] Ahmad M. El-Nagar 2013Intelligent Control For Nonlinear Inverted Pendulum based On Interval Type-2 Fuzzy PD Controller (Department of Industriial Electronics and Control Engineering) (Menof: Elsevier)

[3] Ronzhina M N 2015 Optimal Conditions With Chattering In The Inverted Two-Link Pendulum Control Problem (Lomonosov Moscow State University) (Moscow: Elsevier)

[4] Sorokin V S 2014 Analysis of Motion Of Inverted Pendulum With Vibrating Suspension Axis At Low-Frequency Excitation (Institute of Problems In Mechanical Engineering) (Moscow: Elsevier)

[5] Lal Bahadur Prasad, Barjeev Tyagi, Hari Om Gupta 2012 Modelling & Simulation for Optimal Control of Nonlinear Inverted Pendulum Dynamical System using PID Controller & LQR (Rorkee: Sixth Asia Modelling Symposium) p138-143

[6] Arda M, Kuscu H 2012 Modelling of the Two Dimensional Inverted Pendulum in Matlab/Simulink (Prague Czech Republic: Proceding of 5th International Mechanical Engineering Forum) p 116-126

[7] Bahram Shahian, Micahel Hassul 1993 Control System Design Using Matlab (London: Prentice-Hall International Inc)

[8] Stanislaw H Zak 2003 System and Controls (New York: Oxford Univeristy Press)

[9] Katsuhiko Ogata 2010 Modern Control Engineering Fifth Edition (Boston: Prentice-Hall)

IOP Conf. Series: Materials Science and Engineering **237** (2017) 012006 doi:10.1088/1757-899X/237/1/012006

Hybrid robot climbing system design

Agustinus Purna Irawan*, Agus Halim and Hengky Kurniawan
Department of Mechanical Engineering, Faculty of Engineering, Universitas Tarumanagara, Jakarta, Indonesia 11440

* agustinus@untar.ac.id

Abstract. This research aims to develop a climbing hybrid robot, especially to design the structure of robot that quite strong and how to build an optimal mechanism for transmitting the motor's rotation and torque to generate movement up the pole. In this research we use analytical methods using analysis software, simulation, a prototype, and robot trial. The result showed that robot could climb a pole by with maximum velocity 0.33m/s with a 20 kg load. Based on a weight diversity trial between 10 kg and 20 kg we obtained climb up load factor with value 0.970 ± 0.0223 and climb down load factor with value 0.910 ± 0.0163. Displacement of the frame structure was 7.58 mm. To minimize this displacement, the gate system was used so as to optimize the gripper while gripping the pole. The von Misses stress in the roller was 48.49 MPa, with 0.12 mm of displacement. This result could be a reference for robot development in further research.

1. Introduction

Robots are machines which are designed to help and ease human's jobs, especially to reduce risks of accident. One concept follows and has similarities to, a robot designed to climb palm trees. This robot can climb a palm tree to take coconuts, a job normally done by humans, so it assists in human activities. In this research we will used robot which canmove around on the flat floor by using caster wheels and grip a pole by using a lever mechanism powered by a compressed air, and controlled manually by a wireless joystick controller [1]. The concept dessign was made using Computer Aided Design (CAD), specifically Autodesk Inventor 2016 Student Version, while the pnelumatic dessign component used Festo Fluid SIM V4.2p/1.67. The purpose of this research is to discuss the frame structure analysis for develop a climbing system on a hybrid robot to know if the construction is sufficiently strong and how to build an optimal mechanism so robot can climb up at a certain velocity and how the load factor affect the robot velocity climbing up.

2. Method and materials

Methods that we used in this research were analytical methods using analysis computer's software, simulation, a prototype, and a robot trial. The first stage before actually designing the robot was to have a brainstorming session to determine what mechanism would wark best to grip a pole, climb the pole and also the most effective method to gather accurate data. The second stage ws to figure out the paarts specifications for the robot and then to come up with a design for the robot. At all stages there was constrant evaluation of the robot's system. If a problem occured we would go back and re-evaluated the design and specifications before continuing further with the robot frame structure. The final stage was to test the robot's performance and measure and collect data for comparative analysis purposes.

2.1. Gripping mechanism selection

There are several possible designs for the gripping mechanism of a robot which climbs vertically. One approach considered was the vacuum suction method. Despite being a good method for ascending a flat surface, such as a wall, it would prove in effective for our purposes because it would not be able to grip a cylindrical or conical shape object such as a pole[2]. In terms of locomotion, three alternatives were considered. The first comprised wheel-driven machines, which climb vertical structures by using the wheel motion to ceate upward motion[3], [4], [5], [6]. This mode of movement is very effective and is especially suited to the inspection of long structures. The second locomotion method comprised a legged climbing robot, which usually consists of four or six legs, each of them having either magnets, vacuum pumps or claws attached[7], [8]. This technique provided limited maneuverability and is better suited to rugged or dangerous environments. The final mode of movement was based on the us of arms with grippers or similar devices attached [4], [9], [10]. This provided the robot with greater versatility and range of movements. Several patents exist at present for machines that are capable of tree climbing and pruning [11], [12], [13]. These machines generally completely encircle the trunk and require many actuators and are usually quite heavy[14]. After due consideration, it was decided that we would use a self lock, wheel driven system, especially since it had been proven capable of climbing up a pole. Existing research in area indicates that the major weakness in this method was that the load that a robot could carry was rather light. Our goal was to improve on the system so that the robot could climb to a height of 2.4 m.

2.2. Specification of Hybrid Robot

Parts specification of the robot that we used is shown in the following Table 1.

Table 1. Mechanical and electrical components

Components	Specification
Motor DC	Power window 24V, locked torque: 19.4 Nmm, torque: 5.9 Nmm
Pneumatic Cylinder	Bore diameter: 20 mm and 16 mm, stroke length: 100 mm and 45 mm
Valve	Directional Control Valve 5/2, single solenoid, spring return
Roller	Pitch diameter: 19mm, outer diameter: 50mm, total length: 125mm
Caster wheel	diameter = 50mm
Structure	Aluminium hollow profile (19x19x1mm and 25x3x1mm)
Roller shaft	diameter 10mm [5]
Bearing UP000J	shaft diameter 10mm, outer diameter 35mm, width 10mm
Microcontroller	ATMega 16
Motor Driver	L293D
Motor Driver	EMS 30A H-Bridge
Sensor	Infrared Proximity Sensor

3. Results and discussion

3.1. Hybrid Robot Design

The 3D CAD design of the robot was preceded by a discussion on the ultimate objectives we wanted the robot to obtain, the system of implementation, theoretical calculations, tools and raw materials availability (by arranging a bill of materials), arranging job shifts for the fabrication stages, and final preparations for contructing the robot. This design presents an orthogonal view of the robot, caster wheels were used prevent slipping movement of the robot on a flat floor. We used an aluminium profile for the frame because it is light and provides a sloid structure for the base frame. A passive roller was put in place to keep the robot's position straight while climbing up and down. A pneumatic cylinder was placed in position (Figure 1) to generate grip in the lever mechanism. To grip the pole, the robot deploys the lever mechanism which we added to ensure that the back force created when the grippers grip the pole cause s little displacement to the frame structure as possible, and minimize any bending in the frame components.

1. Motor DC
2. Aluminium hollow25x13x1 mm
3. Aluminiumhollow19x19x1 mm
4. Caster wheel
5. Roller
6. Pneumatic cylinder;
7. Passive roller

Figure 1. Structural design of hybrid robot

3.2. Pneumatic system design on hybrid robot

The pneumatic system design was prepared and put through simulations before we actually built the hardware to minimize the risk of errors (Figure 2). The pneumatic system on the robot uses PET bottles to store pressurized air at a maximum of 0.5 MPa. The flow of the pressurized air is controlled by a directional control valve 5/2 single solenoid spring return which works as the control valve. Under normal conditions, the pressurized air will flow directly from the valve to the cylinder and pressure gauge, so the cylinder piston rod will always be extended causing the grip system to hold firmly to the pole. In the event of a sudden loss of electrical power, this system will allow the robot to maintain its grip on the pole thus preventing the robot from falling down the pole.

IOP Conf. Series: Materials Science and Engineering **237** (2017) 012006 doi:10.1088/1757-899X/237/1/012006

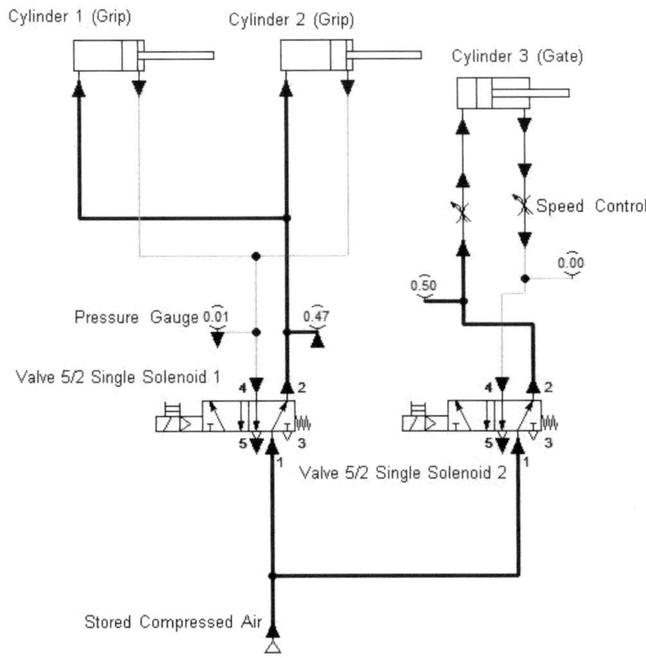

Figure 2. Design of pneumatic system

3.3. The Structural strength simulation results

The structural analysis of the frames occured during the contact force exerted by grip mechanism while gripping the pole and was recorded as 269 N force. By using the frame analysis a displacement of 7.58 mm was measured. This problem was remedied by adding the gate system which decreased the displacement by 4-6 mm.

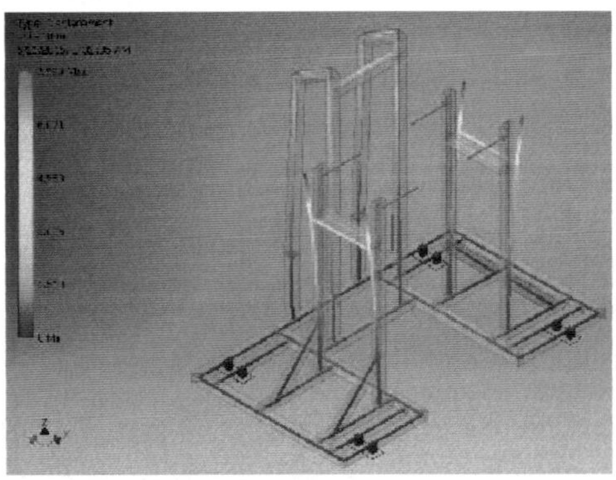

Figure 3. Maximum displacement simulation of frame analysis

The stress analysis simulation was used in order to determine the displacement of the roller structure and the lever that exerted the contact force while gripping the pole. The force applied on the roller is a269 N bearing load, so the values obtained from the simulation are 48.49 MPa of Von Misses stress, 0.12 mm of displacement, and 15 of maximum safety factor value as well as 9 of minimum safety factor value. Safety factor needed to ensure the safety of the robot in lifting weights. The design of the structure must provide a safety factor that is adequate in accordance with the working conditions it receives. The value of the safety factor depends on the working conditions of the product being designed [15] [16].

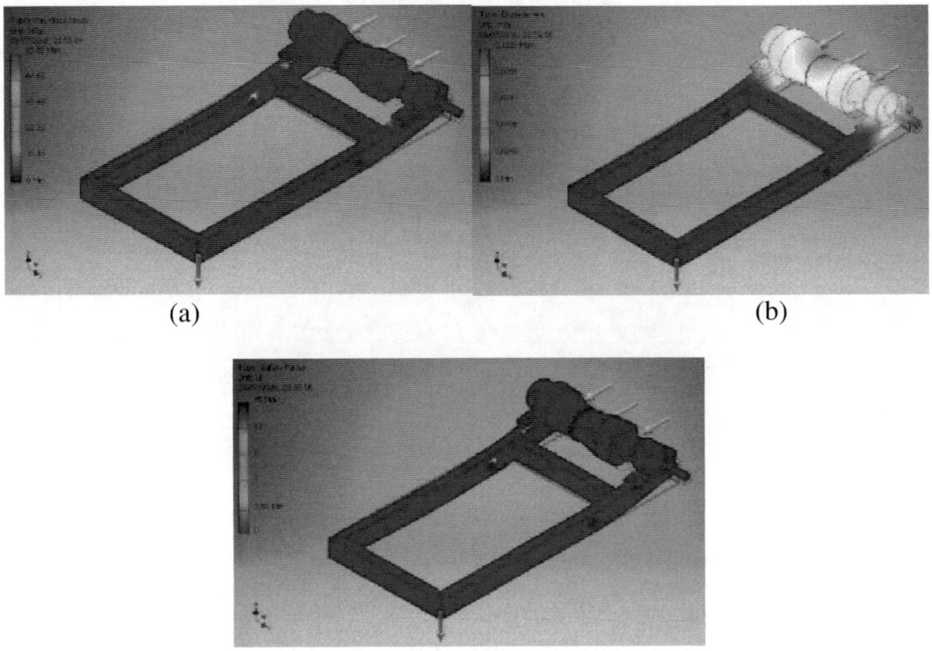

(a) (b) (c)

Figure 4. Von Misses stress simulation (a), maximum displacement simulation (b), safety factor simulation (c)

3.4. Implementation, evaluation, and test result

The implementation result was done based on a 3D CAD design and pneumatic system. Each of the joints of the robot was linked with a riveted joint system, which is used since it is easy to implement, and strong enough to support the structure with a maximum weight limit of 25 kg. The total weight of the robot with supporting equipment such as Li-Po battery was 10 kg (Figure 5). Li-Po battery usage refers to the Omni Climber hybrid robot designed by Tavakoli et al where the robot is operated by a 1000 mAh Li-Po battery [17]. Maksimum climbing speed of robot is 0.33 m/s with a total weight of 20 kg. If compare with the Omni Climber hybrid robot (maximun climbing speed is 0.14 m/s), this robot has climbing speed faster [17].To ensure robot stability while working at altitudes and to safely descend the pole, further research should focus on the influence of elevated wind loads. Safety should be considered for robots to work at altitudes with different wind loads [18].The stability of the robot when climbing is determined by the gripper design. The robot should not climb by swaying, but must be gripped perfectly between the gripper and the pole surface [19].

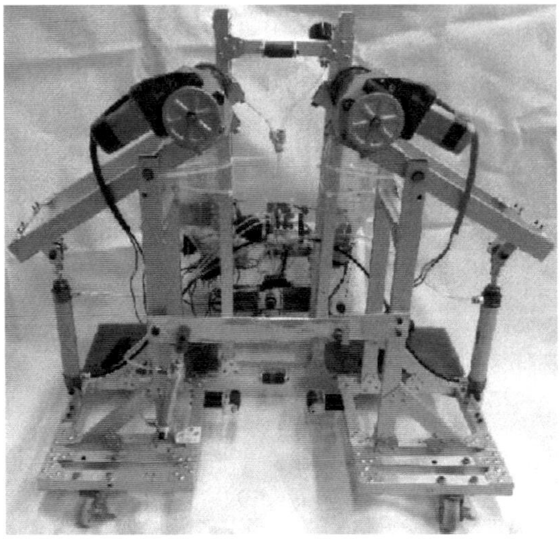

Figure 5. Result of the implementation of the hybrid robot

At the evaluation stage, the robot was tested to climb the pole. A loadwas placed on each left and right side of the robot giving the robot close to maximum weight conditions, which is 20 kg. Result showed that the robot was able to climb the pole without any defects in the frames detected. With a load applied, the robot is tested to climb the pole at various climbing velocities. The result of the tests are shown in the following Figure 6, Figure 7 and Figure 8.

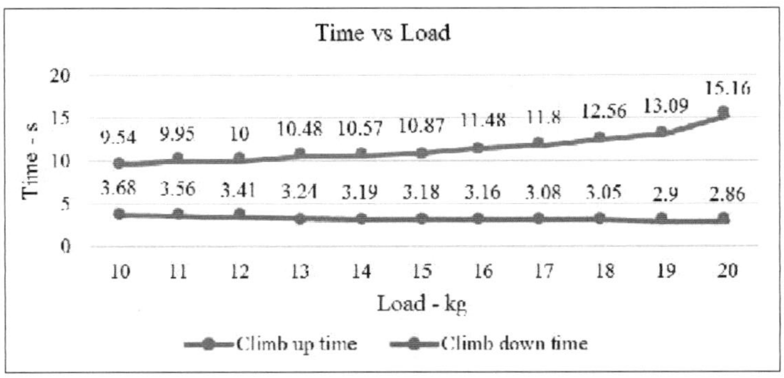

Figure 6. The relation between time and loads with various loads added at the rate of 147.2r/min

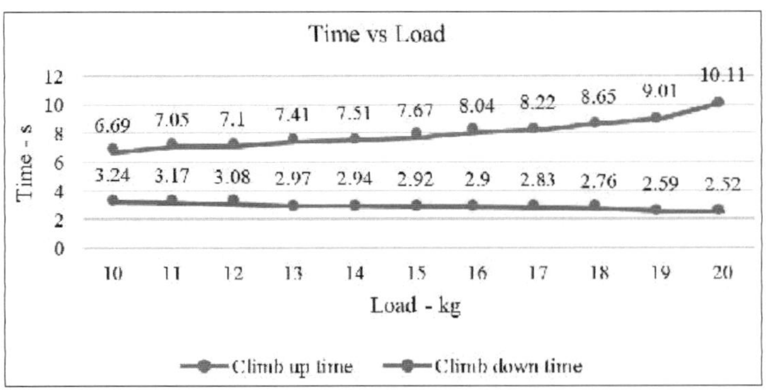

Figure 7. The relation between time and loads with various loads added at the rate of 173.6r/min

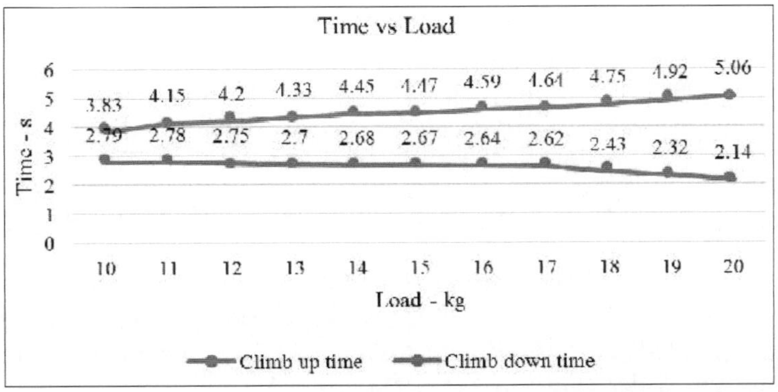

Figure 8. The relation between time and loads with various loads added at the rate of 195 r/min

After collecting the data, the values of the loading factors are shown as follows Table 2 and Table 3.

Table 2. The average loading factor

Rotation (r/min)	Loading Factor	
	Increase	Decrease
147.2	0.978 ± 0.0323	0.972 ± 0.0208
173.6	0.973 ± 0.0267	0.973 ± 0.0127
195	0.959 + 0.0170	0.950 ± 0.0153
Average	0.970 ± 0.0223	0.910 ± 0.0163

4. Conclusion

The hybrid robot is able to climb up and climb down the pole with a total weight of 20 kg to a height of 2.4m. The motor used to climb the pole had enough torque to lift the robot with a maximum velocity of 0.33m/s. The greatest displacement of the frame structure was 7.59 mm. In order to minimize this displacement, the gate system was used so as to optimize the gripper while gripping the pole. The von Misses stress in the roller was 48.49MPa, with 0.12mm of displacement, and 15 of maximum safety factor value as well as 9 of minimum safety factor value. The result test shows that the value of the loading factor when the load was increased is 0.97± 0.0223 and 0.91± 0.0163 when the load was decreased. This result could be a reference for hybrid robot development in further research.

5. Acknowledgements

The authors would like to be obliged to Universitas Tarumanagara for providing laboratory facilities and financial assistance.

6. References

[1] Xu, F.Y., Jingjin, S. and Guo Ping, J. G. 2015*International Journal of Advanced Robotic Systems***12-99** 1
[2] Yoon, Y. and Rus, D. 2007 *IEEE International Conference on Robotics and Automation*, vol 1 pp. 181.
[3] Tavakoli, M., Marques, L., and Almeida, A. T. 2010 *Industrial Robot: An International Journal***37-3** 309
[4] Xu, F. Y., and Wang, X. S. 2011*Journal of Field Robotics***28-3** 441.
[5] Sadeghi, A., Moradi, H. and Ahmadabadi, M. N. 2012 *Journal of Robotica***30-2** 279.

[6] Baghani, A., Ahmadabadi, M. N. and Harati, A. 2005 *Proceedings of the 2005 IEEE International Conference on Robotics and Automation*vol. 1 pp. 2099-2104.
[7] Haynes. G. C., Khripin, A., Lynch, G., Amory, J., Saunders, A., Rizzi, A. A. and Koditschek, D. E. 2009 *Proceedings of the IEEE International Conference on Robotics and Automation*vol 1 pp. 2767–2772.
[8] Xu, F. Y., Wang, X. S. and Jiang, G. P. 2012 *International Journal of Advanced Robotic Systems***9**-1-12.
[9] Lam, T. L. and Xu, Y. S. 2012 *Journal of Field Robotics***29-6** 843
[10] Tavakoli, M., Marjovi, A. and Marques, L. 2008 *Proceedings of the IEEE/RSJ International Conference on Intelligent Robots and Systems*vol. 1 pp. 4130-4135.
[11] Emery, W. and Shuff, H. 1949 *Patent US*2477922.
[12] Whitaker, R. 1949 *Patent US*2482392.
[13] Grasham, C. 1952 *Patent US* 2581479.
[14] Fauroux, J. C. 2010 *Industrial Robot: An International Journal***37**, 287-292.
[15] Irawan, A.P., Fediyanto, Tandi, S. 2006 *Proceedings of Ergo Future* vol. 1 pp. 337-341.
[16] Irawan, A.P., Daywin, F.J., Fanando, Agustino, T. 2016 *International Journal of Engineering and Technology***8-3** 1543
[17] Tavakoli, M., Lourençoa, J., Viegas, C., Neto, P., Almeida, A.T. 2016 *Journal of Mechatronics***36,** 136-146.
[18] Xu,F.Y., Wang, X.S., Wang, L., 2011 *Transactions of the Canadian Society for Mechanical Engineering***35-2** 269
[19] Hariskrishna, T.V., Harshavardhan, P.D. P. R., Pandey V., 2014, *International Journal of Current Engineering and Technology***3,** 85-88.

IOP Conf. Series: Materials Science and Engineering **237** (2017) 012007 doi:10.1088/1757-899X/237/1/012007

Geohydrology, geochemistry, geothermal potency of Rianiate Toba Lake North Sumatera

Juliper Nainggolan[1,*], Cristin Sitepu[1], Sanggam Pardede[1], Markus Diantoro[2]
Department of Physics, Nommensen HKBP University Jl. Sutomo No. 4a Medan, Indonesia
Department of Physics, Faculty of Mathematics and Natural Sciences, State University of Malang, Jl. Semarang 5 Malang 65145, Indonesia

* juliper_n@yahoo.com

Abstract. This research was performed to determine the potency of Rianiate's geothermal an alternative of energy source and determine the types of geothermal that was begun with the position's measurement by using GPS (Global Position System), and then the direct observation of chemical and physical properties such as pH, surface's temperature, color etc. The following steps were taking sample in four different springs indicated by spring 1, spring 2, spring 3, and spring 4. The chemical nature was measured by titrimetry method by using AAS, XRD, and gas Chromathology. The calculating of temperature of subsurface has done by using geothermometer and sequentially followed by geothermal potency's calculation. The position of four springs are located about N: 02° 31,852' and E: 098° 44. 021' where were average height from sea's level is 958 m. The highest surface's temperature is 80 °C and the temperature under soil is about 130.5 °C described the average of geothermal. The calculation of content of chloride, sulfate, bicarbonations revealed that the water can be categorized as chloride type. The trilateral diagram Na/1000 – K/100 - √Mg of hot water is in the regime of *immature water*. From the calculation of Indonesia's Standardized Geotherm, it was obtained that the estimated reservoir potency of Rianiate geotherm is 2,68 MWe.

1. Introduction

The increase of energy demand is strongly relates to the increase of civilization, industries and daily life. It is about 87% of world energy consumption is from fossil. Oil and coal is decreasing in potentially (Gençer and Agrawal, 2016). Another colossal potency of energy source may be coming from geothermal which should be a huge amount available in Indonesia. Geothermal is a source of energy, such as hot water, steam heat, and other gases. Scientists defined that the geothermal energy as the amount of heat that comes from the earth and is close enough to the surface of the earth that can be use economically. Indonesia has abundant geothermal potency, but less explored.

Estimation of geothermal energy potential is base on geological, geophysical and reservoir studies. Geological studies are more emphasize on volcanic systems, geological structures, rock age, type and type of rock changes in relation to geothermal systems. Geochemical studies are emphasizes on the type and level of water mature, the origin of hot water hydrological models and their fluid systems. Geophysical studies produce physical parameters of rocks and subsurface structures of geothermal systems. The reservoir engineering study yields a technical phase that defines the classification of reserves including the physical properties of rocks and fluids as well as the removal of fluids from the reservoir.

The previous of our unpublished work has performed in Dolok area, North Sumatera (2010). We have measured using a geothermeter, than followed by measuring chemical elemental analyses of hot spring. We found that the chemical elemental content of Dolok Marawa area contain of Na, K, Mg, Ca, and SiO_2. We also found that the reservoir temperature was about ± 187 °C(Cristina,2010).

Basically a geothermal system basically was formed as result of heat transfer from a heat source to its surroundings which are occurred by conduction and by convection. The occurrence of geothermal energy sources in Indonesia and its characteristics are briefly described as follows (Hazuardi, 1992): There are three plates interacting in Indonesia, namely the Pacific plate, the Indian-Australian plate and the Eurasian plate. The collision between the Indian-Australian plate to the south and the Eurasian plate in the north produced subduction zones at a depth of 160 - 210 km below Java-Nusa Tenggara Island and at depth of about 100 km under the island of Sumatra (Herdianita, 2006). Based on the magnitude of the temperature, Ralph (1985) distinguishes the geothermal system into three, there are low (<125 °C), moderate (125 °C - 225 °C, and high (>225 °C) temperature heat systems.

It is difficult to get information related geophysics, geology, geochemistry, as well as geothermal condition of Toba lake area, although a preliminary investigation has been performed by a group of research by Ari et al. (2012). Unfortunately, there is insufficient information could be obtained from the previous report. So far, in our knowledge, there is not comprehensive report related to the geochemistry, geophysics, geohydrology, and geothermal conditions of Toba lake area, especially in Rianiate district. In order to obtain the important information to predict the potency of geothermal in Rianiate, we have performed an investigation four different springs, and related rocks and soils in the area of Toba lake. The investigation was comprehensively performed involving geochemistry, geophysics, as well as geothermal experiment.

2. Materials and Method

The research conducted in Rianiate Village, Pangururan sub-district which administratively under governance of Samosir Regency. Geographically, this area is located at position98°42' – 98°47' East Longitude and 2°32 '- 2°45' North latitude. This location has an area of 121.43 km^2, with the boundaries of the region is as follows. North side is Simanindo Sub-district, South side is District Palipi, West side is Sianjur Mulamula District, and East is Ronggur Nihuta District.

2.1 Geophysics Investigation

The principle of conducting a survey of resistance type is by injecting an electric current through electrode current and measuring its response (voltage) to a potential electrode in a particular configuration (Hochstein, 1982). Geoelectric survey was carried out as the following procedure: Determine the 3 paths of data retrieval and simultaneously determine the position of the survey area by using GPS. Measuring distance between electrodes (5 m) on specified path. Data collection using the Geoelectric (Resistivity meter) ARES-G4 v4.7, SN:0609135 (Automatic Resistivity System) schlumberger method on specified path. Processing data obtained from the Geoelectric data using Res2Dinv software to obtain 2D cross-sectional model along the track. The data obtained were presented indifferent colors. Each color indicates resistivity difference in each layer of the 2D cross-sectional model along the path.

2.2 Geohydrology Investigation

Geohydrological investigations in addition to measuring water debit, the number of springs, also investigate water absorption, groundwater runoff, electrical conductivity, and surface flows.

2.3 Geochemistry Investigation

The geochemical investigation focused on chemical content analysis. The existed chemical element may include pH, Hg and CO_2, SiO_2, Al, Fe, Ca, Mg, Cl, SO_4, HCO_3, As, F, Hg, CO, CH_4, H_2, O_2, N_2, NH_3, SO_2, CO_2, H_2S and HCl by titrimetric methods using AAS, Gas Chromatography, and other equipment. These results used to determine the water type of the area (chloride type, sulfate type or carbonate type). Chemical analysis has taken the rock's sample just under spring by means of X-Ray diffraction technique. The instrument used in this experiment was Philips PW 1710 diffractometer of UNIMED Physics Laboratory. The anode or target used is copper (Cu) with the typical wavelength of λ = 1.56054 Å. The measurement was performed under voltage used s 40 KV with a current intensity of 30 mA.

3. Results and Discussions

Firstly, we describe the hot springs in terms their specific geological location, physical as well as chemical properties. The four springs identified as spring 1, spring 2, spring 3, and spring 4. The spring1 is gushing water that has passed through the rock gap away from the source the real. The gushing water is at the coordinates of N: 02^0 31,852' and E: 0980 44, 021' with the elevation from mean sea level (msl): 958 m, there are air temperatures of 26^0 C and hot water surface temperature of 43^0 C with pH of 6.2. Water in acidic conditions is seen from changing of blue litmus paper color to be a red and the red litmus paper unchanged when was immersed into the sample. The Spring 2 is at the coordinates of N: 02^0 31.813 'and E: 098^0 44,014' with the elevation from mean sea level (msl): 946 m, there are air temperatures of 26^0 C and hot water surface temperature of 77^0 C with pH of 6.3. Water in acidic conditions are seen from changing of blue litmus paper color to be a red and red litmus paper unchanged when was immersed into the sample. The spring 3 is at the coordinates of N: 02^0 31, 785 'and E: 098^0 44', 045' with the elevation from mean sea level (msl): 977 m, there are air temperatures of 26.5^0 C and hot water surface temperature of 79^0 C with pH of 6.5. Water in acidic conditions seen from changing of blue litmus paper color to be a red and red litmus paper unchanged when immersed into the sample. The spring 4 is at the coordinates of N: 02^0 31, 759 'and E: 098^0 44', 060" with the elevation from mean sea level (msl): 930 m, there are air temperatures of 27^0C and hot water surface temperature of 80^0 C with pH of 6.4. Water in acidic conditions is seen from changing of blue litmus paper color to be a red and red litmus paper unchanged when was immersed into the sample.

3.1 Geochemistry of springs

Geochemical analysis is very helpful to obtain information about reservoir condition for determining characteristics of geothermal in the study area. Chemical analysis conducted at PT Sucofindo. In the geochemical process of hot spring of the research area, that is known that heat transfer is accompanied by chemical reactions from the magma as a heat source with the medium through which the reservoir and the rock cover up by heat to reach the surface. The indication of heat transfer and the chemical reaction, can be obtain from the content of the elemental/ chemical content (Ca, Mg, K, Na, Cl, NH3, SO4, Hg and HCO3) which is relatively constant in the hot spring sample.

3.2 Type of hot springs

To determine types of hot water, we calculated the ion content in meq/L, the result was compared using trilenear classification diagram. In this case, we focus on the relative content of chloride, sulfate, and bicarbonate ions. The amount of ion content is expressed in units of meq/L (miliequvalent per liter), so it must be converted from ppm to meq/L. The conversion was calculated using Meq/L = ppm/(Rel.atomic mass compound/Valence). From laboratory test data, calculation of chloride ion, sulfate and bicarbonate ions to determine the hot spring type for springs 1, 2, 3 and 4 as in table 1.

Tabel 1. The chloride, sulfate, and bicarbonate ion content of various spring.

content (%)	Spring 1	Spring 2	Spring 3	Spring 4
HCO_3	0.00	0.00	0.00	38.71(9)
SO_4	4.36(0)	7.18(5)	10.02(2)	7.45(3)
Cl	95.63(9)	92.81(4)	89.97(7)	53.82(7)

Based on the percentage of ion content in hot water samples which have been analyzed chemical element content especially Bioarbonate anion content (HCO3-), Clorida (Cl-), and Sulfate (SO42-) that be determined trough Water type. The four of hot springs were included in the chloride type, especially the spring 1, 2, and 3 samples. This is characterized by the high content of chloride ions in hot water compared to the concentration of bicarbonate and sulfate. Also characterized by a relatively acidic pH (pH between 6.3 - 6.5) and surface temperatures between 43 °C to 79 °C. Further anlaysis was converting the three chemical indicators of natrium, potassium and magnesium into a specific ratio of Na/1000, K/100 and √Mg.The results of calculation is depicted in table 2.

Table 2. Elemental ratio of Na/1000, K/100, and √Mg

Springs	Na/1000	K/100	√Mg
1	0.00084	0.001	0.52
2	0.0879	0.2189	0.89
3	0.02978	0.1711	0.94
4	0.0886	0.001	0.74

Based on table 2, the water type was determined by using triangle diagram, generally the water under inspection is in the immature waters area. Although the composition of the most dominant ionic ratio of Na/1000 – K/100 - √Mg relatively difference, the global type are similar. It means that the hot water springs has contaminated by surface water.

3.3 Temperature under soil surface of springs

The calculation of under soil surface performed using Equation 1, Equation 2, and Equation 3 for geothermometer (SiO_2), geothermometer Na-K, and geothermometer Na-K–Ca, respectively. The SiO2 in Equation 1, Na, K in Equation 2 and Na, K, Ca in Equation 3 is the respective content in mg/L. The results of temperature calculation using Equation 1, 2, and 3 can be concluded that the closed reliable value of reservoir temperature are130.5°C. If the average temperature reduced gradient is 3°C per km, it could calculated that the surface temperature is about 69.75 °C and the reservoir temperature is 130.5 °C, the depth of the reservoir is 20.25 km from the surface.

Other important identity is the mineral containing in the underwater springs as well as the crystallinity. A small sample of the rock has just taken and inspected under X-RD machine using Cu-Kα wavelength. The main peaks are found to be at the position of 2θ = 22.56°, 29.9°, 31.46°, 35.94°, 48.61°, and 59.96°are fit with the database AMCSD 0000098 of Markgraf dan Reeder (1985). Those peaks exactly fit with calcite $CaCO_3$. The intensity and the full with at half maximum (FWHM) of the pattern may be described sample preparation or from the rock properties. It obtained that the rock is calcite, while the relative wide of FWHM is describing a relatively low crystallinity of the rocks. This result told us that the rocks may be not naturally built in relatively low pressure.

3.4 Geohydrology

By inspection of the surface temperature of the four samples, samples from spring 2, 3, and 4 are categorized a high temperature regime at about 79°C. This shows that these three hot springs are relatively close to the reservoir compared to the spring 1 sample. The water of spring 1 sample has undergone cooling in the distant stream through the rock gap.So that the surface temperature is very low compared to the other samples. The four samples can be said to have contaminated with surface water. This can be seen from the Na / 1000-K / 100 - $\sqrt{Mg}$ diagram.

The precipitation area is only about 20%. In this area, rainwater precipitates into the earth through the permeability of the rock (feed - zone) then flows into the lake Toba. The area under investigation is measured by the geothermal manifestation of about 2.2 km^2. To determine the potential of geothermal water as an alternative energy source based on Indonesian Geothermal Potential Standardization (DGSM, 1999) is$Q = 0,11585 \times A \times (T_{Res} - T_{cut\text{-}off})$ °C. By putting the cut-off temperature as 120 °C, the potency of expected geothermal of Rianiate is 2.68 Mwe.

3.5 Geophysics

Data retrieval in this research is done by using tools of geoelectric (resistivity meter), ARES G4 v4. 7 SN: 0609135 (Automatic Resistivity System), GPS (Global Position System) and compass on three tracks. Results Measurement and data processing on the three tracks are as follows:First trajectory (crater 2): apparent resistivity (ρa) varies considerably between 1.6 to 444 Ωm. Second trajectory (crater 1): apparent resistivity (ρa) varies considerably between 2.02 to 144.66 Ωm. The third trajectory (side along the highway): The apparent resistivity (ρa) varies greatly between 3.48 to 94.81 Ωm. These prove the underground structure is highly inhomogeneous.

4. Conclusions

Based on the results of data research processing was found that the apparent resistivity (ρ_s) vary from 1.62 Ωm to 444 Ωm. It is revealed that the underground soils are not homogeneous. The three types of geothermometer of reservoir temperature calculation were SiO_2 geothermometer. The expected temperature was 130.5°C. Based on the trilenear diagram of chloride, sulfate, and bicarbonate ions content, the Rianiate's hot water can be categorized ionic chloride's type. Further analyses of element ratio of Na/1000, K/100, and $\sqrt{Mg}$ itcould be concluded that the four hot springs are *immature water*area. It means that the water had contaminated by surface water. The rocks of Rianiate area is mostly calcite under XRD inspection. The low intensities of the XRD pattern indicating that the crystallinity of calcite is not high. We found also that the precipitation area just about 20%. The rains water precipitates into the earth through the feed zone of rock's permeability. The remaining water flows to Toba lake through small river which are laying the three craters. We also found that the reservoir of geothermal potency of Rianiate is 2.68 Mwe.

References

[1]. Kusumasari, B.A. Situmorang, J., Mussofan, W., Herdianita, N.R , Geology and Geochemistry of Hydrothermal System of West Toba (Indonesia): A Preliminary Study, Proceedings of the 2012 Australian Geothermal Energy Conference

[2]. M. Bayrak, Ü. Serpen, O.M. İlkışık, Two-dimensional resistivity imaging in the Kızıldere geothermal field by MT andDC methods, Journal of Volcanology and Geothermal Research 204 (2011) 1–11.

[3]. Gençer, E., Agrawal, R., 2016. A commentary on the US policies for efficient large scale renewable energy storage systems: Focus on carbon storage cycles. Energy Policy 88, 477–484.

[4]. Hazuardi.,(1992), *Pengantar Eksplorasi Panas Bumi,* PPT MIGAS (Pusat Tenaga Perminyakan dan Gas Bumi), Cepu.

[5]. Ralph, P.H., (1985), *Kimia Dasar Prinsip dan Terapan Modern*, Suminar Achmadi, Ph.D. (Penerjemah), Edisi keempat, jilid 2, Erlangga, Jakarta.

[6]. Situmorang, Cristina (2010), Perhitungan Suhu Reservoir Panas Bumi Rianiate Kecamatan Pangururan Menggunakan Persamaan Geothermometer Empiris, Skripsi, FMIPA, Unimed, Medan.

[7]. Markgraf S A, Reeder R J, American Mineralogist 70 (1985) 590-600, High-temperature structure refinements of calcite and magnesite, Sample: T = 24 C, _database_code_amcsd 0000984

Performance comparison of a refrigerator system using R134a and hydrocarbon refrigerant (HCR134a) with different expansion devices

A. Aziz[1], **Izzudin**[1] and **A K Mainil**[2]

[1]Department of Mechanical Engineering, Universitas Riau, Jl.Subrantas km 12,Pekanbaru 28293, Indonesia

[2]Department of Mechanical Engineering, Universitas Bengkulu, Jl. WR Supratman, Kandang Limun, Bengkulu 38371A, Indonesia

azridjal.aziz@lecturer.unri.ac.id

Abstract. The objective of this study is to compare the performance of refrigerator system using working fluid between R134a refrigerant and HCR134a as hydrocarbon refrigerant for substitution of R134a. The use of capillary tube (CT) 1.5 m with HCR134a showed that slightly better COP than among the others, due to the lower pressure of condenser, conversely thermostatic expansion valve (TEV) showed that better COP than among the others with R134a. COP of CT 1.25 m and CT 1.5 m using HCR134a increase about 42.89% and 18.09% compared to R134a, where the electric current of refrigerator system decrease about 11.63% and 10.98%. However, the COP of HCR134a with CT 2.7 m and TEV were obtained lower than R134a about 16.2% and 17.06% and the use of electric current is higher than R134a about 12.98% and 16.5%. The use of HCR134a provides a higher refrigeration effect than R134a about 66.71%-88.27% for various types of expansion devices. The results confirmed that HCR134a could be an alternative refrigerant for replacement of R134a refrigerant.

1. Introduction

The refrigerator systems generally operated using vapour compression refrigeration cycles. The refrigeration cycle has four main components: compressor, condenser, expansion device and evaporator. Refrigerant as the working fluid in the refrigeration cycle will absorb the heat of the evaporator so that the temperature lower than the surrounding temperature, then dissipate this heat of the condenser to surrounding where temperature around condenser higher than the surrounding temperature. The expansion device used to control the refrigerant flow from the high-pressure condensing side in condenser of the system into the low-pressure evaporating side in evaporator. There are two kinds of commonly used expansion device: capillary tubes (CT) and thermostatic expansion valves (TEV).

The pressure drop in the capillary pipe depends on the diameter and length of the pipe. Tube diameters of 0.8–2 mm with lengths of 1–4 m are common. Capillary pipes are usually made of copper with a small pipe diameter. Thermostatic expansion valve (TEV) operated such as embody of mechanism that will detect the superheat of refrigerant gas leaving the evaporator to the suction side of compressor [1]. The TEV used by maintain proper refrigerant flow works automatically depending upon the heat load of the evaporator. If the load on the evaporator increases, refrigerant will evaporate earlier and there will be more superheat at the phial position that charged with the same refrigerant as working fluid in the refrigerator system [1,2].The evaluation on the use of various expansion valves to the performance of refrigeration system have been presented by many researchers [3-6]. Jia and Lee [7] reported the experimental investigations on the use of CT and TEV in storage-enhanced heat recovery room air-conditioner, the results confirmed that TEV performs better performance than CT and preferred for use.

The problem of ozone depletion potential (ODP) and increased global warming potential (GWP) led scientists to evaluate more environmentally friendly refrigerants from chlorofluorocarbons (CFC) or hydro-fluorocarbons (HFC) to maintain environmental sustainability such as hydrocarbons (HC). Under the ASHRAE standard, HC refrigerants have flammable properties (A3) that are a deficiency of HC refrigerants. However, the advantages of HC are zero ODP, lowest GWP, non-toxic, and higher refrigerant performance than other refrigerants, and still in accordance with the existing refrigeration system as drop in substitute. HC refrigerants have been used in many cooling applications. Attention and secure handling of system leak for HC refrigerant is same as other refrigerant that has been used for safety reason of application [8-9].

Many researchers [10-13] have reported the better performance of hydrocarbon as alternative refrigerants to R134a, where the energy consumption of hydrocarbons (R436A and R600a) is reduced about 14.6% and 18.7% [14]. Shikalgar and Sapali [15] reported that COP of refrigeration systems with R290 as hydrocarbon refrigerant is comparable with R134a refrigerant and reducing starting torque of compressor including improvement the life span. Joybari et.al [16] conducted the energy analysis and optimization of R600a as a replacement of R134a in a domestic refrigerator system, the mass charge for R600a 66% lower than R134a and reduces the hydrocarbon refrigerant flammability risk. Chopraet al. [17] investigated the thermodynamic analyses of multiple evaporators vapour compression refrigeration systems for various refrigerants (R410a, R290, R1234YF, R502, R404a, R152a) including R134a and the result shows R152a better performances than other considered refrigerants.The use of additional coil for condenser that operated with CFC refrigerant shows better performance [18], while the use of evaporative cooling [19] can be improved refrigeration system performance.

This study is conducted to obtain the effect of different expansion devices (various length CT and TEV) on the performance comparison of refrigerator using R134a and hydrocarbon refrigerant (HCR134a) as substitute of R134a. The results presented in this research are the performance of refrigerator temperature achievement, pressure of evaporator and condenser, electric current consumption, refrigeration effect and COP. The experimental procedure performed will also be useful for future development of refrigerator system and energy saving.

2. Methods

The experimental method used in this study to conduct the performance comparison of a refrigerator system using R134a and hydrocarbon refrigerant (HCR134a) with different expansion devices (various length of CT and TEV). The experimental study was conducted in a refrigerator system test facility at Thermal Engineering Laboratory, Department of Mechanical Engineering, Universitas Riau. The refrigerator system test facility was fabricated for easy control of expansion device used in this study and for replacement of refrigerant as working fluid. This test facility originally designed for R134a refrigerant with 240 g charge and 120 Watt compressor power, where second test with HCR134a (hydrocarbon refrigerant) using 40% refrigerant charge (96 gram) compared to R134a.

Figure 1(a) shows the schematic diagram of refrigerator system test facility with points of data measurement.The expansion devices used in this test facility are three type of CT with length 1.25m, 1.5m, 2.7 m, and one type of TEV with bulb phial containing R134a refrigerant. Each expansion devices were equipped with a refrigerant valve, if the test was carried out using CT1.25m, the refrigerant valve of the CT 1.25m was opened while the refrigerant valve on the other expansion devices were closed. Figure 1(b) shows the isometric picture of refrigerator system test facility.The components of refrigerator system test facility as shown in figure 1(b)[20] consist of compressor (1), evaporator (2), condenser (3), refrigerated box (4), fan (5), pressure gauge (6), filter drier (7), TEV (8), CT (9), thermometer (10), temperature controller (11), on/off switch (12) and steel frame (13). The data collection was captured every 5 minutes during 2 hours. The data collected from the test using measurement devices are temperature, pressure, and electric current consumption. The calculations performed using the principles of thermodynamics [20] and the thermodynamic properties of the refrigerant were taken using software of REFPROP ver. 8.

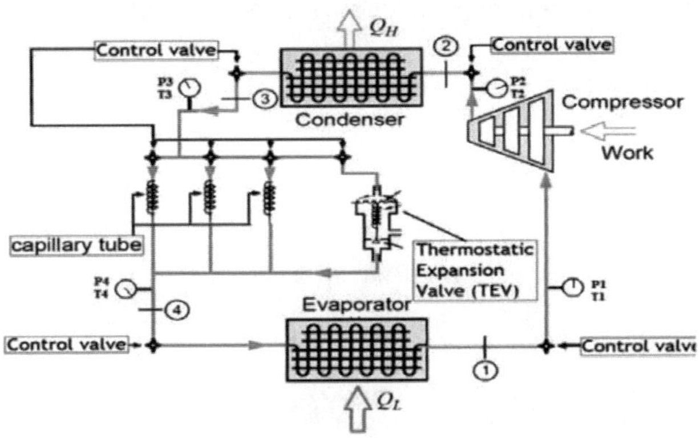

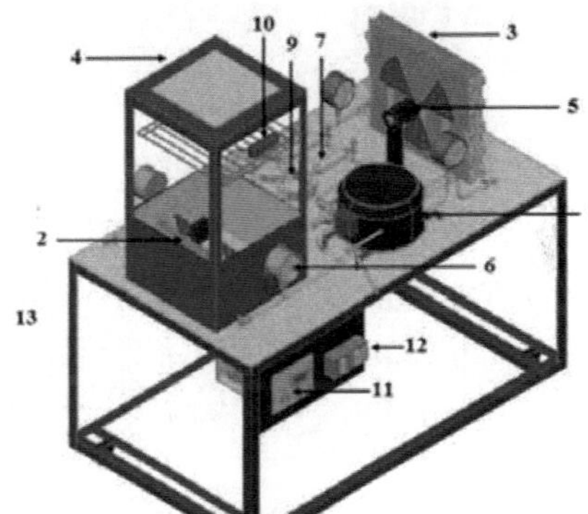

(a) (b)

Figure 1.(a) Schematic diagram of refrigerator system test facility that modified from [20] and **(b)** Isometric picture of refrigerator system test facility [21]

3. Results and Discussion

Figure 2 shows the average box temperature of refrigerator system comparison between HCR134a and R134a at different expansion devices. The average box temperatures with CT 1.25 m and CT 1.5 m using HCR134a compared to R134a tend to be similar with slightly differences as shown in figure 2. This is because the evaporator temperature with CT 1.25 m and CT 1.5 m using HCR134a compared to R134a also slightly lower than the evaporator temperature with CT 1.25 m and CT 1.5 m as shown in figure 2.While the average box temperature with CT 2.7 m and KET using HCR134a are higher than R134a, because the condenser pressure using HCR134a also higher than R134a with CT 2.7 m and KE as shown in figure 4. Base on the proportional relationship between pressure and temperature, when the pressure of compressor increase the temperature of compressor increasing too [20].

As seen in figure 3, the average evaporator pressure with HCR134a for all expansion devices used in this study (CT 1.25 m, CT 1.5 m, CT2.7 m and TEV) higher than average evaporator pressure with the R134a.This is because the latent heat as a part of thermodynamic properties of HCR134a higher than latent heat of R134a, so that the amount of heat that can be absorbed in evaporator by more refrigerants. The higher the amount of heat absorbed in the evaporator the temperature of the refrigerant will be higher too where the evaporator pressure becomes higher too.

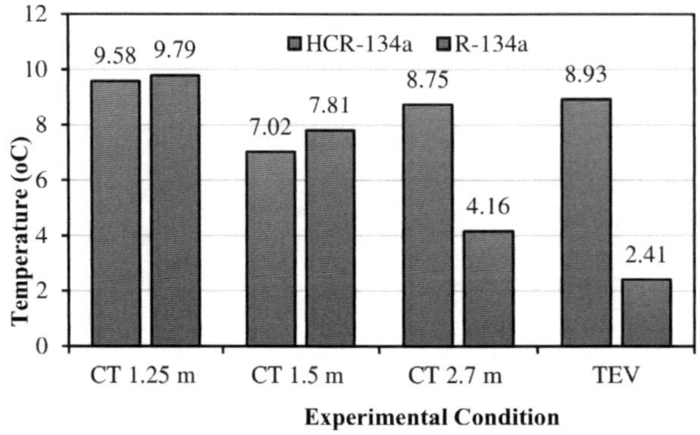

Figure 2. Average box temperature at different expansion devices comparison

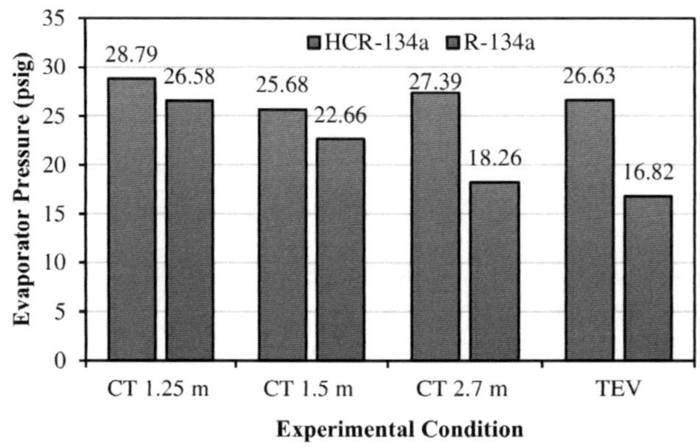

Figure 3. Average evaporator pressure at different expansion devices comparison

Figure 4 shows the average pressure of the condenser with different expansion devices using HCR13a refrigerant and R134a. It can be seen that the condenser pressure using HCR134a is lower than R134a for the capillary tubes of 1.25 m and 1.5 m, while for the CT 2.7 m and KET, condenser pressure using HCR134a is higher than R134a.

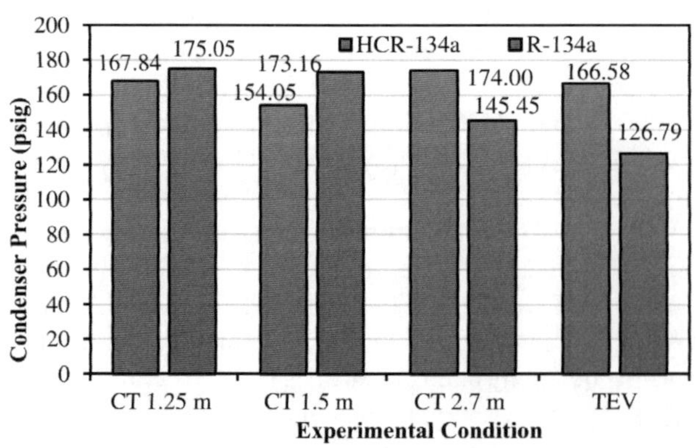

Figure 4. Average condenser pressure at different expansion devices comparison

Figure 5 shows the comparison of average of electric current consumption of compressor usingHCR134a and R134a refrigerant with different expansion devices. The experimental result shows that the electric currentconsumption of compressor using refrigerant HCR134a is lower than R134a about 10.8% and 9.7% for capillary tube 1.25 m and 1.5 m, so that the condenser pressure using HCR34a refrigerant become lower than R134a for testing with CT 1.25 m and 1.5 m where reducing the compressor work due to the lower of electric current consumption using HCR134a. Conversely in the test with CT 2.7 m and KET, the compressor electric current consumption using HCR134a refrigerant are higher about 12.6% and 16.8% respectively than using R134a. This is because the condenser pressure using HCR134a is higher than R134a refrigerant with CT 2.7 m and KET, so the compressor work become heavier and increasing of the electric current consumption using HCR134a refrigerant.

Figure 6 shows a comparison of the refrigeration effect between HCR134a refrigerant and R134a refrigerant with different expansion devices. The analysis of test results obtained that refrigeration effect using HCR134a refrigerant higher than R134a refrigerant with different expansion devices, this is due to the latent heat value of HCR134a refrigerant is higher than refrigerant R134a, which means that amount of heat that can be absorbed by HCR134a refrigerant on evaporator is higher and produce a larger refrigeration effect.

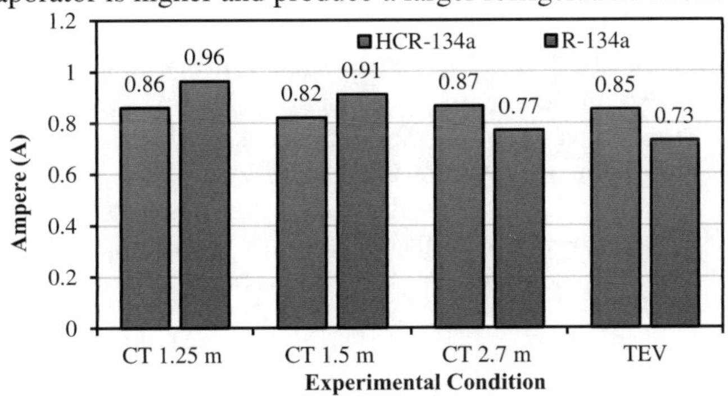

Figure 5. Average electric current consumption at different expansion devices comparison

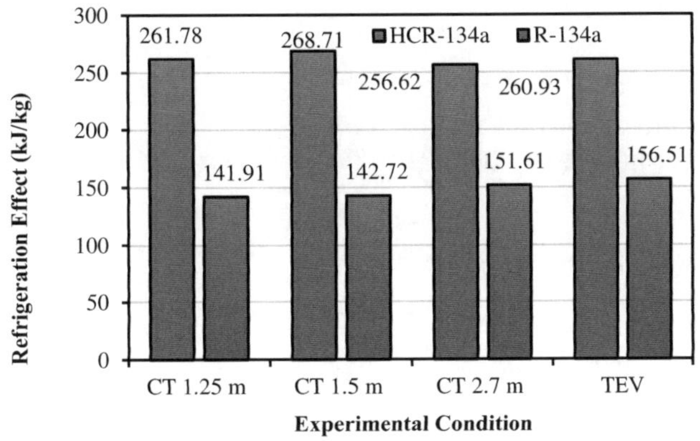

Figure 6. Average refrigeration effect at different expansion devices comparison

Figure 7 compares the COP between HCR134a refrigerant and R134a with different expansion device. It can be seen that COP using HCR134a gives higher result with capillary tube 1.5 m and 1.25 m, this is due to the HCR134a refrigerant has higher refrigeration effect than R134a refrigerant. However, contrary the COP of refrigerator system using HCR134a refrigerant is lower than R134a refrigerant in with CT 2.7 m tube and KET.

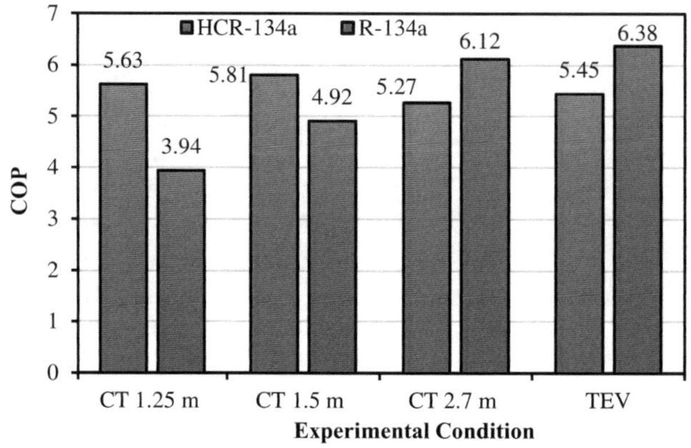

Figure 7. Average COP at different expansion devices comparison

4. Conclusions

The use of different length of CT (1.25 m, 1.5 m, and 2.7m) and TEV in refrigerator system using HCR134a (hydrocarbon refrigerant) compared to R134a refrigerant by experimental investigations. The test result was showed that TEV as expansion device system could better regulate the refrigerant flow than CT where COP of TEV higher than CT. It was found that HCR134a as hydrocarbon refrigerant showed increase the performance with CT 1.25 m and 1.5 m, but decrease the performance with CT 2.7 m and KET. HCR134a hydrocarbon refrigerant could be an alternative refrigerant for replacement the existing R134a refrigerant.

Acknowledgments

The present study was financially supported by Universitas Riau and Ministry of Research, Technology and Higher Education (Penelitian Unggulan Perguruan Tinggi, DRPM, Desentralisasi Research Grant 2017).

References

[1] HundyGF, Trott AR and Welch TC 2008*Refrigeration and Air-Conditioning*4th (London: Butterworth-Heinemann) p 103
[2] Miller R and Miller MR 2006 *Air Conditioning and Refrigeration (*New York: McGraw-Hill) p 300
[3] Sulaimon S, Nasution H, Aziz AA, Abdul-Rahman AH and Darus AN 2014 *J. Eng. Technol. Sci.***46** 141
[4] Choi JM and Kim YC 2002 *Energy***27** 391
[5] Peng JW, Li H and Zhang C L 2016 *Appl. Therm. Eng.***99** 1190
[6] Aziz A, Mainil RI, Mainil AK and Saputra E 2017*AIP Conf. Proc.***1788** 030023
[7] Jia J and Lee WL2015 *Energy Build.***101** 76
[8] Dalkilic AS and Wongwises S 2010 *Int. Commun. Heat Mass Transf.***37** 1340
[9] Harby K 2017 *Renew. Sustain. Energy Rev.* **73** 1247
[10] Saravanakumar R and Selladurai V 2013 *J. Therm. Anal. Calorim.***9** 8

[11] El-Morsi M 2015 *Energy***86** 344
[12] Pendyala S, Prattipati R and Raju AVSR 2017 *Int. J. AirCond.***25**1750019
[13] Thakar SM, Prajapati RP and Solanki DC 2017 *IOSR J. Mech. Civ. Eng.***14** 92
[14] Rasti M, Aghamiri S and Hatamipour MS 2013 *Int. J. Therm. Sci.***7**486
[15] Shikalgar ND and Sapali SN 2017 *Energy Procedia***109** 34
[16] Joybari MM, Hatamipour MS, Rahimi A and Modarres FG 2013 *Int. J. Refrig.***36** 1233
[17] Chopra K, Sahni V and Mishra RS 2014 *Int. J. Air-Conditioning Refrig.***22** 1450003
[18] Aziz A, Satria AB and Mainil RI 2015 *Int. J. Automot. Mech. Eng.***12** 3043
[19] Aziz A, Mainil RI, Mainil AK and Listiono H 2017 *AIP Conf. Proc.***1788**030024
[20] Borgnakke C and Sonntag RE 2009 *Fundamentals of Thermodynamics* 7th(New Jersey: John Wiley &Sons) p 450
[21] Hartanto BH 2014 *Expansion Device Effect on The Performance of Vapor Compression Refrigeration System*(in Indonesian) Bachelor thesis (Pekanbaru: Universitas Riau) p 29

Mechanical behaviour of selected bulk oilseeds under compression loading

Č. Mizera[1]*, D. Herák[1], P. Hrabě[2], Z. Aleš[3] and J. Pavlů[3]
[1]Department of Mechanical Engineering, Faculty of Engineering, Czech University of Life Sciences Prague, Prague, Czech Republic
[2]Department of Material Science and Manufacturing Technology, Faculty of Engineering, Czech University of Life Sciences Prague, Czech Republic
[3]Department for Quality and Dependability of Machines, Faculty of Engineering, Czech University of Life Sciences Prague, Kamýcká 129, CZ 16500 Prague, Czech Republic

* mizera@tf.czu.cz

Abstract. Pressing of vegetable oils plays an important role in modern agriculture. This study was focused on the linear pressing of soybean seeds (*Glycine max L.*), Jatropha seeds (*Jatropha curcas L.*) and palm kernel (*Elaeisguineensis*). For pressing test the compressive device (ZDM, model 50, Germany) was used. The maximum pressing force of 100 kN with a compression speed of 1 mm s^{-1} was used to record the force–deformation characteristics. The pressing vessel with diameter 60 mm and initial height of seeds 80 mm were used. The specific energy per gram of oil of soybean, palm kernel and Jatropha was 158.92 ± 7.21, 128.78 ± 8.36 and 68.26 ± 5.94 J.goil-1, respectively. The oil content of soybean, palm kernel and Jatropha was 20.4 ± 1.23, 44.7 ± 2.27 and 34.2 ± 1.75 %, respectively. Water concentration, dynamic and kinematic viscosity of obtained oils was also determined.

1. Introduction

Production of vegetable oil is one of the most important part of modern agriculture. Vegetable oils have been used for centauries for food and non-food applications [1]. A large part of obtained oil is edible oils, which are used directly in food industry as follows: mayonnaise, margarine, cooking fats etc. [2]. Separation of oil from oilseeds is an important processing operation. The process has a significant effect on the quality and quantity of protein and oil obtained from the oilseeds [3]. One of the most common methods of obtaining oil is mechanical pressing. This method is based on mechanical compression of oleaginous materials. The vegetable oil obtained via this method is characterized by high quality and can be used for special applications [4]. The most commonly used equipment for mechanical pressing is screw press. This method is the most popular oil extraction as the process is simple, flexible, continues and safe [5]. For designing screw pressing technology is necessary to know the basic mechanical behaviour of bulk oilseeds by linear pressing. For this study soybean (*Glycine max L.*), Jatropha (*Jatropha curcas L.*) and palm kernel (*Elaeisguineensis*) were chosen, because this seeds represent the main oilseeds in Indonesia. Some of the pressing parameters of this seeds have already been determined [2; 3; 6; 7; 8; 9]. The aim of this study was to determine the pressing curves, specific energy, amount of oil and initial slope of deformation curve of selected oilseeds. The content of water in oil and viscosity of the obtained oil were also determined.

2. Materials and Methods

Whole seeds of soybean (*Glycine max L.*), Jatropha (*Jatropha curcas L.*) and palm kernel (*Elaeisguineensis*) obtained from North Sumatra, Indonesia were used for the compression test. The moisture content, which was determined using standard oven method, ASAE method (ASAE S410.1 DEC97, ASAE, 1998), for soybean, Jatropha and palm kernel was 7, 8; 8,9 and 6,2 %, respectively. The maximum quantities of oil present in each batch of the oilseeds used for this study were determined using the Soxhlet extraction technique. This was done according to the CSN method (CSN EN ISO 659 – 46 1034, CSN, 2011).

2.1. Compression test

To determine the relationship between the pressing force and deformation, the compressive device (ZDM, model 50, Germany) was used to record the course of deformation function. The pressing vessel and plunger (Fig. 1) of diameter 60 mm were used. The bulk seeds of initial height 80 mm were measured and pressed at the rate of 1 mm.s-1 under room temperature. The measuring range of force was between 0 kN and 100 kN. The experiment was repeated three times for each type of seeds.

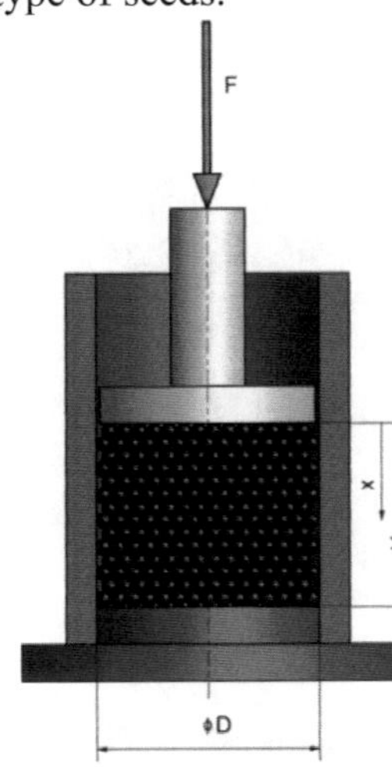

Figure1. Scheme of pressing vessel

The volume energy is the area under the compressive force-deformation curve and was calculated using Eq. 1.

$$E_x = \sum_{n=0}^{n=i-1}\left[\left(\frac{F_{n+1}+F_n}{2}\right)\cdot(X_{n+1}-X_n)\right] \quad (1)$$

where: E_x – deformation energy, J; F_{n+1} and F_n–values of compressive force, N; X_{n+1} and X_n– values of seed deformation, m;

The initial slope of deformation curve was determined as the slope of the tangent at zero time.

2.2. Oil analysis

The Soxhlet apparatus was used to determine the oil content of seeds. For Soxhlet extraction 10 g of find ground seeds was extracted with petroleum in a Soxhlet apparatus for 6 h at 60°C.

The Coulometer WTD developed to perform routine analysis of moisture, was used to measure the content of water in given oil samples. Employing the coulometric method allows detection of small or even trace amounts of water in a variety of solutions; including, but not limited to, solutions based on organic acids, alcohols, esters, hydrocarbons and many other organic solvents.

Stabinger Viscometer was used for determination of viscosity. This analyser covers a wide measuring range with one measuring cell and provides precision comparable to gravimetric capillary viscometers.

3. Results and discussion

Measured data of individual pressing curve for soybean, Jatropha seed and palm kernel are shown in Fig. 2. The compression test was repeated three times for each type of seeds.

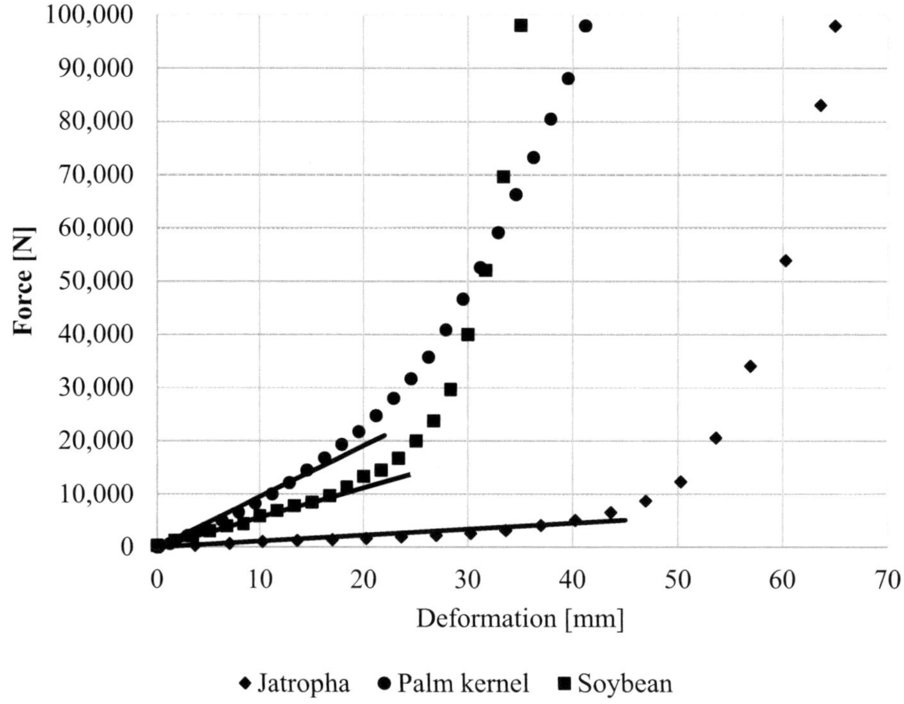

Figure 2. Measured amounts of pressing characteristic of different type of oilseeds

The oil content of seeds was determined by Soxhlet apparatus and is shown in Table 1. The parameters of mechanical pressing of the 3 oilseeds are presented in Table 1. Deformation at peak load 100 kN ranged between 34-35, 41-43 and 64-66 mm in soybean, palm kernel and Jatropha seeds. The deformation was significantly higher in Jatropha seeds than in the other seeds (Table. 1).

Table 1. Mechanical parameters of the bulk oilseeds loaded compressively to 100 kN

	Soybean	**Palm kernel**	**Jatropha**
Oil content [%]	20.4 ± 1.23	44.7 ± 2.27	34.2 ± 1.75
Initial mass [g]	167.1 ± 8.87	152.3 ± 6.52	77.2 ± 3.15
Gained oil [g]	4.2 ± 0.22	10.5 ± 0.69	10.65 ± 0.56
Initial slope of deformation [$N.m^{-1}$]	588.57 ± 28.38	736.67 ± 44.63	99.96 ± 4.58
Strain [-]	0.44 ± 0.03	0.53 ± 0.02	0.81 ± 0.04
Deformation energy [J]	670.56 ± 32.22	1237.64 ± 88.48	726.99 ± 45.51
Specific energy [$J.g_{oil}^{-1}$]	158.92 ± 7.21	128.78 ± 8.36	68.26 ± 5.94

(mean ± standard deviation)

Strain in bulk seeds of soybean, palm kernel and Jatropha seeds at a peak compressive force of 100 kN was 0.44, 0.53 and 0.81, respectively. Mean deformation energy at peak compression was 670.56, 1237.64 and 726.99 J, for 80 mm initial height of soybean, palm kernel and Jatropha seeds, respectively. The energy demands for the deformation of soybean and Jatropha seeds at peak load were significantly less than the energy requirement for the deformation of palm kernel at the same load (Table 1).Similar mechanical properties of the Jatropha seeds were also found by other authors [7; 8; 10].The oil obtained from individual seeds is also shown in Table 1. From these values the specific energy per gram of oil was determined. From Table 1 is evident that the energy demand to obtain 1 g of oil in the case of Jatropha seeds was significantly less than the energy requirement for the other seeds.Mechanical properties of soybean were also determined by other authors [11; 12]. Very similar results of pressing palm were also found by other authors [13, 14].

The measured values of water concentration and viscosity of oils are shown in Fig. 3 and Fig. 4.

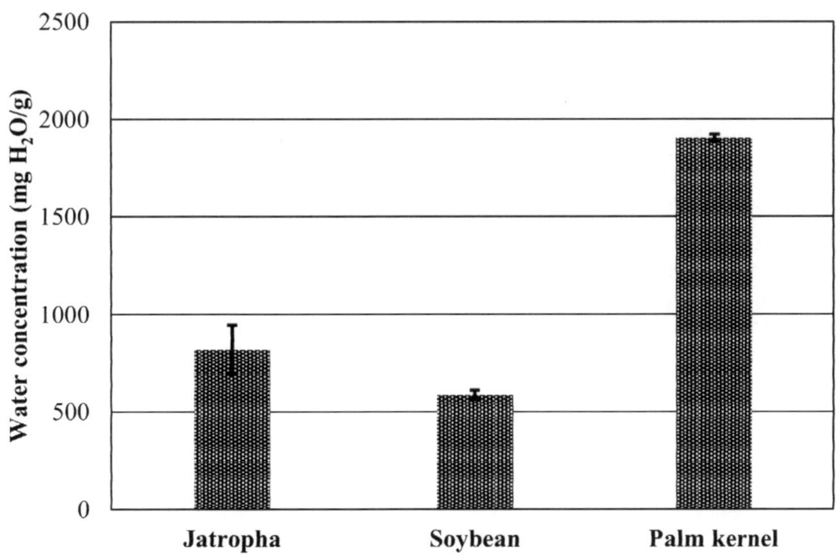

Figure 3. Water concentration of analysed oils

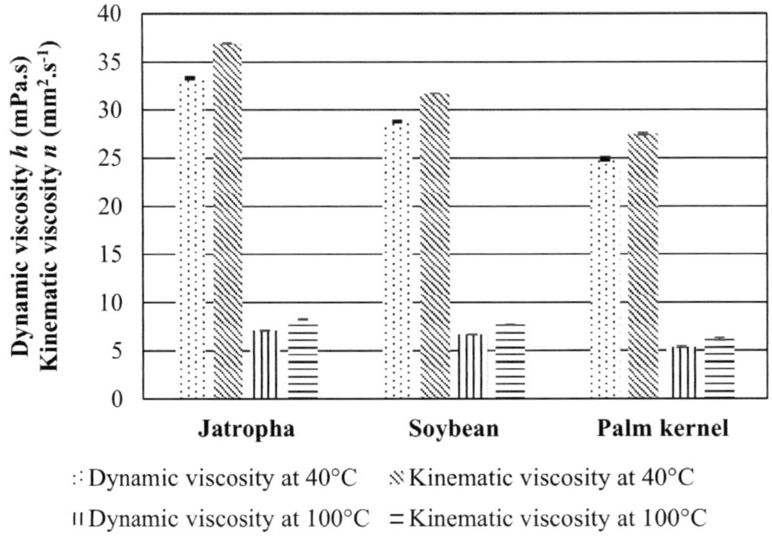

Figure 4. Dynamic and kinematic viscosity at 40°C and 100°C of analysed oils

It is obvious that the highest concentration of water was measured in palm kernel oil sample in Figure 3. Water content has also impact on value of viscosity. The lowest value of viscosity was measured within palm kernel shown in Figure 4. Some results of soybean oil has been already determined.

4. Conclusion

This study was focussed on the Mechanical behaviour of soybean, palm kernel and Jatropha seeds under compression loading. Individual pressing curves of seeds were determined and analysed. Results indicate that the specific energy per gram of oil at Jatropha seeds was approximately two times lower than the energy requirements for other seeds. The oil content of soybean, palm kernel and Jatropha seeds by Soxhlet extraction was also determined. Oil analysis of obtained oils such as water concentration, dynamic and kinematic viscosity was also determined. The results obtained in this study can help with the design of the pressing technology.

Acknowledgment

This paper has been supported by Internal Grant Agency of Faculty of Engineering – Czech University of Life Sciences Prague – IGA 2017: 31130/1312/3111.

References

[1] Hasenhuettl G L 1991Fats and fatty oils, in *Kirk-Othmer's Encyclopedia of Chemical Technology*, J.I. Kroschwitz and M. Howe-Grant (Eds.), pp. 282-285. New York: Wiley & Sons. 10: 252 – 287.

[2] Bargale P C 1997Mechanical oil expression from selected oilseeds under uniaxial compression, Ph.D. Thesis, University of Saskatchewan, Canada

[3] Bargale P C and Patil R T 2007 Effect of pretreatments on oil recovery and oil-point

pressure of soybean under uniaxial pressing. Journal of Food Science and Technology 44(5) pp 478-482

[4] Willems P, Kuipers N J M and de Haan A 2008 Hydraulic pressingof oilseeds: experimental determination and modeling of yield andpressing rates. *Journal of Food Engineering* 89(1) pp 8-16

[5] Beerens P, 2007 Screw-pressing of Jatropha seeds for fuelling purposes in less developed countries, Ph.D. Thesis, Eindhoven University of Technology,Netherlands

[6] Ajibola O O., Okunade D A and Owolarafe O K 2002 Oil Point Pressure of Soybean. *Journal Food Process Engineering* 25 pp 407-416

[7] Herak D, Kabutey A, Divisova M and Simanjuntak S 2013 Mathematical model of mechanical behaviour of Jatropha curcas L. seeds under compression loading. *Biosystems Engineering* 114 pp 279-288

[8] Kabutey A., Herák D., Chotěborský R., Sigalingging R. and Mizera C 2015 Effect of compression speed on energy requirement and oil yield of Jatropha curcas L. bulk seeds under linear compression. *Biosystems Engineering* 136 pp 8-13

[9] Herak D, Kabutey A, Choteborsky R, Petru M and Sigalingging R 2015 Mathematical models describing the relaxation behaviour of Jatropha curcas L. bulk seeds under axial compression. *Biosystems Engineering* 131 pp 77-83

[10] Sigalingging R, Herák D, Kabutey A, Mizera Č and Divišová M 2014 Tangent curve function description of mechanical behaviour of bulk oilseeds. *Scientia Agriculturae Bohemica* 4 (45) pp 259-264.

[11] Dijkstra A J., 2015 Soybean Oil. *Encyclopedia of Food and Health* 18 pp 58-63

[12] Lee H, Min S C 2014 Development of antimicrobial defatted soybean meal-based edible films incorporating the lactoperoxidase system by heat pressing. Journal of Food Engineering 120 pp 183-190

[13] Akinoso R., Raji A O., Igbeka J C 2009 Effects of compressive stress, feeding rate and speed of rotation on palm kernel oil yield. *Journal of Food Engineering* 93 pp 427-430

[14] Owolarafe O K., Osunleke A S., Odejobi O A., Ajadi S O., Faborode M O 2008 Mathematical modelling and simulation of the hydraulic expression of oil from oil palm fruit. *Biosystems Engineering* 101 pp 331-340

IOP Conf. Series: Materials Science and Engineering **237** (2017) 012010 doi:10.1088/1757-899X/237/1/012010

Effect of heat recovery water heater system on the performance of residential split air conditioner using hydrocarbon refrigerant(HCR22)

AAziz[1], Thalal[1], I Amri[2], Herisiswanto[1]and A K Mainil[3]
[1]Department of Mechanical Engineering, Universitas Riau,
[2]Department of Chemical Engineering, Universitas Riau,
Jl.Subrantas km 12,Pekanbaru 28293, Indonesia
[3]Department of Mechanical Engineering, Universitas Bengkulu,
Jl. WR Supratman, KandangLimun, Bengkulu 38371A, Indonesia

azridjal.aziz@lecturer.unri.ac.id

Abstract. This This paper presents the performance of residential split air conditioner (RSAC) using hydrocarbon refrigerant (HCR22) as the effect on the use of heat recovery water heater system (HRWHS). In this study, RSAC was modified with addition of dummy condenser (trombone coil type) as heat recovery water heater system (HRWHS). This HRWHS is installed between a compressor and a condenser by absorbing a part of condenser waste heat. The results show that RSAC with HRWHS is adequate to generate hot water with the temperature range about 46.58°C - 48.81°C when compared to without HRWHS and the use of dummy condenser does not give significant effect to the split air conditioner performance. When the use of HRWHS, the refrigerant charge has increase about 19.05%, the compressor power consumption has slightly increase about 1.42% where cooling capacity almost the same with slightly different about 0.39%. The condenser heat rejection is lower about 2.68% and the COP has slightly increased about 1.05% when compared to without HRWHS. The use of HRWHS provide free hot water, it means there is energy saving for heating water without negative impact to the system performance of RSAC.

1. Introduction

One of the ways to improve energy efficiency is recovery or reuse of waste heat energy from the energy conversion device like residential split air conditioner (RSAC). The most commonly reason was used in an attempt to modify the refrigeration system is energy efficiency. Refrigeration system that serves as air conditioning machine is commonly used to achieve comfortable zone by utilizing the cooling effect of the evaporator for human activity in home, office or another building. Refrigeration system is one type of energy conversion machines, where the amount of energy is needed to produce a cooling effect. Vice versa, the amount of energy as heat rejected by the system into the surrounding tomeet the principles of thermodynamics. In thermodynamics energy balance analysis, the heat energy rejected (heating effect) from the condenser are combination of compressor work and cooling effect energyin the evaporator. It means that heat energy rejection in the condenser is bigger than cooling effect in evaporator, so that some of this energy has the potential to be recovered for heating water. There is saving energy by using it to produce hot water [1-4].

The objective of refrigerator is to maintain refrigerated room at low temperature by remove heat from it. Rejected heat to higher temperature medium is only part of the operation to meet the principles of thermodynamics equilibrium, not the purpose. Vice versa, the objective of heat pump is to maintain heated room at high temperature by absorbing heat from low-temperature sources such as water or ambient air in winter, and this heat is supplied to a warmer zone such as a house [5]. The objective of hybrid refrigeration device is to maintain a refrigerated room at low temperature by remove heat (Q_L) from it and at the same time maintain heated zone of high temperature and rejected some of heat to higher temperature to meet the principles of thermodynamics equilibrium (Q_H).

The ozone depleting potential (ODP) and global warming potential (GWP) have become the most important criteria for choosing a refrigerant that will be used in the refrigeration system [6]. The most of the refrigeration systems operate with the refrigerant R22. The long-term alternatives for R22 include hydro fluorocarbon (HFC) refrigerants [7]. HFC have a significant GWP and could be replaced by hydrocarbon (HC) refrigerants in most of refrigeration systems. The use of HC refrigerants has an environmental advantage of a greatly reduced GWP when it is compared with HFC refrigerants [8]. Many researcher focused in the research intosubstitutes for R22 with hydrocarbon refrigerant (HCR22) at low and medium condensing temperatures [9-11].

Several studies of waste heat recovery from condenser in air conditioning for water heating has been performed by many researchers, including applying heat recovery process[12-14], heat recovery technique and economics analysis [15-16]. The use of evaporative cooling [17] can be applied to improve RSAC performance using HRWHS.Wang et al. [18] conductedabout experimental investigation on a conventional airconditionerworking as air-water heat pump. The results are shown that COP got 10% improvement and the average temperature of tank water affects COP of the equipment working as air-water heatpump significantly.

The previous literature shows that most of studies have been performed on residential air-conditioning integrated with water heater system using R22 as the refrigerant and their components. But there has little reports presented about the performance of HCR22 refrigerant used in residential air conditioning integrated water heater system with trombone coil dummy condenser as heat exchangers which are widely applied in present residential air-conditioning water heater system or heat pump.In this paper, a performance comparison of residential split air conditioner (RSAC) with and without heat recovery system using hydrocarbon refrigerant (HCR22) were investigated.Some of the parameters beinganalyzed and discussed include: the cooling capacity, COP, heat rejection, compressor power, condensing pressure, evaporatingpressure, test room temperature and hot water temperature.

2. Methods

The performance comparison as effect of heat recovery water heater system is evaluated using the similar experimental apparatus performed byAzridjal et al. [14]. The schematic diagram of the experimental apparatus is shown in figure 1[19]. The experimental apparatus is modified from RSAC with cooling capacity 2.6 kW. The trombone coil type dummy condenser as heat exchanger for heating water was made up from 3/8 inch diameter of copper pipe, with 5 windings and 5.5 meters length [14]. The modified RSAC has two mode of operation: AC original mode without HRWHS or cooling mode from original AC and RSAC with HRWHS mode(simultaneously cooling and heating mode). This experimental apparatus is equipped with 3 valves (valve 2, 2a and 2b), so that the apparatus can serve as AC original mode or ACWH mode by set up the valves. The working fluid in this experimental apparatus is HCR22. The HCR22 refrigerant mass circulating in original AC mode (without HRWHS) and RSAC with HRWHS mode are 252 gram and 300 gram respectively.

The test room facility with size 2.26 x 1.75 x 2 m (length x width x height) is equipped with 30 pieces of 100 Watt incandescent lamp as the cooling load to indoor unit (evaporator) with variation of 0 W, 1000 W, 2000 W, and 3000 W at 19°C room temperature reference setting. The first experimental test was started with original AC mode by opening valve 2 and closing valves 2a and 2b. Then, the second test was conducted by closing valve 2 and opening the valves 2a and 2b as RSAC with HRWHS mode. The data werecollected every 5 minutes during 2 hours of testing for cooling load variation of 0 W, 1000 W, 2000 W, and 3000 W. The temperatures from apparatus were measured by K-type thermocouple using data acquisition module with accuracy 0.2% (± 0.5°C) and has a resolution 0.1°C. The pressures were measured by bourdon type pressure gauge in condenser side (high pressure) and evaporator side (low pressure) with accuracy ±5 psi and ±1 psi, respectively. The electric current input and voltage input were measured by ampere-meter (accuracy ± 2.0 percent and 3 digits) and voltmeter (accuracy ± 1.0 percent and 3 digits), respectively. The charge of refrigerant mass was measured by digital weight scale with accuracy ±10 gram. The calculation of system performance was done by assuming ideal vapor compression cycle, where the enthalpy is calculated based on the high pressure and low pressure [5]. The thermodynamics properties calculation was carried out by software of REFPROP ver. 6.01 base on pressure data collection. Data without heat recovery as comparison in this study was used data from previous publication [20].

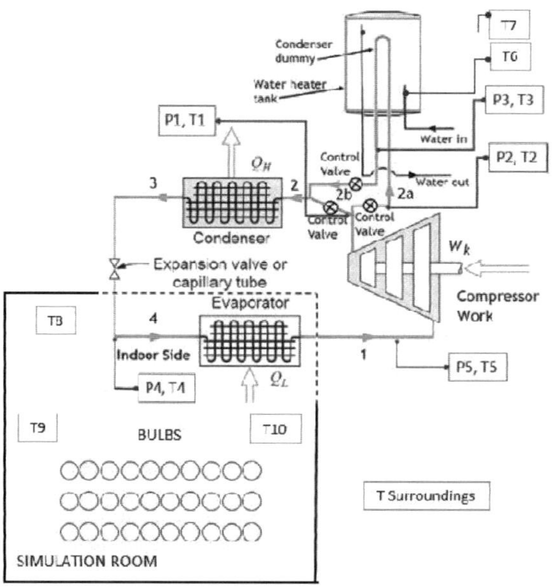

Figure 1. The schematic of the experimental apparatus [20].

3. Results and Discussion

The compressor power at different cooling load with and without HRWHSis shown in figure 2(a). While figure 2(b)shows the average pressure of the compressor inlet (low pressure) and outlet (high pressure) with and without HRWHSat different the cooling loads from 0W, 1000W, 2000W and 3000W. It can be seen from figure 2(b) that the compressor pressure differences with and without HRWHS about 8.9% and 6.8% at the compressor inlet and outlet respectively.

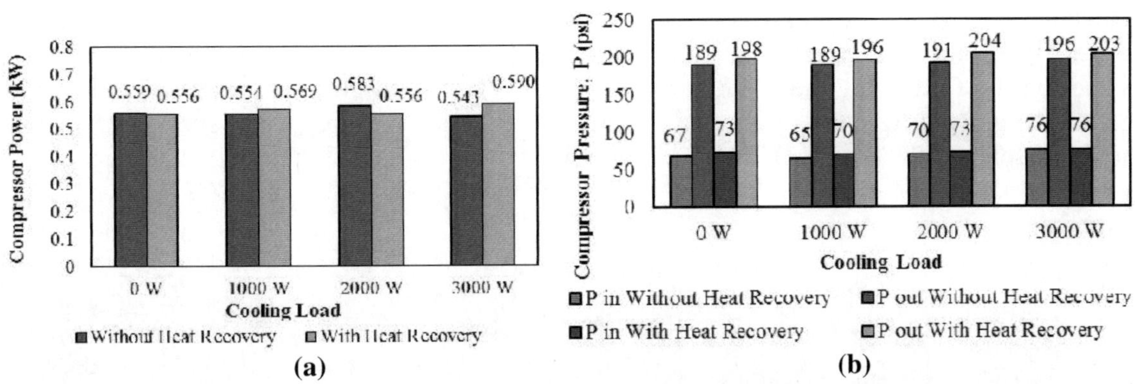

Figure 2. Compressor power at different cooling loads with and without HRWHS (a) and compressor pressure at different cooling loads with and without HRWHS (b).

Figure 3(a) shows the cooling capacity for two hours operation of system with and without HRWHS in different cooling load. In general, there was no significant difference in the variation of the cooling load against the cooling capacity with or without heat recovery HRWHS as seen in figure 3(a). It is because, the cooling capacity with HRWHS decrease only about 0.39% (very small portion) in different of cooling load from low to high. It can be stated that the use of heat recovery system for utilization hot water not affecting the cooling capacity of RSAC. At the simultaneously operation for cooling and heating, the free hot water was obtained from the use of HRWHS.

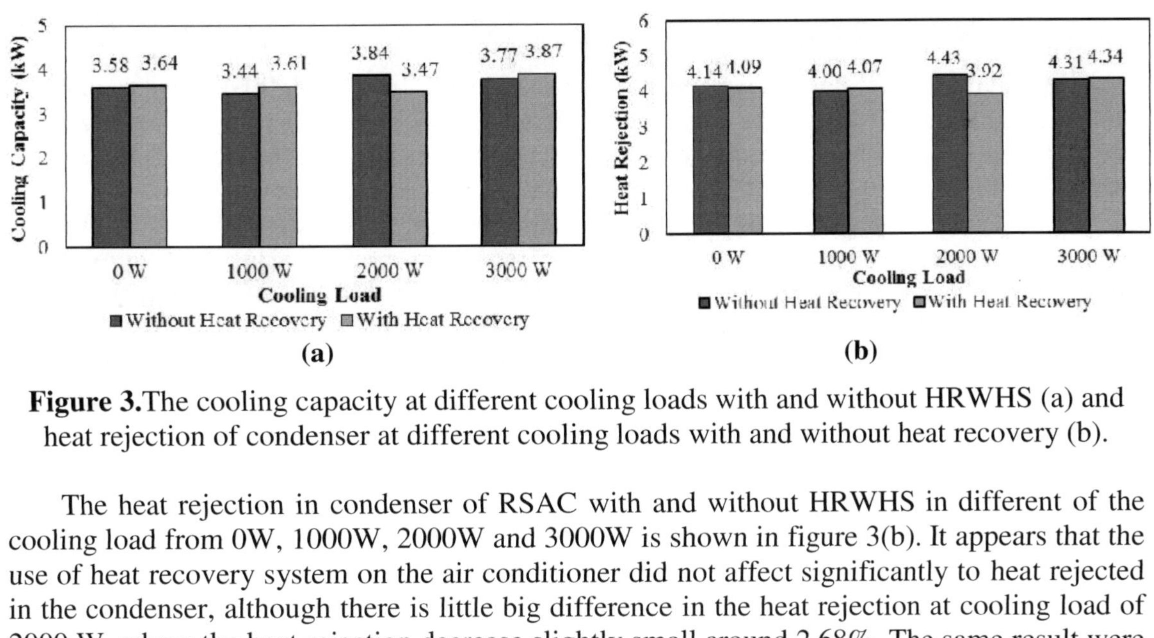

Figure 3. The cooling capacity at different cooling loads with and without HRWHS (a) and heat rejection of condenser at different cooling loads with and without heat recovery (b).

The heat rejection in condenser of RSAC with and without HRWHS in different of the cooling load from 0W, 1000W, 2000W and 3000W is shown in figure 3(b). It appears that the use of heat recovery system on the air conditioner did not affect significantly to heat rejected in the condenser, although there is little big difference in the heat rejection at cooling load of 2000 W, where the heat rejection decrease slightly small around 2.68%. The same result were also obtained for compressorpower,compressorpressure and cooling capacity as shown in figure 2(a), figure 2(b) and figure 3(a) respectively.

The COP in condenser of RSAC with and without HRWHS in different of the cooling load from 0W, 1000W, 2000W and 3000W is shown in figure 4(a), while COP_R(COP as refrigerator/cooling) and COP_{HP}(COP as heat pump/heating) are COP for cooling and heating, respectively. The similar results are also shown in figure 4(a) when it compared with another figure (figure 2(a), figure 2(b) and figure 3(a) and figure 3(b)), where there is no significant impact on the COP with and without recovery HRWHS at different cooling load. Although there is slightly difference of the COP with and without heat recovery system, the COP increase around 1.05% compared to without heat recovery system (dummy condenser), but in general the difference is very small so there is no significant impact on COP of the RSAC system.

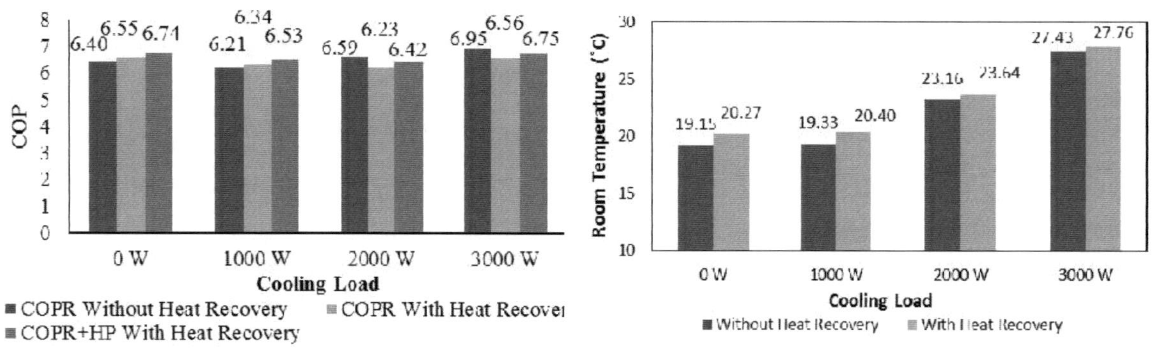

(a) (b)

Figure 4. COP of RSAC system at different cooling loads with and without HRWHS (a) and the room temperature (indoor) of RSAC system at different cooling loads with and without HRWHS (b).

The room temperature (indoor) of RSAC with and without heat recovery system in different cooling load from 0W, 1000W, 2000W and 3000W is shown in figure 4(b). As shown in figure 4(b), the increasing of cooling loads from 0 W to 3000 W causes that the room temperatures increases. It is because the higher the cooling load applied, the longer the time required to reach the reference temperature, so for cooling time 2 hours, then only the smallest cooling load (0 W and 1000 W) which can reach temperatures of reference (19°C).

The hot water temperature of RSAC with and without heat recovery system in variation of the cooling load from 0W, 1000W, 2000W and 3000W is shown in figure5. The hot water temperature with HRWHS obtained in the differentcooling load from 0 W to 3000W tend to be similar, so the hot water temperature was not affected by the cooling load different. The temperature of the hot water is only obtained in the RSAC system with HRWHS. The RSAC system without HRWHS is original mode of RSAC system without dummy condenser that does not produce hot water as shown in figure 5.

The main advantageof RSACwith heat recovery system is free energy for hot water. This is one form of heat recovery in RSAC, where most of the heat released into the environment is absorbed for heating 50 liters of water. This hot water can be used for bathing and washing. As seen in figure 5, RSAC with heat recovery (dummy condenser) is adequate to generate hot water with the temperature range around 46.58°C - 48.8°C compared without heat recovery, and the use of dummy condenser does not give significant effect to the RSAC performance. The heat value which is not great but if heat accumulated for 120 minutes the hot water can be obtained, as shown in figure 5 while the room temperature is maintained for the similar evaporator cooling capacity as shown in figure 4(b).

It can be seen from figure 2, figure 3 and figure 4(a) there are no significant difference in the performance characteristics of the RSAC original without HRWHS (condenser dummy) when compared with HRWHS (the condenser dummy). The design of the condenser dummy is expected no impact to the performance of air conditioning, so the RSAC original has been turned into hybrid RSAC system as the air conditioning system for space cooling and water heater.

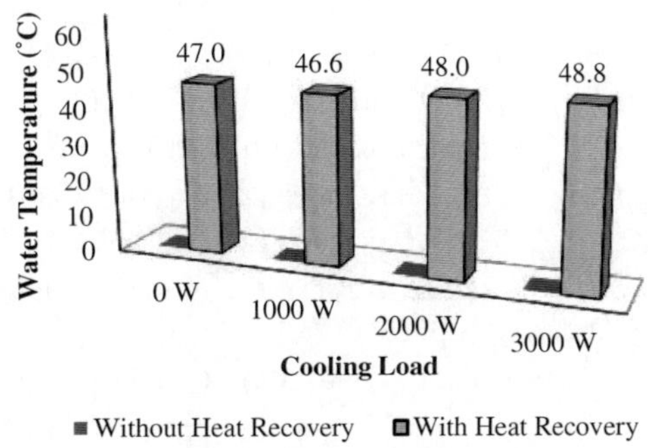

Figure 5. The hot water temperature of RSAC system at different cooling loads with and without HRWHS

4. Conclusions

The effect of heat recovery water heater system (HRWHS) on the performance of residential split air conditioner using hydrocarbon refrigerant (HCR22) was presented by different in cooling load as actual cooling load simulation. It is found that RSAC system with HRWHS (dummy condenser) capable to generate hot water in temperature range 46.58-48.81°C within 2 hours. The results show the use of dummy condenser does not have a big impact on the performance of RSAC because of the increase or decrease in RASC system performance that occurs in slightly small difference. The compressor power increase 1.42%, while cooling capacity decrease 0.39%, where heat capacity decrease 2.68% and COP increase 1.05%. The use of the heat recovery system on the RSAC as the water heater only requires additional refrigerant 19.05%. In general the use of heat recovery does not affect the performance of the RSAC and can be applied as air conditioner and water heater simultaneously to provide better performance.

5. Acknowledgments

The present study was financially supported by Universitas Riau and Ministry of Research, Technology and Higher Education (PenelitianUnggulanPerguruanTinggi, DRPM, Desentralisasi Research Grant 2017).

6. References

[1] Joudi KA and Al-Amir QR 2014 *Energy Convers. Manag.***86** 496
[2] Jia J and Lee WL 2014*Energy Build.***78**132
[3] Ji L, Pei G, Chow TT, He W, Zhang A, Dong J and Yi H 2005 *Appl. Energy.***80** 307
[4] Yang L, Yuan H,Peng JW, Zhang CL 2016 *Renew. Energy.***87** 1067
[5] Ceng YA and Boles MA 2011*Thermodynamics an Engineering Approach* 7th (New York:McGraw-Hill)
[6] Wongwises S, Kamboon A and Orachon B 2006 *Energy Convers. Manag.***47**1644
[7] Harby K 2017 *Renew. Sustain. Energy Rev.***73** 1247
[8] Garland NP and Hadfield M 2005 *Tribol. Int.***38** 732

[9] Bayrakci HC and Ozgur AE 2009 *Int. J. Energy Res.***33** 1070
[10] Dalkilic AS and Wongwises S 2010 *Int. Commun. Heat Mass Transf.***37**1340
[11] Cleland DJ, Keedwell RW and Adams SR 2009 *Int. J. Refrig.* **32** 1403
[12] Çomaklı K, Çakır U, Sahin E and Kus AC 2013*Int. J. Automot. Mech. Eng.***8** 1108
[13] Chaiwongsa P and Duangthongsuk W 2011 *Procedia Eng.* **8**165
[14] Aziz A, Satria AB and Mainil RI 2015 *Int. J. Automot. Mech. Eng.***12** 3043
[15] Aziz A, Mainil RI, Mainil AK and Saputra E 2017 *AIP Conf. Proc.***1788**030023
[16] Hepbasli A and Kalinci Y 2009 *Renew. Sustain. Energy Rev.***13** 1211
[17] Aziz A, Mainil RI, Mainil AK and Listiono H 2017 *AIP Conf. Proc.***1788**030024
[18] Wang Y, You Y and Zhang Z 2011 *Procedia Eng.***23** 493
[19] Aziz A and Satria AB 2014 *Proceeding 1st Int. Conf. Ocean. Mech. Aerosp. -Science Eng.*vol 1(Pekanbaru: ISOMASE) p 150
[20] Aziz A, Thalal and Mainil AK 2017 Effect of Cooling Load on The Performance of R22Residential Split Air Conditioner When Reftrofitted With Hydrocarbon Refrigerant (HCR22)*Adv. Sci. Lett.* (submitted publication).

IOP Conf. Series: Materials Science and Engineering **237** (2017) 012011 doi:10.1088/1757-899X/237/1/012011

Performance of a small compression ignition engine fuelled by liquified petroleum gas

Himsar Ambarita[1,*], Eko Yohanes Setyawan[1,2], Sibuk Ginting[3], Waldemar Naibaho[3]

[1]Sustainable Energy and Biomaterial Centre of Excellent, Faculty of Engineering, University of Sumatera Utara, Jl. Almamater Kampus USU Medan 20155, Indonesia

[2]Mechanical Engineering, Faculty of Engineering , National Institute of Technology Malang, jl. Bendungan Sigura-gura No.2 Malang 65152, Indonesia

[3] Mechanical Engineering, Faculty of Engineering , Nommensen HKBP University, Jl. Sutomo No. 4a, Medan 20234, Indonesia

[*] himsar@usu.ac.id

Abstract. In this work, a small air cooled single cylinder of diesel engine with a rated power of 2.5 kW at 3000 rpm is tested in two different modes. In the first mode, the CI engines run on diesel fuel mode. In the second mode, the CI engine run on liquified petroleum gas (LPG) mode. In order to simulate the load, a generator is employed. The load is fixed at 800 W and engine speed varies from 2400 rpm to 3400 rpm. The out power, specific fuel consumption, and brake thermal efficiency resulted from the engine in both modes are compared. The results show that the output power of the CI engine run on LPG fuel is comparable with the engine run on diesel fuel. However, the specific fuel consumption of the CI engine with LPG fuel is higher 17.53% in average in comparison with the CI engine run on diesel fuel. The efficiency of the CI engine with LPG fuel is lower 21.43% in average in comparison with the CI engine run on diesel fuel.

1. Introduction

Green House Gases (GHGs) emission is a big problem for the world in the next decades. To avoid the catastrophe, many countries have committed on reducing their GHGs emission. The Government of Indonesia (GoI) has released its target on reducing GHG emissions by 26% from level business as usual (BAU) by 2020 and it can be increased up to 41% by international aid [1]. Recently, the GoI revised the target emissions reduction to 29% of BAU scenario by 2030. The source of emissions can be divided into 5 sectors, they are energy, waste, industrial and process, agricultural, and forestry. According to inventory in 2010, total Indonesian GHGs emission was 1,334 MTon CO_2e. The BAU scenario in 2030 is 2,869 MTon CO_2e. The emission target by 2030 is 2034 MTon CO_2e or reduction of 834 MTon CO_2e. Energy sector, individually, has a target of 314 MTon CO_2e. It is 11% of the total reduction target. The energy sector includes all activities of burning fossil fuel and fugitive emissions. One of the strategies to achieve this target is promoting renewable energy fuel and or converting liquid fuel into gas fuel which has a lower CO_2 emission [2]. In Indonesia, the main consumption of fossil fuel is diesel oil which is used in compression ignition (CI) engines [3]. The CI engines are mainly used in heavy machineries such as truck, agricultural engines, marine, and power plants. Thus, reducing diesel oil consumption will reduce GHGs emissions significantly. On the other hand, in Indonesia fossil fuel (gasoline, diesel and kerosene) is still subsidized and it becomes a load for the GoI budget. These facts suggest that reducing diesel oil consumption in CI engines will give a significant impact on mitigation GHG emissions and it will help the GoI in reducing fossil fuel subsidy.

There are two methodologies that can be used to reduce diesel fuel in the CI engines. The first method is dual-fuel mode. Here, the diesel fuel is reduced and substituted by gas. The second method is gas fuel mode. In this method, the original CI engine is converted into spark ignition engine and fuelled by gas fuel such as biogas and liquified petroleum gas (LPG). In the CI engine with dual-fuel mode, after compression stroke, the compressed of fuel gas and fresh air, a number of diesel fuel, called the pilot, is injected to the combustion chamber [4,5]. In this work, we focus on the study of a CI engine run on pure gas. The CI engine is converted into spark ignition engine and fuelled with LPG. Several studies related to a CI engine fuelled with gas fuel have been found in literature. Chandra et al. [6] reported a study on the performance evaluation of a 5.9 kW CI stationary engine which is converted to spark ignition. The CI engine run on three different gas fuels. They are compressed natural gas (CNG), methane enriched biogas (Bio-CNG) and biogas produced from bio-methanation of Jatropha and Pongamia oil seed cakes. The composition of the methane-enriched biogas was 95% CH_4- 3% CO_2 and the biogas is 65% CH_4- 32% CO_2. The results showed that the methane-enriched biogas showed almost similar engine performance as compressed natural gas in term of brake power output, specific gas consumption and thermal efficiency. Lee et al. [7] reported an experimental study on a 30kW biogas power generation enhanced by using waste heat to preheat inlet gasses. The effect of preheating the inlet gas to different temperature was investigated by employing a waste-heat recovery system. The results showed that power generation increases with increasing biogas methane concentration, except when excess air ratio less than 0.85. However, thermal efficiency increased with increasing methane concentration only when excess air higher than 0.95, although, on the relatively rich side, there was no benefit. The improved generator performance obtained by preheating the inlet gas was apparent when the excess air ratio is relatively high, such as bigger than 1.3. Homdoung and Tippayawong [8] reported a study onan agricultural CI engine, single cylinder, four strokes, indirect injection engine, 598 cc, and compression ratio of 22. The CI engine was converted into spark ignition engine and operated 100% on producer gas. The results showed that maximum brake thermal efficiency of 23.9% was obtained. The smoke density of the engine was lower than the diesel engine; however, CO emission was higher with similar HC emission. The same strategy was also used by [9, 10], converting an original CI engine into spark ignition engine and fuelled with biogas.

The above literatures show that study on CI engine run on gas fuel in order to decrease the diesel oil and GHGs emission has come under scrutiny. However, study on CI engine fueled with LPG has receive less attention. In this study, the performance of a small CI engine converted into spark ignition engine and fueled with LPG is investigated experimentally. The main objective here is to explore the performance characteristics of the CI engine if it is fueled with LPG. The performance of the CI engines run on LPG and diesel fuels are compared. The results are expected to supply the necessary information on development of alternative solutions for reducing GHGs emission in energy sector.

2. Method and equipment

In this study, a CI engine purchased from Indonesian market is used in the experiment. The commercial name of the engine is KAMA Model YL170F with rated power of 2.5 kW and

2.8 kW at engine speed 3000 and 3600 rpm, respectively. The type of the engine is an air cooled single cylinder four-strokes with dimension of stroke and bore is 70 mm and 55 mm, respectively. It is categorized as a small CI engine with a weight of 27 kg and typically found it is used by Indonesia small farmer. The specifications of the tested CI engine are presented in Table 1.

Table 1. Specification of the CI engine

No	Parameter	Value
1	Commercial name/model	KAMA Diesel Engine YL170F
2	Number of cylinder/stroke	Single-cylinder/4 strokes and Vertical
3	Cooling system	Air cooled
4	Bore × Stroke	70 mm × 55 mm
5	Maximum power	3.8 kW at 3600 rpm
6	Rated power	2.5 kW at 3000 rpm and 2.8 kW at 3600 rpm
7	Engine weight	27 kg
8	Overall dimension	330 × 376 × 415

2.1. Engine modification

In order to perform the study, an experimental apparatus has been designed and developed as shown in Figure 1. It consists of a unit of the tested CI engine, generator, series of lamps, LPG tank, gas mixer, and measurements apparatus. As stated in the previous section, the CI engine is tested in two different operation modes. In the first mode, the CI engine is tested without modification. The diesel fuel is used. In the second mode, since the gas fuel is used, several modifications are made. Cylinder head of the engine is modified by changing the nozzle and replaced it with spark plug. A gas mixer has been designed and developed in order to mix the fresh air with biogas. Here, an intake manifold is employed to inject the mixture into the engine.

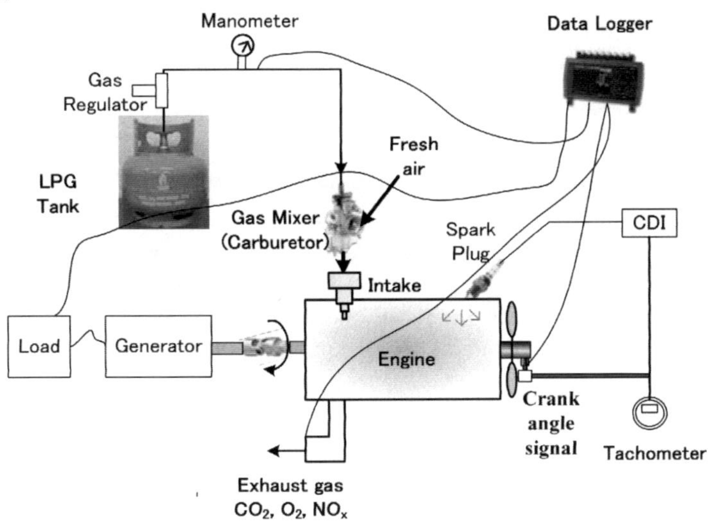

Figure 1. Experimental apparatus

A crank angle signal is employed to detect the engine speed signal. The signal is used to regulate the combustion timing. In order to regulate the pressure of the LPG fuel before

IOP Conf. Series: Materials Science and Engineering **237** (2017) 012011 doi:10.1088/1757-899X/237/1/012011

injected to the engine, a gas regulator is used. Pressure of the LPG when it is mixed with fresh air is measured using a manometer. The engine speed is measured using a tachometer. An ignition coil connected with a power supply (battery) and crank angle signal is used to determine the combustion time.

In order to simulate the load, a generator with commercial name Brushless is employed. The generator is a self-exciting, 2-poles, single phase. Specifications of the generator are as follows. The rating frequency, rating voltage, rating power, and power factor are 50/60 Hz, 110-240 V, 0.8 – 1.1 kVA, and 1.0, respectively. The electricity from the generator is used to light a series of lamps which can be operated at load 400W and 800W. The diesel oil is purchased from PERTAMINA the government oil retailer in Indonesia. Before used, the LHV of the diesel oil is tested. The LHV and density of the diesel fuel is 42.64 MJ/kg and 0.82 kg/m^3, respectively. The LPG used in this study is produced by PERTAMINA and retailed in the Indonesian market. It consists of 30% Propane and 70% Butane. The heating value and density at atmosphere are 48.8 MJ/kg and 1.71 kg/m^3, respectively. The temperature is measured using thermometer type of KW 06-278 Krishbow with an accuracy range of $\pm 0.5\%$ $\pm$ 1°C. The load of electricity is measured using a Multi tester Meter CD800A. The revolution speed of the CI engine is measured using Tachometer.

The experiments are performed in two modes. In the first mode the CI engine is run on pure diesel. Here the load is fixed at 800 W and engine speed varied from 2600 rpm to 3400 rpm. For every engine speed, when the CI engine is stable, the measurement is carried out for 5 minutes. Electric power, fuel, temperature, and engine speed are measured. In the second mode, the CI engine is converted into spark ignition mode, the LPG from the tank is mixed with the fresh air in the mixer. The pressure of the LPG from the tank is decreased to 1.8 bar by using gas regulator. The same measurements with the pure diesel are performed. Every test is replied for three times and the measurement is averaged.

2.2. Problem formulation

To carry out the analysis, performance parameters are formulated. The used performance of the CI engine is discussed in term of electric power output, efficiency and specific fuel consumption. These parameters are elaborated in the following.

The resulted power P_E (Watt) of the CI engine is calculated by using thevoltage and electric current resulted by the generator. The below equation is used.

$$P_E = V \times I \qquad \text{........ (1)}$$

where V [Volt] and I [Ampere] are voltage and current resulted by the generator, respectively. The next parameter is efficiency. Itis defined as electric power resulted by the generator divided by total energy from the fuel burn in the CI engine. For the first mode, the CI with pure diesel fuel,it is calculated by using equation (2).

$$\eta = \frac{P_E}{\dot{m}_{diesel} \times H_{diesel}} \qquad \text{......... (2)}$$

where H_{diesel} [kJ/kg] is the heating value of the diesel fuel. On the other hand, in the second mode, converted CI engine with LPG fuel, the efficiency is given by

$$\eta = \frac{P_E}{\dot{m}_{LPG} \times H_{LPG}} \qquad \text{......... (3)}$$

where H_{LPG} [kJ/kg]and $\dot{m}_{LPG}$ [kg/s] is the heating value and mass flow rate of the LPG, respectively. The next comparison parameter is specific fuel consumption (sfc [g/kWh]). It is a ratio of fuel consumption to the useful energy. Here, it can be viewed as how many gram of fuel is needed to produce 1 kWh of electrical energy. For diesel operation mode, it is given by:

$$sfc = \frac{\dot{m}_{diesel} \times 10^3}{P_E} \qquad \text{......... (4)}$$

While for LPG fuel mode it is given by:

$$sfc = \frac{\dot{m}_{LPG} \times 10^3}{P_E} \qquad \text{......... (5)}$$

Using the above formulated parameters, the performance of the CI engine is analyzed.

3. Results and Discussions

The results will be discussed in 4 subsections, they are output power, total efficiency, specific fuel consumption, and after test visualization of the combustion chamber.

3.1. Output power

The output power is calculated using the measured voltage and the current of electricity as formulated in equation (1). The electric power as a function of engine rotation speed for both diesel fuel and LPG fuel are shown in Figure 2. As a note, the load is fixed at 800 W. It can be seen clearly that, for both modes, the output power increases with increasing engine speed increases. The output power of the CI engine run on pure diesel is similar with the CI engine run on LPG fuel at the same speed. This figure reveals that the CI engine can be run with LPG fuel with output power comparable with the same engine run on pure diesel mode.

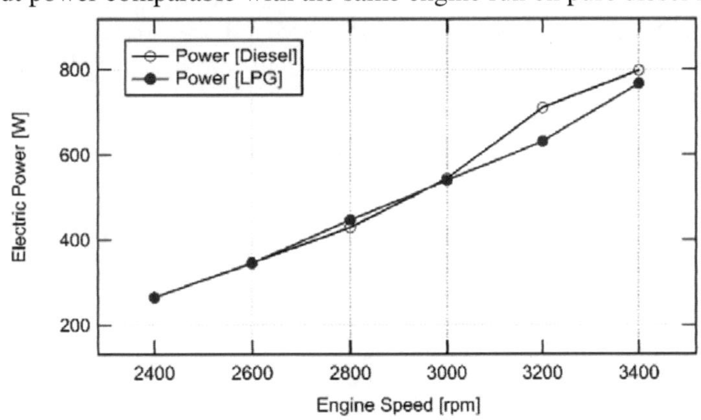

Figure 2. Output power of the CI engine at load 1500 W

3.2. Specific fuel consumption

Figure 3 shows the specific fuel consumption (*sfc*) as a function of engine speed. The specific fuel consumption can be viewed as a parameter to show how effective the heat energy in the fuel can be converted into power. Here, it is the ratio of fuel to electrical energy. The lower *sfc* is the better. The figure shows that *sfc* decreases with increasing engine speed. In the first mode, the highest and the lowest *sfc* is 0.59 and 0.98 kg/kWh, respectively. On the other hand, in the second mode, highest and the lowest *sfc* is 1.16 and 0.69 kg/kWh, respectively. These values suggest that *sfc* in the first mode is better than in the second mode. As expected, the CI engine run in pure diesel mode is better than the engine converted into spark ignition engine. This is because the present CI engine is originally designed for being operated with diesel fuel. All of the system is designed for optimum operation for diesel fuel. Thus, operating the CI engine without any modification to improve the combustion process made the engine less effective.

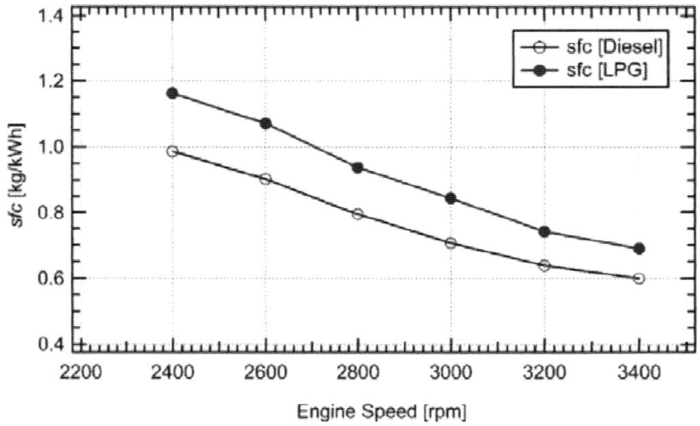

Figure 3. Engine speed vs. specific fuel consumption

3.3. Efficiency

In this study, the calculated efficiency is different from thermal efficiency. In the thermal efficiency, the output will be measured as brake power. Here, the efficiency is calculated using the electric output power from the generator. The efficiency for both cases are shown in Figure 4.

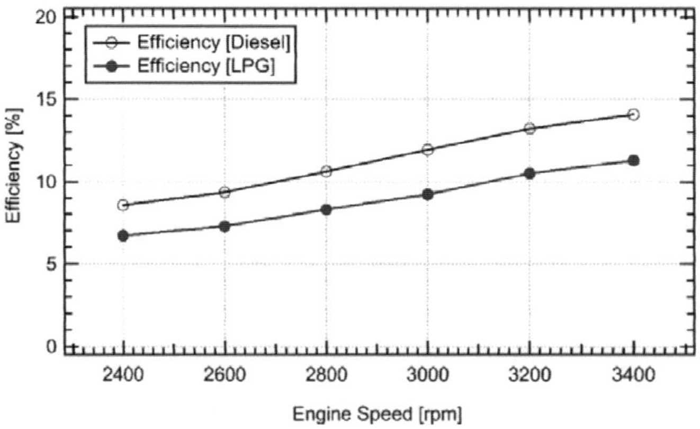

Figure 4. Engine speed vs. Efficiency

It can be seen in the figure that efficiency increases with increasing engine speed. As a note, the present engine has rated power of 2.5 kW and 2.8 kW at engine speed 3000 and 3600 rpm, respectively. In the experiment, the used load is only 800 W. It is far below the rated power. Thus, the present CI engine is operated far from its optimum efficiency. Thus, the efficiency still far from its maximum value. The comparison of the CI engine with pure diesel and LPG shows that efficiency of the CI engine run on diesel mode is higher than efficiency of the LPG fuel. In the diesel fuel, the efficiency varies from 8.5% to 14.1%. In the LPG fuel, the efficiency varies from 6.7% to 11.2%. These values suggest that efficiency of the CI run on diesel fuel is higher than LPG fuel. The average difference is 21.5%. It's because the CI engine is originally designed for diesel fuel. All of the system is designed for optimum operation for diesel fuel. Thus, operating the CI engine without any modification to improve the combustion process made the engine less efficient. The similar trend also shows by the specific fuel consumption.

3.4. After test visualization

The last parameter compared here is after test visualization. The photograph of the cylinder head of the engine before tested, after tested with diesel fuel and after tested with LPG fuel are shown in Figure 5. It is shown clearly, that the cylinder head of the engine is very clean before it is tested. After tested with diesel fuel, there was significant carbon deposition on the cylinder head. This fact clearly shown in Figure 5(b). However, the carbon reside is significantly lower in the cylinder head when the engine tested with LPG fuel, as shown in Figure 5(c). These carbon deposition reveals that operating the CI engine with LPG fuel make the engine cleaner and it can be operated longer. Thus, it needs less maintenance cost.

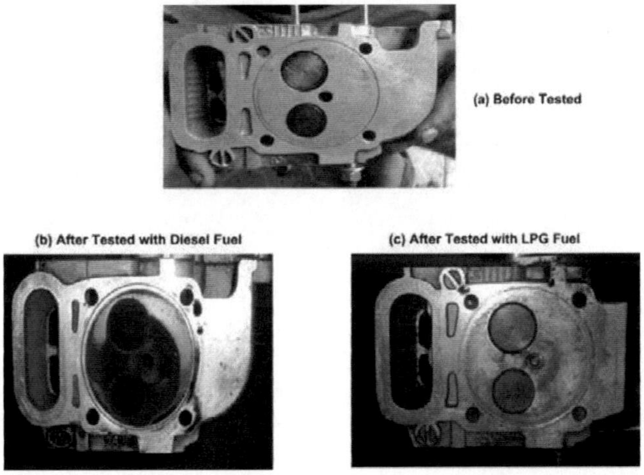

Figure 5. Visualization of cylinder head before and after tested

4. Conclusions

The performance of a small CI engine has been tested in two difference fuels. In the first mode, the diesel fuel is used and in the second mode LPG fuel is used. In the second mode, the CI engine is converted into spark ignition engine by replacing the oil pump into spark plug. The conclusions are as follows. The output power of the CI engine run on LPG fuel is comparable with the engine run on diesel fuel. However, the specific fuel consumption of the CI engine with LPG fuel is higher 17.53% in average in comparison with the CI engine run on diesel fuel. The efficiency of the CI engine with LPG fuel is lower 21.43% in average in comparison with the CI engine run on diesel fuel. It is suggested to increase the efficiency of the CI engine run on LPG fuel by further modification in compression ratio and fuel injection system.

Acknowledgments
The authors gratefully acknowledge that the present research is supported by Ministry of Research and Technology and Higher Education Republic of Indonesia. The support is under the research grant BP-PTN USU of Year 2016.

References

[1] Gov. of Indonesia Presidential Decree No No 61 Year 2011, National Action Plan For ReducingGreenhouse Gas Emissions.
[2] Gov. of Indonesia, 2016, First Nationally Determined Contribution Republic of Indonesia.
[3] Min. of Energy and Mineral Res., 2011, Handbook of Energy and Economic Statistics of Indonesia.
[4] H. Ambarita, 2017 *Case Studies in Thermal Engineering* (in Review)
[5] H. Ambarita, E.P. Sinulingga, M. KM. Nasution, H. Kawai, 2017 *IOP Conference Series: Materials Science and Engineering***180**, 012025.
[6] R. Chandra, V.K. Vijay, P.M.K. Subbarao, T.K. Khura, 2011 *Applied Energy***88**, 3969 - 3977.
[7] T.H. Lee, S.R. Huang, C.H. Chen, 2013 *Renewable Energy***50**, 342 - 347.
[8] N. Homdoung, N. Tippayawong, N. Dussadee, 2015 *Energy Conversion and Management***94,** 286-292.
[9] C. Poompipatpon, K. Cheenkachorn, 2011 *Energy***36**, 6862-6866.
[10] Bedoya ID, Saxena S, Cadavid FJ, Dibble RW, Wissink M 2012 *Applied Energy* 618-629.

Field tests of a natural vacuum solar desalination system using hybrid solar collector

Eko Y. Setyawan[1,2], Richard A.M. Napitupulu[3], Parulian Siagian[1,3], Himsar Ambarita[1,*]

[1]Sustainable Energy and Biomaterial Centre of Excellent, Universitas Sumatera Utara,
Jl. Almamater Kampus USU Medan 20155, Indonesia
[2]Mechanical Engineering, National Institute of Technology Malang, Indonesia
[3]Mechanical Engineering, Universitas HKBP Nommensen, Medan, Indonesia

[*] himsar@usu.ac.id

Abstract. This study deals with field test of a natural vacuum solar desalination system using hybrid type solar collector. In order to perform the tests, a natural vacuum solar desalination has been designed and fabricated. The dimension of evaporator is 1000 mm $\times$1000 mm $\times$ 200 mm, while dimension of solar collector is 1000 mm $\times$1500 mm. The system is tested by exposing to solar radiation in Medan city of Indonesia for five days. The solar radiations during test are 8.79 MJ/m^2, 10.14 MJ/m^2, 6.88 MJ/m^2, 11.05 MJ/m^2, and 11.36 MJ/m^2, respectively. The produced fresh waters are 160 ml, 180 ml, 118 ml, 206 ml, 220 ml, respectively. The conclusions are as follows. The produced fresh water is still very low due to the heat from the solar collector is not transferred perfectly to the evaporator. There produced fresh water is strongly affected by solar irradiation. It is recommended to minimize the heat loss from the evaporator and the transfer fluid.

1. Introduction

Fresh water needs are increasing day bay day due to modernization activities such as industrialization, motorization, life standard of human being. The ability of the world to support the demands of natural fresh water is limited. According to U.S. Geological Survey, there are 332,519,000 cubic miles of water present in this planet. However, 97% of this water is found in the ocean, about 2% is frozen up in glaciers and ice caps and it is hard to imagine only less than 1% of the water on earth is fresh. Based on the estimation by United Nations Organization, around 1800 million people in the world will be under severe water scarcity [1]. This situation can be avoided if human being found technologies to produce fresh water from seawater. There are several technologies are developing by researchers around the world. One of the most promising technologies is solar powered desalination system, here after named as solar desalination systems. The solar desalination system can be divided into direct and in-direct solar desalination systems. The in-direct solar desalination systems consist of membrane based and non-membrane based. The non-membrane based solar desalination systems can be divided into Multi-stage flash, Multi-effect distillation, Vapor compression, Freese Desalination, Adsorption Desalination, and Natural Vacuum Desalination. Due to its simplicity, in this study, we focus on natural vacuum desalination.

The main objective of natural vacuum solar desalination systems is that can be operated with low heat sources. In the system seawater is filled in evaporator (which is a container which) is vacuumed naturally by using the gravity of the water. The advantage of using the

IOP Conf. Series: Materials Science and Engineering **237** (2017) 012012 doi:10.1088/1757-899X/237/1/012012

vacuum is that low grade heat source such as solar energy can be used efficiently. Several researches of natural vacuum desalination have been reported in literature. Al-Kharabsheh and Goswani [2, 3] reported the research on preliminary experimental and theoretical analysis of a natural vacuum water desalination system using low-grade solar heat. The effects of several various operating conditions were studied. Simulation for Gainesville, Florida, the daily output from a system of 1 m^2 of evaporator area with 1 m^2 of solar collector area could reach 6.5 kg fresh water. Ambarita [4,5] performed experimental and numerical study on the natural vacuum solar desalination and using electrical heater to simulated the solar heat source. There are several researchers [6,7,8] have been reported their studies on natural vacuum solar desalination system. However, those studies focus on the simulations and the heat sources are simulated using electric heater.

In this study, a field test of a natural vacuum solar desalination system is carried out. The system is coupled with hybrid solar collector. The system is field tested in Medan city of Indonesia. The objectives are to explore the characteristics of the solar desalination system in term of temperature, pressure and fresh water produced. The results are expected to supply the necessary information on developing effective and efficient solar desalination system.

2. Method

In order to perform the study, a natural vacuum solar desalination unit has been designed and fabricated. The system consists of evaporator, condenser, solar collector, hot water circulation unit, seawater, brine and fresh water unit. The schematic diagram and photograph of the experimental apparatus are shown in Figure 1. The evaporator is made of stainless steel 304 with a thickness of 2 mm. It is a square container with a dimension of 1000 mm × 1000 mm × 200 mm and the top is a conical with bottom diameter and height of 400 mm and 200 mm, respectively. The upper part of the evaporator is connected with a condenser. The condenser is a horizontal tube with diameter and length of 150 mm and 500 mm, respectively. It is made of Stainless steel 304 with a thickness of 2.54 mm. In this case, ambient air is used to cool the condenser and to absorb latent heat of the freshwater. In order to provide a better heat transfer rate, 20 fins are installed to the condenser. The fins are made of aluminum with a diameter and thickness of 254 mm and 0.6 mm, respectively. The distance between fins is 40 mm. The evaporator and condenser are placed at a location 10.33 m above the ground. The reason for this is to provide natural vacuum inside the evaporator.

To provide evaporation inside the evaporator, an amount of heat must be supplied to overcome heat evaporation. To do so, the heat from the solar is collected by a solar collector and it is transferred into the evaporator by using a water as a transfer fluid flowing inside the heating coil. The diameter and total length of the heating coils is 10 mm and 3000 mm, respectively. A pump is used to circulate the transfer fluid inside the heating between solar collector and evaporator. The pump is powered using a solar cell with a power of 100 Wp. The solar collector used in this study is a flat plate type solar collector with a dimension of 1500 mm × 1000 mm. In order to reduce the heat loss from the top of the collector, a double glasses cover is used.

A data acquisition system to measure temperature, pressure, and ambient is installed to the experimental apparatus. Temperatures are measured by using *J*-type thermocouples with uncertainly equal to 0.1°C. There are 20 points are measured in the system, they are four thermocouples placed in the evaporator, four in the condenser, four in the solar collector and the rests in pipe and water containers. A multi channels data logger (Agilent 34972A) is used to record temperatures with interval of 1 minute. Solar radiation is measured using HOBO pyranometer smart sensor. The ambient temperature and relative humidity (RH) is measured using HOBO temperature RH smart sensor with an accuracy of 0.2°C and ± 2.5%RH, respectively. The wind speed around the experimental apparatus is measured with HOBO wind speed smart sensor with accuracy ± 1.1 m/s. The schematic diagram of the system and data acquisition system are shown in Figure 1(b). To start up the system, it will be filled completely with water initially, and the water will then be allowed to fall under the influence of gravity, in order to establish the vacuum. The extraction rate of the brine water from the evaporator is controlled at a rate of 0.36 Liter per hour by using a valve.

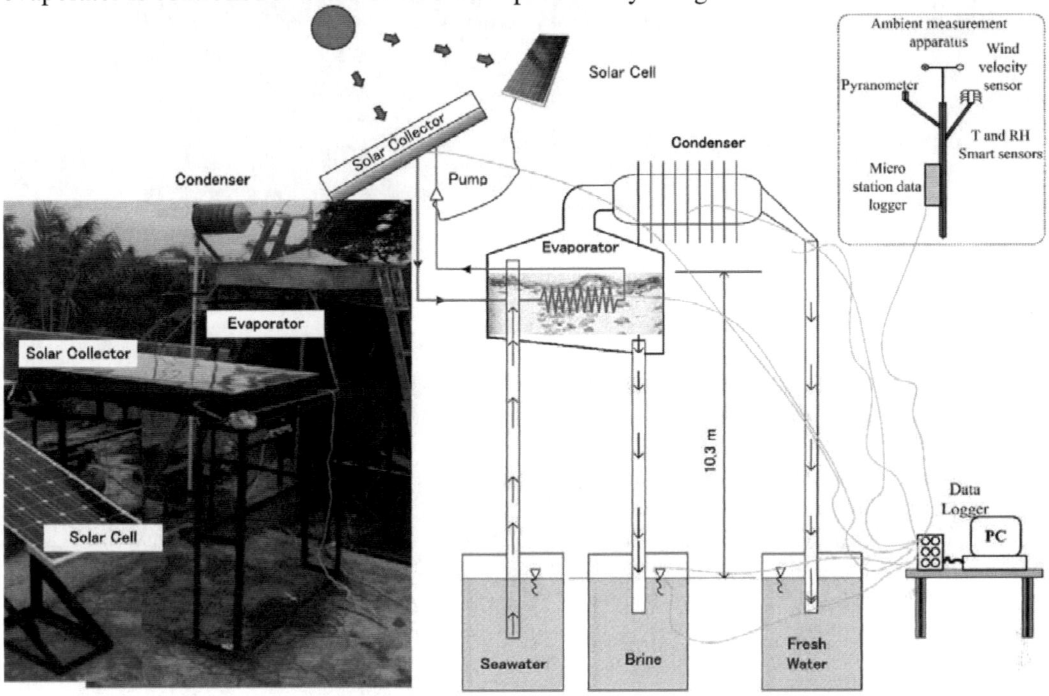

(a) Photograph (b) Schematic Diagram

Figure 1. Solar radiation measurement apparatus

3. Results and Discussions

In this study five days of field tests (experiments) have been performed. The experiments were carried by exposing the solar desalination unit on a top roof of a building in Medan city of Indonesia. The location of the tests is 20 m above the sea level. The experiments are from 22 May to 26 May, 2017. The results are discussed below.

IOP Conf. Series: Materials Science and Engineering **237** (2017) 012012 doi:10.1088/1757-899X/237/1/012012

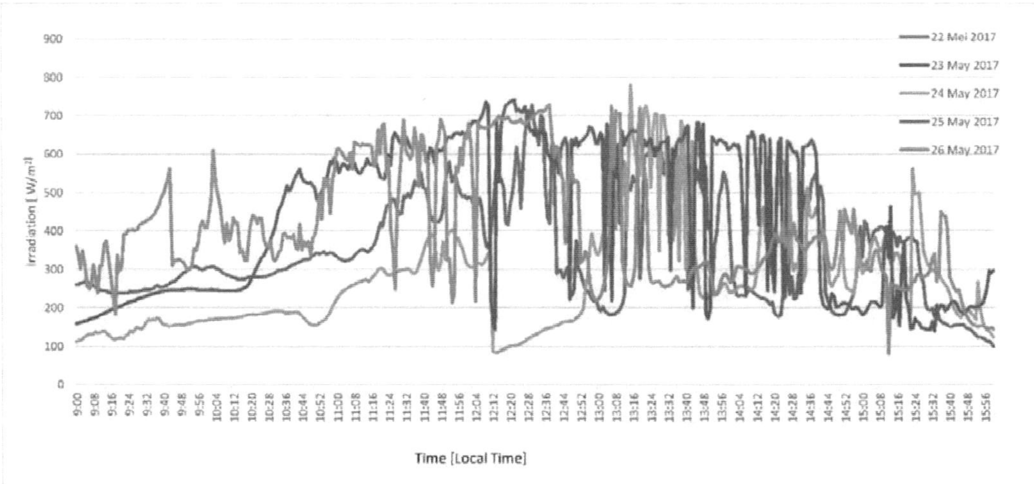

Figure 2. Solar radiation during field tests

3.1. Solar radiation during field tests

During the field tests, the solar irradiation is shown in Figure 2. Every test started from 9.00 and finished at 16.00 of local time. The figure shows that during the field tests, the solar irradiation varies from about 100 W/m^2 to 800 W/m^2. In general, by the noon solar irradiation increases with increasing time and after noon, it decreases with increasing time. The solar irradiation can be falling down suddenly due to blocking of the clouds. Solar energy is calculated by multiplying solar irradiation with time in second. The solar energy during the field tests at 22 May, 23 May, 24 May, 25 May, and 26 May are 8.79 MJ/m^2, 10.14 MJ/m^2, 6.88 MJ/m^2, 11.05 MJ/m^2, and 11.36MJ/m^2, respectively. These values suggest that the field test at day 3 (May 24) is categorised as cloudy day and the shows the lowest solar energy. The highest solar energy is captured at day 6 (26 May).

3.2 Temperature of the evaporator

In order to explore the characteristics of the evaporator, the measured temperatures are discussed. In this paper, temperatures of the fourth day (May 25) are selected and shown in Figure 3. As expected, temperatures of evaporator show similar trend with solar irradiation. In general, by the noon the temperature increases with increasing time and after noon, it decreases with increasing time. The temperature can be falling down suddenly as solar irradiation falling down. The figure clearly shows that the highest temperature is captured at the top part of the evaporator. This is because of two reasons. The first reason is that the hot vapor is collected at the top and the second reason is the top is heated by solar irradiation. The temperature of the coil in (transfer fluid from the solar collector flow in to evaporator) is lower than top part of the evaporator. The temperatures of the coil out, temperature of evaporator bottom and evaporator wall are almost similar. This is due to high heat loss to the ambient.

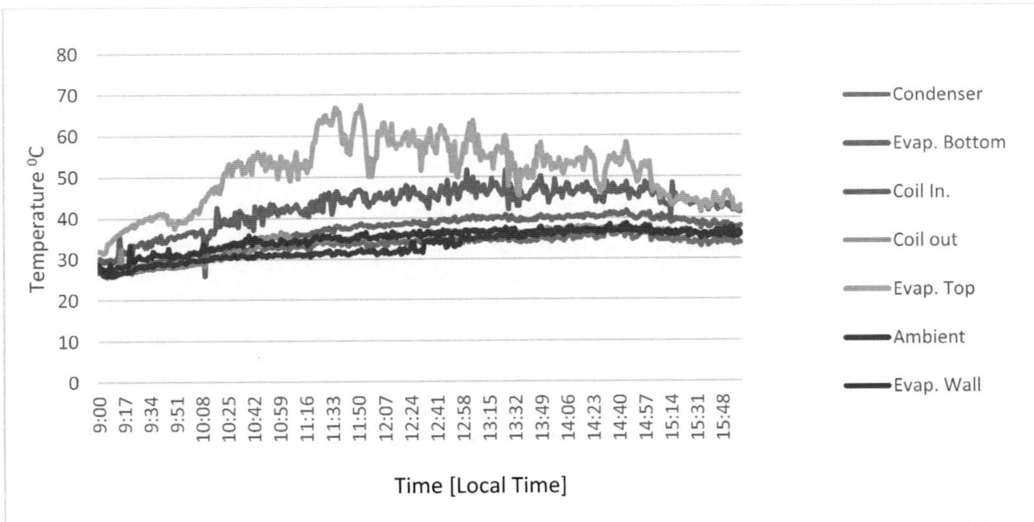

Figure 3. Temperature on the evaporator and condenser during 25 May 2017

3.3 Temperature of the solar collector

Temperatures of the solar collector during the fourth day of field test are shown in Figure 4. The figure shows that temperature of the collector varies from 30°C to 95°C. The highest temperature is shown by absorber plate. This is because the solar irradiation is converted into heat by solar absorber. Since the lower glass cover close to absorber plate, its temperature is similar to temperature of the absorber plate. The heat from the solar collector is used to heat the transfer fluid which come from coil in and it out from coil out. The figure shows that there is no significant temperature difference of the fluid in and out from the solar collector. The examination shows that the maximum temperature difference is only 0.95°C.

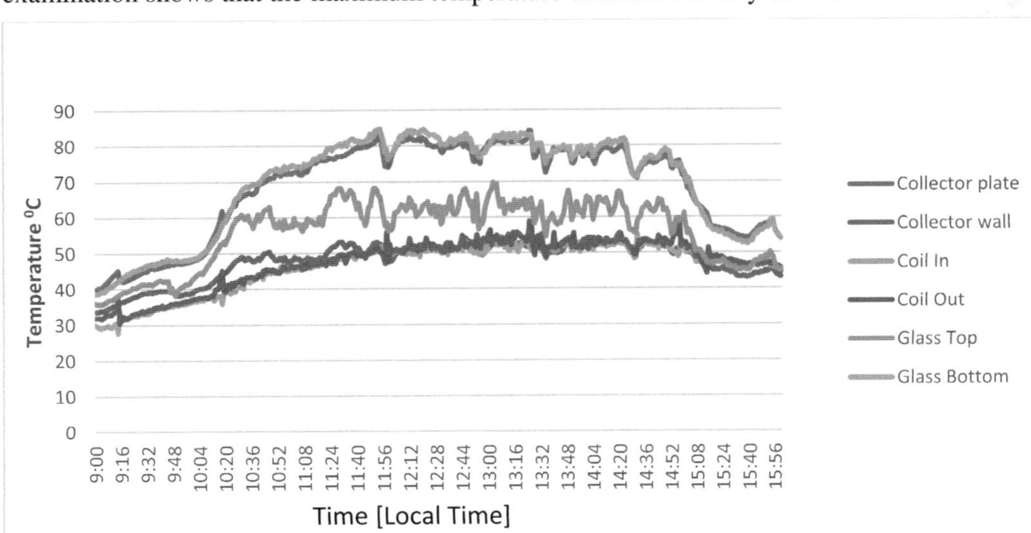

Figure 4. Temperatures of the solar collector during 25 May 2017

3.4 Pressure inside the evaporator

The objective of installing the system 10.3 m above the ground is to capture the vacuum pressures inside the evaporator. The pressure inside the evaporator during the field tests are shown in Figure 5. The figure clearly shows that pressure inside the evaporator is close to vacuum. When the system is started, the pressures inside the evaporator decrease suddenly close to vacuum. However, the figure shows that the absolute can't be reached. As temperature increase pressure inside the evaporator increase or lower than vacuum pressure. This is because steam from evaporation fills the evaporator and this makes pressure will increase.

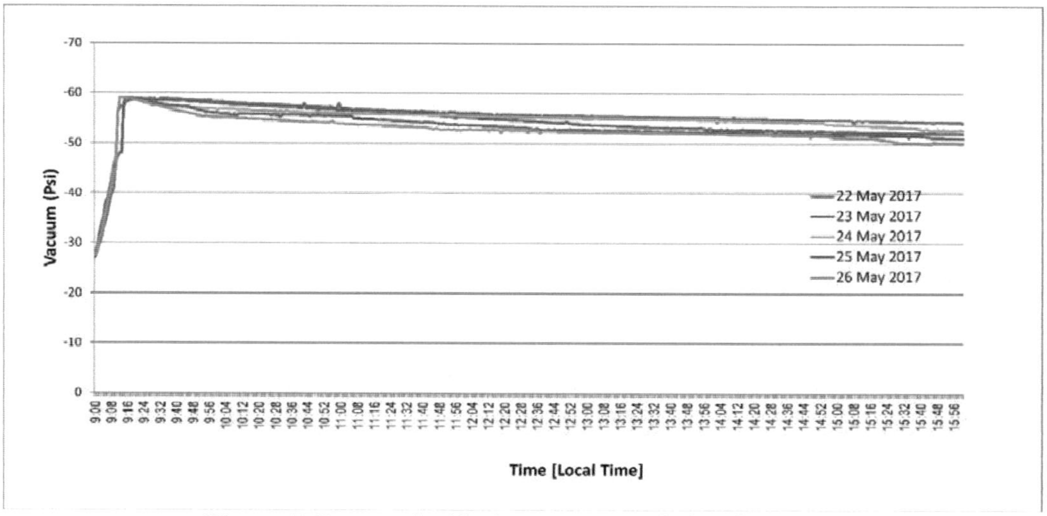

Figure 5. Pressure inside the evaporator during field tests

3.5 Fresh water produced

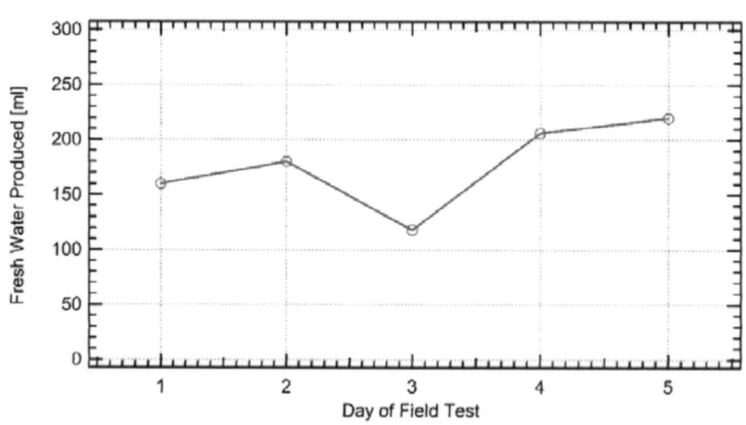

Figure 6. Daily Fresh water produced

Production rate for all field tests are shown in Figure 6. The figure shows that the fresh water produced varies from 116 ml to 220 ml. The fresh water produced in these experiments are very low. This is because the heat from the solar collector is not transferred perfectly into the evaporator. However, the figure shows a clear correlation of solar energy and fresh water produced. As calculated in the previous section that the lowest solar energy is captured at the third day of field test. The figure clearly shows that the lowest water fresh water produced is at the third day. The highest solar energy is captured at the fifth day. As expected the highest fresh water produced is at the fifth day.

4. Conclusions

In this study, a natural vacuum solar desalination with hybrid type solar collector has been designed and tested by exposing to solar radiation in Medan city of Indonesia. The system has been texted for five days. The solar radiations during tests are 8.79 MJ/m^2, 10.14 MJ/m^2, 6.88 MJ/m^2, 11.05 MJ/m^2, and 11.36MJ/m^2, respectively. The produced fresh waters are 160 ml, 180 ml, 118 ml, 206 ml, 220 ml, respectively. The conclusions are as follows. The produced fresh water is still very low due to the heat from the solar collector is not transferred perfectly to the evaporator. There produced fresh water is strongly affected by solar irradiation. It is recommended to minimize the heat loss from the evaporator and the transfer fluid.

References

[1] H. Sharon, K.S. Reddy, 2015, Sustain Energy, 41, 1080–1118.
[2] Al-Kharabsheh, A. and Goswami, D. Y., 2004, Journal of Solar EnergyEngineering, 126, 774-780.
[3] S. Al-Kharabsheh, D.Y. Goswami, 2003, Desalination, 156, 323–325.
[4] Ambarita H, 2016, Case Studies in Thermal Engineering, 8, 346–358.
[5] Ambarita H, 2016, AASEC, 180, 169-171.
[6] V.G. Gude, N. Nirmalakandan, S. Deng, A. Maganti, 2012, Appl Energy 91-1, 466–74.
[7] V.G. Gude, N. Nirmalakhadan, S. Deng, A. Maganti, 2012, Energy Convers, 192–198.

IOP Conf. Series: Materials Science and Engineering **237** (2017) 012013 doi:10.1088/1757-899X/237/1/012013

Internal combustion engine run on biogas is a potential solution to meet Indonesia emission target

Himsar Ambarita
[1]Sustainable Energy and Biomaterial Centre of Excellent, Faculty of Engineering, University of Sumatera Utara, Jl. Almamater Kampus USU Medan 20155, Indonesia

* himsar@usu.ac.id

Abstract. Indonesia has released two different Greenhouse Gas (GHG) emissions reduction targets. The first target, released in 2009, is reduction GHG emissions 26% from Business-as-Usual (BAU) level using own budget and up 41% if supported international aids by 2020. The second target is reduction 29% and 41% from BAU by 2030 using own budget and with international support, respectively. In this paper, the BAU emissions and emissions reduction target of these two targets are elaborated. In addition, the characteristics of emissions from transportation sector are discussed. One of the potential mitigation actions is switching fuel in transportation sector. The results the most promising mitigation action in the transportation is switching oil fuel with biofuel. The Government of Indonesia (GoI) focuses on using biodiesel and bioethanol to run internal combustion engine in transportation sector and biogas is aimed to fuel power plant unit. However, there is very limited of success stories on using biogas in the power plant. The barriers and challenges will be discussed here. It is suggested to run internal combustion engine with biogas.

1. Introduction

Global warming or climate change is the rise of the average temperature of the earth's climate system. The global warming is caused by GHG emission from all activities in the world. To avoid the catastrophe, mostcountries coordinated by United Nations Framework Convention (UNFCC), have committed on reducing their GHG emissions. Parties of UNFCC have agreed that deep cuts in emissions are required and that global warming should be limited to well below 2.0°C compared to pre-industrial levels. The Government of Indonesia (GoI) has released two targets in the last 10 years. The first target is released in 2009. Here, GoI has committed on reduction GHG emissions 26% from Business-as-Usual (BAU) level using its own budget and up 41% if supported international aids by 2020 [1]. The second target is released in 2016. The emission reduction is 29% and 41% from BAU by 2030 using own budget and with international support, respectively [2].According to Intergovernmental Panel on Climate Change (IPCC), the GHG emissions can be divided into four sectors. They are Energy, Industrial Process and Product Use (IPPU), Agriculture, Forestry, and other Land Use, and Waste sectors [3]. Since the targets have been released, the main issue in Indonesian development includes emission reductions or mitigation actions. Several studies on mitigation actions to meet Indonesia target have been found in literature.

Dutu [4] reported the study on the challenges and policies in Indonesia's energy sectors. The study showed that Indonesia's energy sector faces many regulatory, environmental and infrastructure hurdles. Indonesia's energy policy can be improved through greater use of renewables, especially geothermal. The sector should be further developed until more

IOP Conf. Series: Materials Science and Engineering **237** (2017) 012013 doi:10.1088/1757-899X/237/1/012013

renewable energy come on line. In order to boost investment on renewable energy development, the government control over the oil industry should be reduced. Clarifying and simplifying regulations is keys to attracting foreign companies and protecting the environment. Jafari et al [5] examined the long run Granger causality relationship between economic growth, carbon dioxide emissions and energy consumption from 1971 to 2007 in Indonesia. It was found that there was no relationship between these parameters except the causality effect that runs from urban population to energy consumption. The finding suggested that energy conservation strategies in Indonesia may not procedure desirable effect on emission target and Indonesia does not have to relinquish economic growth. Alam et al [6] examined the impacts of income, energy consumption and population growth on GHG emissions by using a yearly time series data from 1970 to 2012 of India, Indonesia, China, and Brazil. The results revealed that GHG emissions have increased significantly with increase in income and energy consumption. In addition, population growth has been statistically insignificant for Indonesia and GHG emissions will decrease over time when income increase. It was suggested that Indonesia should not develop any mitigation actions or its policy that might have conservative impacts on income. Mujiyanto and Tiess [7] reported the study on secure energy supply in 2025 of Indonesia energy policy. The results showed that investment incentive for new technologies, exploration and efficient production are necessary. It was suggested to provide clear and comprehensive energy policy strategy and regulatory framework. Deendarlianto [8] presented energy mix model for transportation sector in Indonesia. Several strategies have been examined such as introduction compressed natural gas (CNG) technology, vehicle retirement program, and hybrid vehicles. The results revealed that the most effective strategy in mitigation action is retirement program of old vehicles. On the other hand, the CNG program does not give significant impact on emission reduction, subsidy cost and fuel consumption. Sukarno et al [9] investigated the effects of household lifestyle on residential energy consumption. The study showed that power consumption (operation and standby mode), number of appliances and operating hours are the three aspects that play an important role on decreasing GHG emission.

The above reviewed studies reveled that study on the Indonesian mitigation actions, especially in energy sector, has come under scrutiny in recent years. However, the study on effect of fuel switching strategy on the emission reduction target has not found yet. In this study, the potency of employing biogas to run internal combustion engine will be examined. The main objective is to supply the necessary information for GoI to meet the emission reduction target.

2. Methods

This study is a combination of literature review and analysis. The policy released by the GoI relate to GHG emissions are reviewed. The history of energy consumption by using statistic data is presented and based on the history, energy consumption is projected to develop BAU emission. The potency of biogas from selected sector is examined. Several research papers on combustion engine fueled by biogas are reviewed to draw fossil fuel replacement ratio.

3. Results and Discussions

3.1. Policy and Energy Consumption

The milestone Indonesian policies related to GHG emissions have been reviewed and shown on Figure 1. As a first response to climate change, Indonesia has released a national development plan reinforced with a policy umbrella which than derived into Long-term

National Development Plan (RPJP) and Mid-term National Development Plan (RPJM) in 2007. This is known as a starting for Indonesian commitment on low emission development. Since then, many regulations to support the commitment have been released as shown in the figure. There are three significant milestones that received many attentions in national and international levels. The first regulation is President Regulation No 61 Year 2011 on National Action Plan on reduction GHG emissions (RAN-GRK). In this regulation, GoI has designed several mitigations action that must be implemented in national level to reach Indonesia emission targets. In the regulation, it is emphasized that local government should develop the similar regulation to support the RAN-GRK. Thus, in 2012 there are 33 government regulation in provincial levels have been released. The second significant regulation is President Regulation No 71 Year 2011 regarding on GHG inventory. The third regulation is President Regulation No 22 Year 2017 on National Energy Planning. In this regulation, a planning to meet energy consumption in Indonesia until 2050 has been formulated. In addition, actions to meet the new emission target 29% from BAU by 2030 have been formulated.

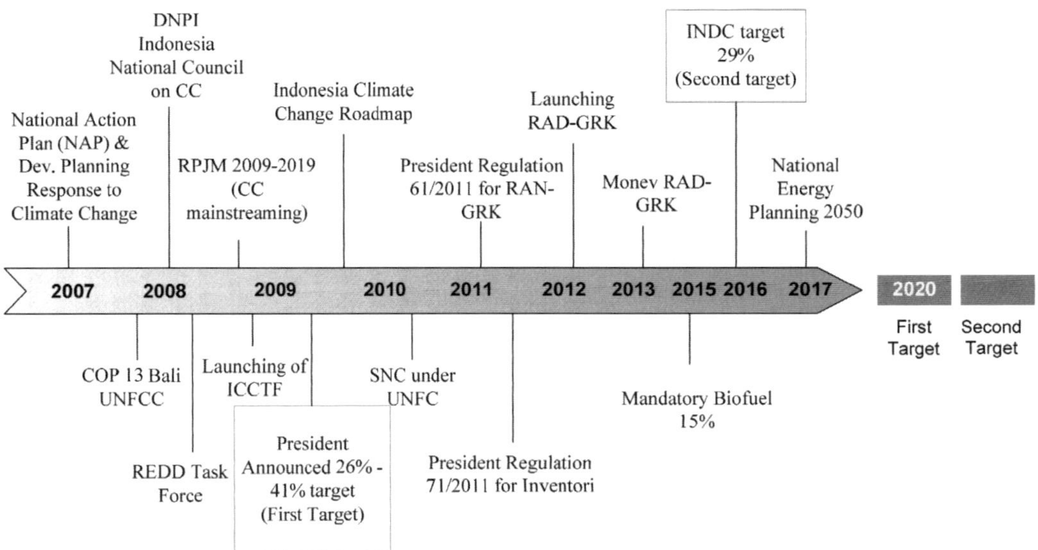

Figure 1. Milestone Indonesian Policies related to GHG emission

As mentioned in those policies, there are two targets emissions reduction have been released by the GoI. In each target, there are two number are stated, they are target by using own resource and target by additional international support.Both targets are shown in Table 1, the first and second targets have time limit by 2020 and by 2030, respectively. In the first target, the total emission reductions are 767 MT CO_{2eq} and 1,189 MT CO_{2eq} for 26% and 41 targets, respectively. It can be seen that the biggest emission target is come from Forest and Pet Land sector. In specific, the target from Energy and Transportation sector is only 36 MT CO_{2eq} and 56 MT CO_{2eq}. However, this is related to energy which is strongly affects economy growth. Thus, mitigation action to meet this target must be treated properly. In the second target, there is no explanation of target division in each sector.

Table 1 Emission target in two Indonesian targets

Sector	Emission Reduction Target in Million Ton (MT) CO_{2eq}			
	By 2020		By 2030	
	26%	41%	29%	41%
Forest and Pet Land	672	1039	-	-
Agriculture	8	11	-	-
Energy and Transport	36	56	-	-
Industry	1	5	-	-
Waste	48	78	-	-
TOTAL	767	1,189	848	1,191

In Indonesia, the energy consumption typically divided into several subsectors, they are household, industrial, transportation, commercial and others. In2015, the total energy consumption in Indonesia was 1,040,677,025 Barrel Oil Equivalent (BOE). The final energy consumption and energy share in Indonesia in 2015 are shown in Figure 2. The figure shows that transportation sector is the biggest consumer with percentage of 36% of this energy and followed by Industrial and Household sectors with percentage of 31% and 15%, respectively.

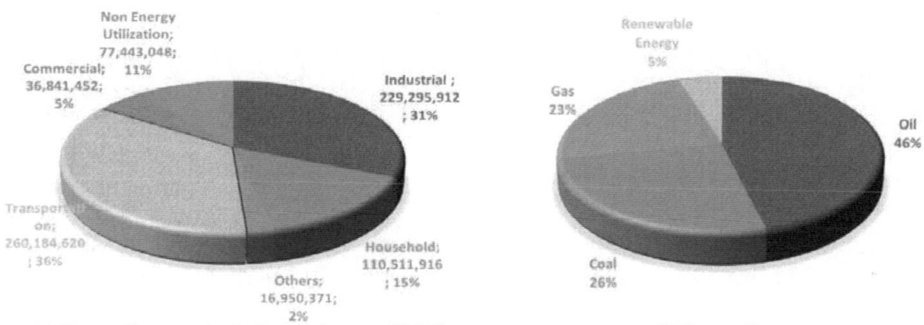

(a) Energy Consumption in Each Sub-sector [BOE] (b) Energy Share

Figure 2. Final energy consumption and energy share in 2015

The analysis reveals that the intensity of energy per capita in Indonesia in 2015 was 2.86 BOE/capita. The energy consumption per capita in the last 5 years from 2011 to 2015 are 3.5, 3.55, 3.18, 3.19, and 2.86 BOE/capita, respectively. These number shows that the energy consumption decrease with increasing time. This is might be because the increasing in energy price and energy conservation and energy efficiency programs. The energy share of the final energy consumption is also shown in the figure. The figure shows that 46% of energy consumption comes from oil and followed by coal and gas with share of 26% and 23%.

As a note, Indonesia has abundant new and renewable energy resources. The potency and installed capacity of the present Indonesian new and renewable energy are shown in Table 2. As a comparison in year 2016, the total capacity of generated national wide electricity is 51,915 MW. In 2015, total Indonesian natural gas reserve is 150.39 TSCF and oil production is 690 kilo barrels per day (bpd). These facts suggest that the new and renewable energy potencies can filled these electricity, oil consumption and oil reserves. However, only 5% of the final energy consumption is supplied by renewable energy resources. The above reviewed policies revealed that there is a strong commitment from GoI to utilize new and renewable energy resources.

Table 2 New and Renewable Resources

New and Renewable Energy Resource	Potency	Installed Capacity	Utilized (Percent)
Water	94,476 MW	5,024 MW	5.3%
Geothermal	29,544 MW	1,403.5 MW	4.8%
Bioenergy	32,000 MW and 200,000 bpd of Biofuel	1,740.4 MW	5.4%
Solar Energy	1.8 kWh/m^2/day 207.9 GW	78.5 MW	
Wind and Hybrid	3 – 6 m/s Up to 60 GW		
Ocean Energy	61 GW (Wave: 1,995 MW, OTEC: 41,001 MW, Stream: 17,989 MW)		
Shale Gas	574 TSCF		
Coal Bed Methane	456.7 TSCF		

Note: TSCF: Trillions of standard cubic feet

Figure 3 shows the BAU projection of energy consumption and the alternative projections. The alternative projection is defined as energy consumption if all of mitigation actions and utilization of renewable energy resources are implemented. In the figure, the BAU energy consumption is made based on the assumption that the growth of Gross Domestic Products (GDP) of Indonesia is 5.6% yearly during 2016 -2050.The population growth is 1.19% in 2015, 1.00% in 2020, 0.8% in 2025, and 0.62% from 2030 to 2050. In 2015 the population is 255 million and it will be 335 million in 2050. During 2016 to 2050, it is assumed that fuel price increases as 1.2%, 0.9%, 2.3% per year for coal, natural gas, and oil, respectively. In the projection of alternative 1 (Alt 1), the same assumptions in BAU are used but, all mitigation actions and renew able energy programs are implemented perfectly. The share energy in 2015 is 5%, the projection of renewable energy share in 2025, 2030, and 2050 are 22%, 26%, and 31%, respectively. The penetration of the present proven technologies is assumed to be 75%. In the alternative 2, all of the assumption in alternative 1 are employed. But, the difference is the GDP growth. In alternative 2, the GDP growth is assumed to be 7.1% yearly.

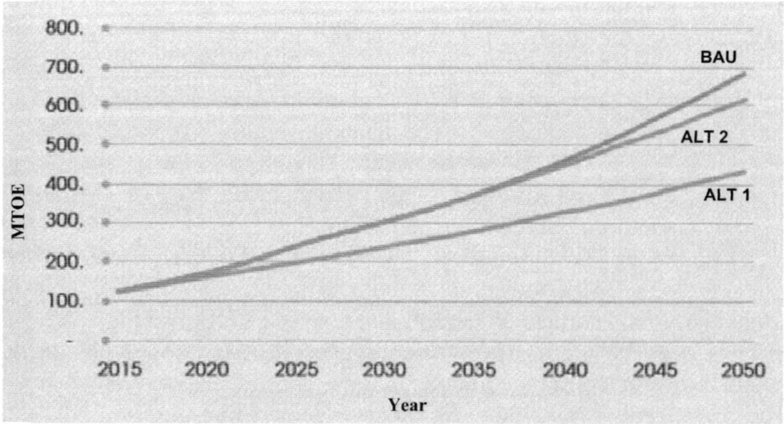

Figure 3. Projection of final energy consumption and alternative projections [10]

By using the above assumptions, the BAU scenario reveals that the final energy consumption will grow from 128.8 million tons oil equivalent (MTOE) in 2015 to 238.8 MTOE in 2025 and 682.3 TOE in 2050. In the same way, the GHG emissions will grow from 484 $MTCO_{2eq}$ in year 2015 to 948 $MTCO_{2eq}$ in year 2025 and 2,886 $MTCO_{2eq}$ in year 2050. These values show that the final energy consumption and GHG emission growing 85% and 95%, respectively in year 2025 in comparison with the energy consumption and emission in year 2015. On the other than in year 2050 the growth of energy consumption and emission are 430% and 496%, respectively. If all of the technologies are implemented, the final energy consumption in 2025 and 2050 are 201.5 MTOE and 430.3 MTOE, respectively. In the same way, the emission will grow to 687 $MTCO_{2eq}$ in year 2025 and to 1,360 $MTCO_{2eq}$ in year 2050. On the other hand, in the Alt2 scenario the final energy consumption in 2025 and 2050 are 244 MTOE and 621 MTOE, respectively. In this scenario, the emission will grow to 888 $MTCO_{2eq}$and 1,928 $MTCO_{2eq}$ in year 2025 and 2050, respectively.By using the alternative 1, the emission reduction that can be reached in 2020 and 2030 are 11% and 33%, respectively. These values, shows that the target of emission reduction can be reached in energy sector by implementing the alternative perfectly.

3.2. Combustion Engine run on Biogas

The main challenge in the implementation of alternative 1 is in the transportation subsector. This subsector is the main energy consumer in the energy sector. The transportation sector includes all of vehicles for public use and for agricultural use. In this subsector, the proposed mitigation actions mainly consist of penetration of technologies efficient combustion engine, hybrid cars, electric cars and fuel switching. The fuel switching program is supported by Indonesian Ministry of Energy and Mineral Resources regulation no 25 year 2013 on the revision of regulation no 32 year 2008 concerning the provision, utilization and commerce of biofuel as alternative fuels. The target of fuel switching in gasoline and diesel oils has shown in Table 3.

Table 3 Mandatory biofuel content for subsidized public service obligation fuels

Fuel	2013	2014	2015	2016	2020	2025
Mandatory Gasoline mixture bioethanol content	0%	0.5%	1%	2%	5%	20%
Mandatory Diesel fuel mixture biodiesel content	10%	10%	10%	20%	20%	25%

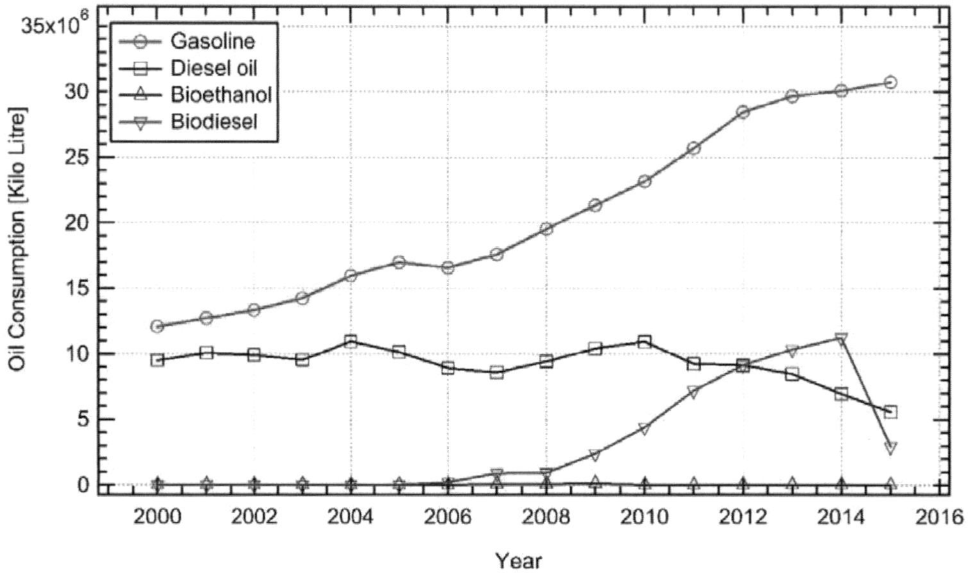

Figure 4. History fuel consumption in transportation sector in Indonesia

The effect of that regulation can be seen in the national oil consumption as shown in Figure 4. The figure shows the gasoline, diesel oil, bioethanol and biodiesel consumption in Indonesia from 2000 to 2015. The figure shows that, in the transportation subsector the consumption of fuel increase with increasing time. There is a strong effect of oil price to fuel consumption. The effect of mandatory biofuel regulation can be seen clearly in the figure. However, only biodiesel shows the significant growth. In several years, it can be over than diesel fuel. This is because Indonesia has abundant biodiesel resources from Palm Oil industry. On the other hand, the growth of bioethanol is very low. The highest bioethanol consumption was 126,048 Kilo liter in 2006 and after that it is zero. Even though, according to regulation the mandatory bioethanol consumption in 2016 must be up to 2%, however, the statistics data show that the bioethanol consumption in 2016 is zero.

The other solution to this situation is to promote technology of fueling internal combustion (IC) engines with biogas. According to studies reported by Ambarita et al [11, 12, 13] there are several technologies can be applied as solution. The IC engines presently in service can be divided into spark ignition (SI) engine and compression ignition (CI) engine. The SI and CI engines are fueledby gasoline and diesel oil, respectively. The SI engine can be run on biogas without any significant modification. However, there are exists several challenges such as biogas container, converter kit, knocking, and performances of the SI engine. Overall, biogas is a perfect fuel for switching gasoline fuel into biofuel to run SI engine. On the other hand, for CI engine there are two methods, can be employed. They are dual-fuel mode and full biogas mode. In the dual-fuel mode, the biogas is injected to the combustion chamber from intake manifold and a certain amount of diesel oil is used to start the ignition. In this method, there is no significant modification is needed. When the biogas is absence or no sufficient biogas is present, the CI engine can easily be converted into pure diesel mode. In the full biogas mode, the CI engine is converted into SI engine. Here the fuel injector is replaced with spark ignition engine. The main weakness of this mode is that the modified CI engine cannot be easily converted into pure diesel mode. The study by Ambarita [11] showed that for a small CI engine run on dual-fuel mode, the diesel oil replacement ratio varies from 15.3 % to 87.5%. This means, this technology is potential to reduce diel oil consumption as well as GHG emission.

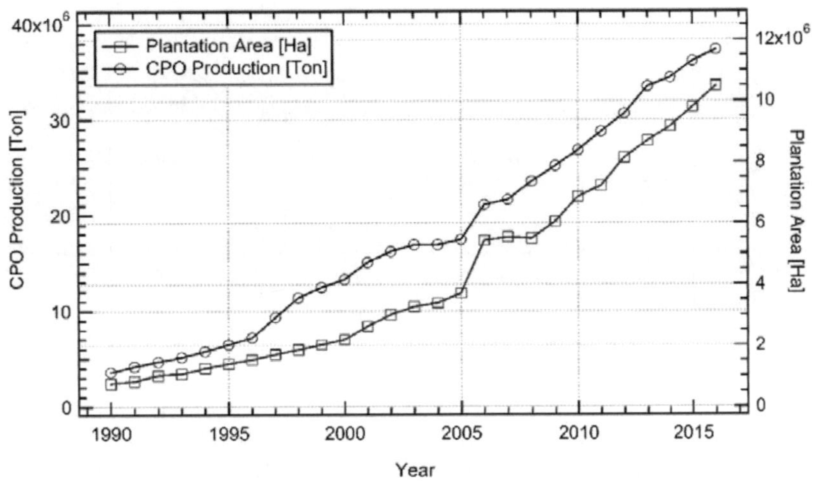

Figure 5. History of Plantation Area and CPO production of Indonesia

As mentioned above, the potential biofuel for fuel switching application is biogas. The biogas is an important source of biofuel produced from anaerobic biodegradation of organic material. It consists of methane typically range from 40 - 70% and its low heating value is between 15 and 30 MJ/Nm^3. Indonesia, as an agricultural country, has an enormous potency of biogas. One of the potential biogas resources in Indonesia is Palm Oil Mill Effluent (POME). The traditional treatment of POME is by collecting it in a pool. After a certain time, it will release methane gas into atmosphere. As a note, methane gas is also a greenhouse gas that results in climate change. According to IPCC, methane gas 21 time more dangers than CO_2. An analysis shows that 70% of GHG emissions from Crude Palm Oil (CPO) production process resulted by traditional POME production. Thus, strategy to capture POME and use it as fuel to replace oil for transportation and machineries has multiple effects on GHG emission reduction. This is because, it will reduce methane and CO_2 both from as GHGs. Indonesia is known as the biggest CPO producer in the world and followed by Malaysia. In order to estimate the potency of GHG emission from Palm Oil Industry of Indonesia, the area and CPO production in the last 25 years are shown in Figure 5. The figure shows that the CPO production and plantation area increase with increasing time. The average growth of plantation area and CPO growths in the last 5 years are 5.9% and 6.8%, respectively. The average production of Palm Oil plantation varies from 1.7 to 2.8 ton CPO/Ha/year.

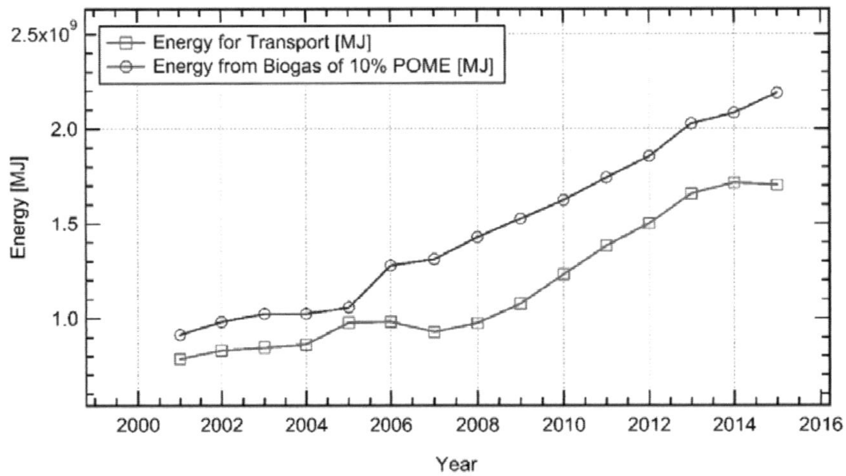

Figure 6. Energy for Transportation and Biogas from POME

Harsono et al [14] reported a study on the anaerobic treatment of POME to produce biogas as an energy source and to avoid GHG emissions. The study and data collections were carry out in a Palm Oil industry at a Palm Oil company PT Asam Jawa in Sumatera Utara province of Indonesia. The industry has plantation area of 130,000 Ha. The results revealed that total energy input and energy output of POME treatments are 84 MJ/ha/year and 2021 MJ/ha/year, respectively. The net energy yield is 1937 MJ/ha/year. Since the Palm Oil industry in Indonesia is well developed, this energy yield can be used to predict energy produced from converting POME into biogas. If only 10% of the energy produced in Biogas can be used in transportation sector, the energy produced from biogas and energy used for transportation are presented in Figure 6. The figure shows that it is possible to employ the biogas energy to replace oil consumption in transportation sector.

4. Conclusions

In this work, regulations related to climate changes in Indonesia have been reviewed. The strategies to meet the emission targets have been elaborated. The conclusions can be drawn here are as follows. The GoI has shown strong commitment on reducing its GHG emissions. Several regulations have been released to support the commitment. In the energy sector, there are two types of strategies are employed. Technologies intervention of energy efficient and switching to renewable energy resources. It was shown that Indonesia has a very big potency of new and renewable energy resources. However, in the condition of 2015, the utilized renewable energy resource is only about 5%. The GoI has a plan to increase its present energy share 5% in 2015 to 22%, 26%, and 31% in year 2025, 2030, and 2050, respectively. In order to meet these targets, biodiesel and bioethanol mandatory targets are introduced. However, the target especially bioethanol, seem to be failed. Thus, it is suggested to employ Biogas resulted from POME to be biofuel for transportation sector.

Acknowledgments

The authors gratefully acknowledge that the present research is supported by Ministry of Research and Technology and Higher Education Republic of Indonesia. The support is under the research grant BP-PTN USU of Year 2016.

References

[1] Gov. of Indonesia Presidential Decree No No 61 Year 2011, National Action Plan For Reducing Greenhouse Gas Emissions.
[2] Gov. of Indonesia, 2016, First Nationally Determined Contribution Republic of Indonesia.
[3] IPCC, 2006, Guidelines for National Greenhouse Gas Inventories.
[4] R. Dutu, 2016 *Energy Policy***98**, 513-519
[5] Y. Jafari, J. Othman, AHS Mohd Nor, 2012 *Journal of Policy Modeling***34**, 879-889
[6] Md. M. Alam, Md. W. Murad, AH Md. Noman, I. Ozturk, 2016 *Ecological Indicators***70**, 466-479
[7] S. Mujiyanto, G. Tiess, 2013 *Energy Policy***61**, 31-41
[8] Deendarlianto, A. Widyaparaga, BM. Sopha, A. Budiman, I. Muthohar, IC. Setiawan, A. Lindasista, J. Soemardjito, K. Oka, 2017 *Renewable and Sustainable Energy Reviews***70**, 12-23
[9] I. Sukarno, H. Matsumoto, L. Susanti, 2017 *Urban Climate***20**, 20-32
[10] National Energy Council of Indonesia, 2016 *Outlook Energy Indonesia*
[11] H. Ambarita, 2017 *Case Studies in Thermal Engineering***10**, 179 - 191
[12] H. Ambarita, E.P. Sinulingga, M. KM. Nasution, H. Kawai, 2017 *IOP Conference Series: Materials Science and Engineering***180**, 012025.
[13] H. Ambarita, TI. Widodo, DM. Nasution, 2017 *Journal of Physics: Conference Series***801**, 012095
[14] SS. Harsono, P. Grundmann, S. Soebronto, 2014 *Journal of Cleaner Production***64**, 619-627

IOP Conf. Series: Materials Science and Engineering **237** (2017) 012014 doi:10.1088/1757-899X/237/1/012014

Experimental study on a prototype solar water heater using refrigerant R141b as a transfer fluid

Himsar Ambarita[1*] and Tekad Sitepu[2]
[1] Sustainable Energy and Biomaterial Centre of Excellent, Universitas Sumatera Utara, Medan, Indonesia 20155
[2] Department of Mechanical Engineering, Faculty of Engineering, Universitas Sumatera Utara, Medan, Indonesia 20155

[*] himsar@usu.ac.id

Abstract. A prototype of a heat pipe type solar water heater by using refrigerant as a heat transfer fluid is investigated experimentally. The objective is to explore the performance and characteristics of the prototype when it is filled with R141b. A prototype of the solar water heater with flat plate collector is designed and fabricated. In the experiments, two different refrigerants, R141b and R718, are employed, respectively. The initial pressure of transfer fluid is varied from 10 psi to 55 psi. The prototype is exposed to solar irradiation in a location in Medan city. Solar collector temperatures, solar irradiation, water temperature, and ambient temperature are measured. The efficiency of the system is analyzed. The results show that at the same initial pressure, the solar water heater filled with R141b is better than R718. The optimum initial pressure of the solar water heater filled with R141b is 30 psi. Thermal efficiency of the solar water heater at pressure 30 psi can be up to 34%. The main conclusion can be drawn here is that the solar water heater using refrigerant R141b as a transfer fluid results in a better performance in comparison with conventional water heater.

1. Introduction

Hot water is one of the needs of human kind due to weather, production process, industry and or life style. In spite of the fact that Indonesia is a humid tropical country, the need of energy to provide water heater is significant and it increases recently. The need of water heater typically founds in hotel, hospital, restaurant and residential. The energy sources to provide water heater typically from gas, oil, coal, and electric energy. These energy resources result in Greenhouse (GHG) gas emission and burden Indonesian economy due to subsidy for oil price and electricity [1,2,3]. An alternative energy resource to provide hot water is extremely needed. The alternative energy resource to provide water heater is solar energy because it was clean and sustainable. Thus, studies on solar water have received many attentions from researchers.

There are many types of solar water heaters found in literature, such as evacuated tube, simple box water heater, and heat pipe type solar water heater. In this study, we focus on heat pipe type solar water heater. Jaisankar et al [4] reported a comprehensive review on solar water heater. It is stated that heat loss from double covers can be reduced by employing an aero design profile. This is because the proposed design lessens air movement in the above covers. Chun [5] et al reported an experimental study on employing heat pipe for solar water heater. The result show that the performance of the solar water heater strongly affected by type of transfer fluid, present of center, and other design parameters. It was recommended that using black-chrome with high technology manufacturer will improve the performance significantly. Arab et al [6] proposed vibrated heat pipe solar water heater. In the experimental work, it was shown that the best performance of the proposed model resulted by pipe filled with 70% transfer fluid. This model is stable and show longer durability. The comparison of thermosyphon and vibrated heat pipe solar water heater are made. The average temperatures of hot water resulted by thermosyphon and vibrated heat pipe type are 35.5°C and 34.5°C, respectively.

Studies of the performance of different fluid in heat pipe type solar water heater are found in literature. Esent and Hesen [7] compared the performance of R410a, R407C, and R134a as a transfer fluid. The results showed that higher performance is shown by R410a. Josep and Yakasai [8] reported the performance comparison of R12, R134a, and ethanol as a transfer fluid in heat pipe type solar water heater. The results revealed that R134a show a better efficiency and higher hot water temperature.

The above studies showed that organic fluid that usually used for refrigeration unit can be used as a transfer fluid in heat pipe type solar water heater. There is a strong effect of refrigerant type that usually used as a transfer fluid. In this study, we focus on using R141b as a transfer fluid in a heat pipe type solar water heater. However, the performance and characteristics need to be explored. The objective of this study is to explore the performance of R141b as a transfer fluid in a heat pipe solar water heater. The optimum operational condition, in term of initial pressure, is investigated. The results are expected to supply the necessary information on development high performance solar water heater.

2. Methods

In order to perform the study, a prototype as an experimental set up has been designed and fabricated. Figure 1 shows the schematic diagram of the experimental set up.

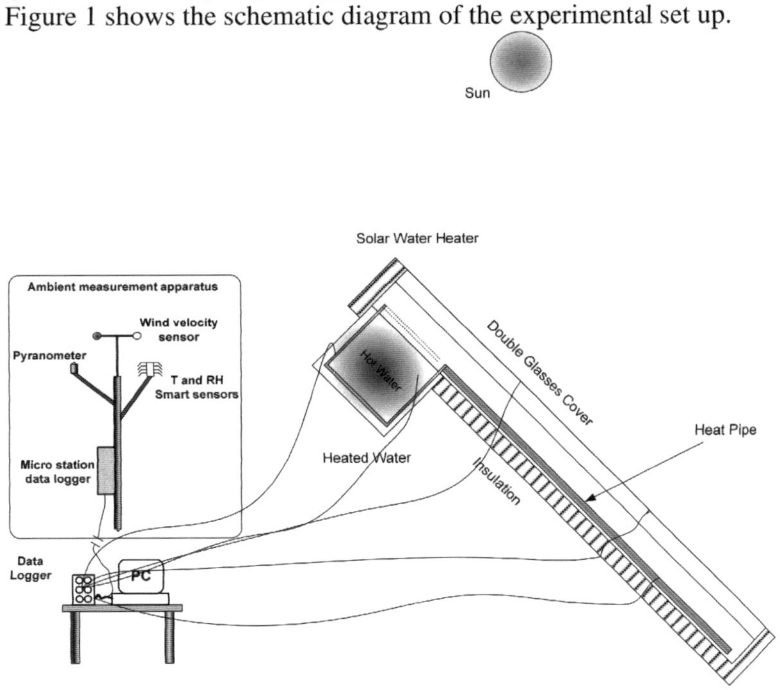

Figure 1. Experimental Set up

It consists of heat pipe type solar water heater and data acquisition system. The solar water heater consists of a flat plate type solar collector, heat pipe, and hot water tank. The dimension of collector area is 1000 mm ×200 mm and the thickness is 200 mm. In order to reduce the heat transfer loss from the collector, it is covered with insulation made of 100 mm of rock wool. The distance of glass cover to plat absorber is 100 mm. The heat pipe is placed on the absorber plate with a distance of 100 mm.Several parts of the heat pipe are passed through a water tank. The water tank is placed beneath the top part of the solar collector. The volume of the water tank is 5 liters. In order to perform experiment, a data acquisition system is used. The system consists of HOBO micro station datalogger and temperature measurement system. There are 20 thermocouples are used to measure temperatures. The sensors are made of *J*-type thermocouples with uncertainly equal to 0.1°C. Agilent 34972, multi channels data logger is used to record temperatures with interval of 1 minute. The solar radiation is measured using HOBO pyranometer smart sensor. The ambient temperature and relative humidity (RH) is measured using HOBO temperature RH smart sensor with an accuracy of 0.2°C and ±2.5% RH, respectively. The wind speed around the experimental apparatus is measured with HOBO wind speed smart sensor with accuracy ± 1.1 m/s.

In order to make performance comparison of the solar water heater, a similar solar water heater is also made but filled with different transfer fluid. In this experiment, the transfer fluid fills the heat pipe and it is evacuated at a certain pressure.

3. Results and Discussions

The working mechanism of a heat pipe solar water heater was explained as follows. The refrigerant as a transfer fluid filled the pipe. When the solar water heater is exposed to solar irradiation, the received heat will evaporate the refrigerant. The refrigerant vapor flows up to the hot water tank. In the tank, the water cools the refrigerant vapor and converted it into liquid refrigerant. The liquidrefrigerant will flow down and cycle repeats. In the experiments, temperatures and solar irradiation are measured and the performance are analysed. The results are discussed below.

3.1. Characteristics

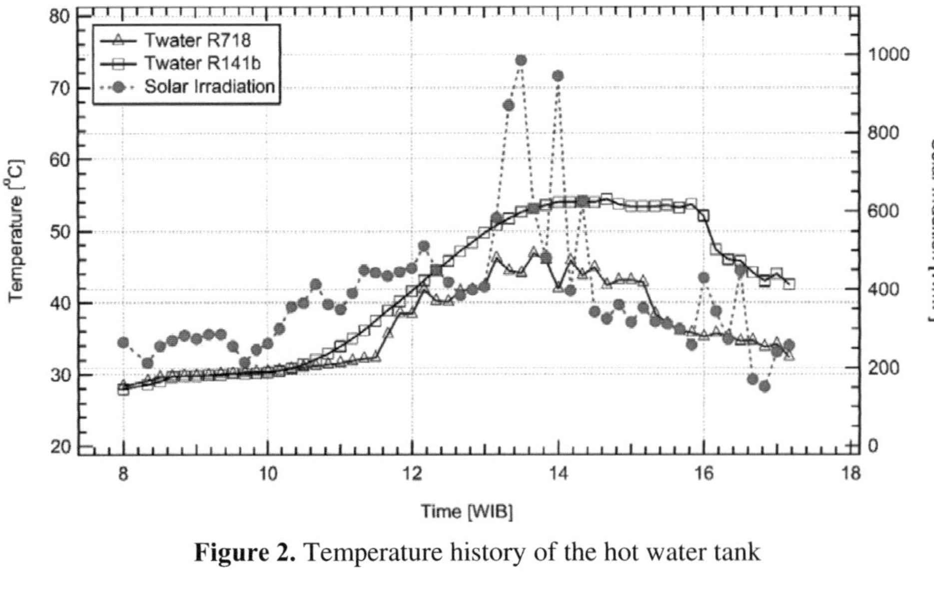

Figure 2. Temperature history of the hot water tank

Figure 2 shows temperatures history of the hot water inside the tank for solar water heater with transfer fluid R718 and R141b. Solar irradiation during the experiment is also shown in the figure. The figure shows that solar irradiation increases as time increase. However, this only for time before 14.00 local time. After that, solar irradiation decreases as time increase. The maximum solar irradiation is 1000 W/m^2 and it was captured between 13.00 and 14.00 local time. It can be seen that at the present experiment, the weather condition is categorized cloudy. The measured solar irradiations are below the theoretical clear sky solar radiation. When the R718 is used as transfer fluid, temperature of the hot water reaching its maximum value of 46.88°C at 13.40 local time. After reaching the maximum value, the temperature maximum decrease with increasing time. This is because solar irradiation and ambient temperature are decreased. The heat loss to ambient is significant. The same trends are also shown by solar water heater when the heat pipe filled with R141b as transfer fluid. In this case the temperature maximum is 54.4°C and captured at 14.34 local time. These values suggest that temperature of the hot water from R141b is higher than R718.

It can be seen clearly that there is a relation of solar irradiation with temperature of the hot water in the tank. When solar irradiation increases, by noon, the temperature of the hot water increases significantly. On the other hand, after the maximum temperature is reached, the solar irradiation not effectively enhances the hot water temperature. There are three reasons for this. The first is that temperature of the hot water is already high. The second ambient temperature is already went down and the third, solar irradiation is not sufficient enough to enhance the temperature of the water.

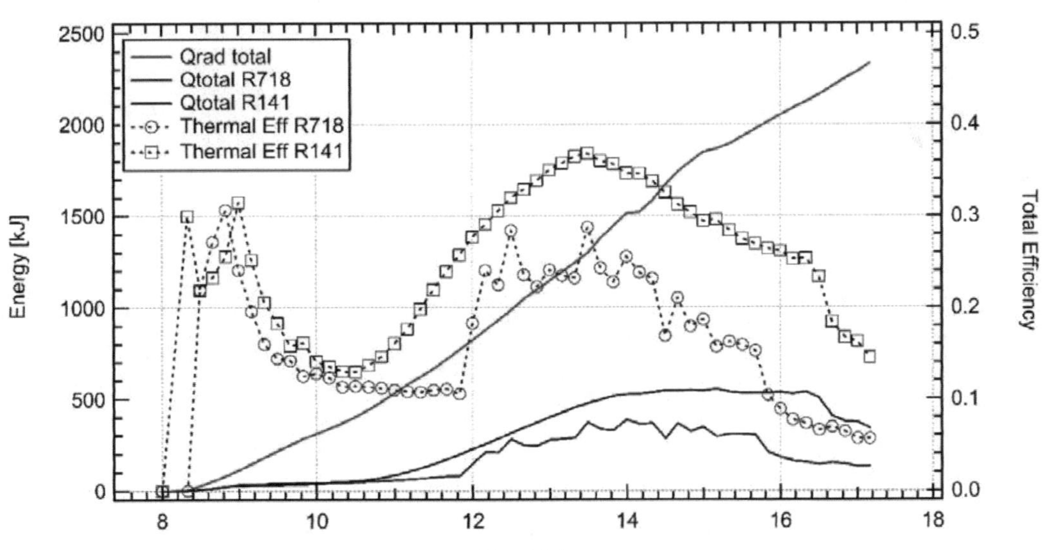

Figure 3. Total heat and thermal efficiency of the system

Figure 3 has shows total (cumulative) energy received by solar collector and total energy utilized by each hot water tank. Thermal efficiency of the solar water heater is also shown in the figure. The figure shows that total energy received by solar collector increases as time increases for all experiments. This is because the solar collector receives solar energy from the sun continuously. In the water heater case, if the solar irradiation sufficient enough the total heat utilized increase with increasing time. On the other hand, if the solar irradiation is not sufficient enough to overcome the heat loss, the total utilized decrease with increasing time. This is because temperature of the hot water in the tank is higher than ambient temperature. These facts can be seen clearly in the figure. The total heat utilized by solar water heater filled with R141b increases as time increase and reaching maximum value of 552 kJ at 14:37 local time. After this, the utilized energy decreases with increasing time. This is due to heat loss to ambient. For the solar water heater filled with R718, the maximum total heat utilized is 388 kJ at 13:39 local time. Similar with the previous water heater, the total utilized heat decreases with increasing time. This is because the heat loss to ambient. This value suggests that solar water heater filled with R141b is better than filed with R718.

3.2. Effects of the initial pressure

Based on the above experiments, the performance comparison of the solar water heater with R718 and R141b as transfer fluid has been carried out. The results show that R141b show the better performance. In the next experiment, the operational parameter is investigated. The parameter is initial pressure of the heat pipe. There are six groups of experiments have been carried out. The initial pressure varies from 10 psi to 45 psi. The initial and maximum temperatures of the hot water tank for all experiments are shown in Table 1.

Table 1 Effects of the initial pressure on the performance of solar water heater filled with R141b

Experiment	Initial Pressure	Initial Temperature	Temperature Maximum
I	30 psi	27.91°C	51.18°C
	45 psi	27.30°C	49.18°C
	55 psi	27.29°C	39.76°C
II	10 psi	29.02°C	39.34°C
	30 psi	28.35°C	40.0°C
III	15 psi	25.82°C	31.89°C
	30 psi	25.61°C	40.69°C
	45 psi	25.56°C	30.51°C
IV	30 psi	24.59°C	60.41°C
V	30 psi	26.34°C	62.18°C

In the first experiment, for the same radiation, three different initial pressure 30 psi, 45 psi and 55 psi shows different performance. The highest maximum temperature is 51.18°C shown by the solar water heater with initial pressure 30 psi. Higher initial pressures show lower maximum temperature. This suggest that lower initial pressure is better. In order to compare the performance of the water heater at much lower initial pressure, two different experiments were carried out, experiments II and experiment III. Here the initial pressure 30 psi is compared with initial pressure 10 psi and 15 psi, respectively. The results show that initial pressure 30 psi is better. In the third experiment, three different initial pressures are compared they are 15 psi, 30 psi and 45 psi. The results show that solar water heater at initial pressure of 30 psi is also better. In the last two experiments, the fourth and the fifth, the initial pressure 30 psi is show high temperature of hot water. They are above 60°C. These facts reveal that the best operational condition for the solar water heater filled with R141b is initial pressure 30 psi.

4. Conclusions

In this work, a prototype heat pipe type solar water heater has been designed and tested by exposing under solar irradiation in Medan city of Indonesia. The performances of two different refrigerants are compared. The results show that the performance R141b is better that R781. The effect of initial pressure of the solar water heater with refrigerant R141b shows that the initial pressure of 30 psi is the optimum initial pressure. It is suggested to develop heat pipe type solar water heater by employing R141b as a transfer fluid. For the present prototype with initial pressure of 30 psi, thermal efficiency and maximum hot water resulted by the solar water heater can be up to 34% and 62.18°C, respectively.

Acknowledgments

The authors gratefully acknowledge that the present research is supported by Ministry of Research and Technology and Higher Education Republic of Indonesia. The support is under the research grant TALENTA-USU of Year 2017.

References

[1] Gov. of Indonesia Presidential Y Decree No No 61 Year 2011, National Action Plan For Reducing Greenhouse Gas Emissions.
[2] Gov. of Indonesia, 2016, First Nationally Determined Contribution Republic of Indonesia.
[3] IPCC, 2006, Guidelines for National Greenhouse Gas Inventories.
[4] S. Jaisankar, J. Ananth, S. Thulasi, S.T., Jayasuthakar, K.N. Sheeba, 2011 *Renewable and Sustainable Energy Reviews* **15(6)**, 3045-50.
[5] W. Chun, Y.H. Kang, H.Y. Kwak, Y.S.Lee, 1998 *Applied Thermal Engineering* **19**, 807-17
[6] M. Arab, M. Soltanieh, M.B. Shafii, 2012 *Experimental Thermal and Fluid Science* **42**, 6-15
[7] M. Esent, H. Hesen, 2005 *Solar Energy* **79**, 459 - 468
[8] J. Eneburekhan, U.T. Yakasai, 2008 *Desalination* **243**, 208-217

IOP Conf. Series: Materials Science and Engineering **237** (2017) 012015 doi:10.1088/1757-899X/237/1/012015

Active noise control in a duct to cancel broadband noise

Kuan-Chun Chen, Cheng-Yuan Chang, Sen M. Kuo
Active Noise & Vibration Control Development Center,
Department of Electrical Engineering, Chung Yuan Christian University, Taiwan

ccy@cycu.edu.tw

Abstract. The paper presents cancelling duct noises by using the active noise control (ANC) techniques. We use the single channel feed forward algorithm with feedback neutralization to realize ANC. Several kinds of ducts noises including tonal noises, sweep tonal signals, and white noise had investigated. Experimental results show that the proposed ANC system can cancel these noises in a PVC duct very well. The noise reduction of white noise can be up to 20 dB.

1. Introduction

Exhaust duct is commonly use for ventilation or cooling system. However, when the fresh air has transmitted by the duct, the noise happens accordingly. This paper studies the noise problem in a 1-dimensional duct. Traditionally, based on acoustic theory, noise propagation in a 1-dimensional can keep its same acoustic property at its cross section [1]. In general, the conventional method to cancel duct noise only is effective in higher frequency. However, for the ventilation or cooling system, the noise has its main power at low frequency, so it is suitable to use active noise control (ANC) technique as an alternative.

There are some papers discuss the performance of ANC system in a duct. But most of them only cancel narrowband noise or use simulation to verify performance and lack implementation [2],[3]. Actually, there are random noise component in mechanical system such as ventilation or cooling system. In order to cancel random noise, online modeling for the error path and secondary path is necessary. Also, the filter length of the adaptive filter has to be increased to fit the acoustic plant. In [3], an ANC system with filtered-X least mean square (FXLMS) algorithm is developed. The adaptive filter has updated using the FXLMS algorithm [5]. In order to realize the ANC system, we have to use a microphone to get the noise source signal as the reference signal. This techniques used a signal at upstream and control the noise at downstream. It is called the feed forward ANC algorithm, which is proven to effective to cancel broadband noise [6]. In [7], the noise for the heating, ventilation and air conditioning (HVAC) system is studied. Several ANC systems had also developed for applications [8]-[10].

This paper studies the feed forward ANC system with FXLMS algorithm. A feedback neutralization algorithm has also proposed to overcome the acoustic feedback problem. Besides, we also studied the distance between the reference microphone to the secondary speaker to make the PVC duct short, more practical for applications. Several experiments are presented to show the effectiveness of the ANC system to reduce noises in ducts.

2. Feedforward ANC system with FXLMS algorithm

Active noise control uses an artificial noise to destructive interfere the undesired noise. In order to cancel the noise effectively, several ANC approaches has been developing. The most popular way is feedback and feed forward structures. The feedback ANC uses only one error microphone to synthesize the original noise, hence, only periodic noise signal can be cancel.

However, the feed forward ANC structure applies a microphone upstream to measure the undesired noise as a reference signal of ANC system, and uses an error microphone downstream to measure the residual noise. This method is effective to cancel broadband noise under the constraint of causality issue.

2.1. Single channel feed forward ANC system

Figure1 shows the single channel feed forward ANC system, where $x(n)$ denotes the reference signal, $e(n)$ represents the error signal. The $P(z)$ is the primary path between reference microphone and error microphone and the $W(z)$ is the adaptive filter that will estimate the primary path. After deriving the anti-noise $y(n)$, we send it to destructively interfere the undesired noise $d(n)$. The $S(z)$ is the secondary path that is the electrical and acoustical information of the transfer functions from the output of the adaptive filter to the error sensor, including the digital to analogue converter, low pass filter, power amplifier, loudspeaker, error microphone, pre-amplifier, anti-aliasing filter and analogue to digital converter. The $\hat{S}(z)$ is the estimated model of $S(z)$.In the single channel feed forward ANC system, we design finite impulse response (FIR) filter for $W(z)$ and $\hat{S}(z)$, an adaptive filtered-X least mean square (FXLMS) algorithm is applied to update the weights, so we have

$$\mathrm{y(n)} = \sum_{l=0}^{L-1} w_l(n)x(n-l) \quad (1)$$

$$x'(\mathrm{n}) = \sum_{m=0}^{M-1} \hat{S}_m x(n-m) \quad (2)$$

$$w_l(n+1) = w_l(n) + \mu e(n)x'(n-1), L = l-1, \quad (3)$$

The most widely used FIR, or transversal, the digital filter computes its output in a linear fashion. That is, given a set of L filter coefficients, $w(n)$, l=0,1,...,L-1, and a data sequence,{ $x(n)\, x(n-1 \ldots x(n-L+1)$ }, the output signal is computed as (1).

The $x'(\mathrm{n})$ is a filtered version of the reference input signal. Where $\hat{S}_m$ is an Mth-order estimate of the secondary-path impulse response. A further simplification can be achieved by replacing $\hat{S}_m(z)$ with a simple delay, as in Ziegler's method.(2)

Where a one-sample delay was used to produce $x(n-1)$. Both weights are update by the FXLMS algorithm as (3).

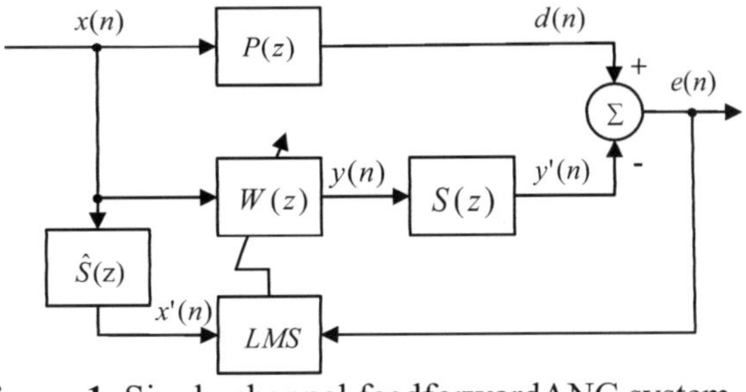

Figure 1. Single channel feedforwardANC system

2.2. Feedback neutralization

In general, the ANC system uses a microphone at upstream to measure the noise source as a reference signal and uses a loudspeaker to play the antinoise. However, when the loudspeaker is playing the antinoise, the reference microphone also takes some antinoise signal besides the noise source. This situation has illustrated in Fig. 2. In order to solve this problem, a feedback neutralization algorithm has proposed. In Fig. 2, the $u(n)$ is the reference signal picked up by a reference microphone. It also takes some component from the antinoise$y(n)$. Assuming that the transfer function from the secondary speaker to the reference microphone is $F(z)$, and the$\hat{F}(z)$ is its estimated filter. Therefore, we applies a negative feedback signal to the $u(n)$ and get the true reference signal.

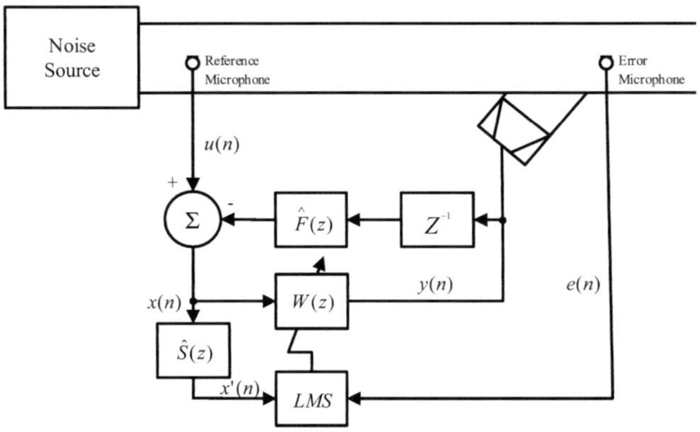

Figure 2. Feed forward ANC with feedback neutralization.

The models $\hat{S}(z)$ and $\hat{F}(z)$ can be estimated simultaneously by using the off-line modeling technique illustrated in Fig.3. The filters $\hat{S}(z)$ and $\hat{F}(z)$ must have enough taps to successfully model S(z) and F(z). The impulse response length is determined by the propagation delay associated with the secondary source and signal-conditioning low pass filters, $\hat{W}(z)$ is able to model the primary path and generate the proper signal to attenuate the primary noise .

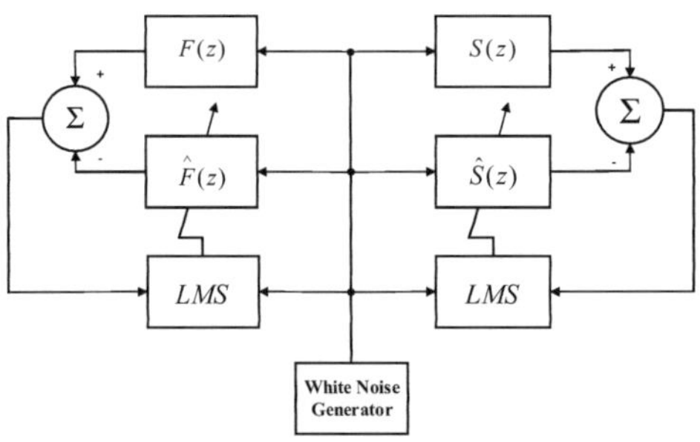

Figure 3. Simultaneous modeling of both feedback and secondary paths.

3. Experimental Results

3.1. Experimental setup

A duct made by polyvinyl chloride (PVC) is applied to implement all the duct noises cancellation experiments. One speaker (primary speaker) has mounted at one end of the duct to play the noise as the noise source. Another speaker (secondary speaker) has mounted at the side to play the anti-noise signal. One micro electromechanical systems (MEMS) microphone is inserted near the primary speaker to measure the noise source as the reference noise for the ANC system and another MEMS microphone is put at the end to measure the residual noise signal and acted as feedback signal to the ANC system. The apparatus of the duct system is shown in Fig. 4. The distance from the reference microphone to the secondary speaker is 135cm, and the diameter of the duct is 6 inches. The core for the ANC system is Texas Instrument (TI) TMS320C6713DSK and an Alexis Microtube Duo microphone preamp is used for both the reference and error microphones. Two SMSL SA-98E power amplifiers are used to drive the speakers. We also use several analog circuits such as ADC/DAC and low pass filters for necessary data processing and conversion.

Figure 4. The duct system.

3.2. Results

The first experiment shows the single tonal noise reduction. The primary speaker plays a pure tone with frequency 350 Hz and use the ANC system to cancel it. Experimental result has taken from the error microphone and has shown in Fig. 5(a). The horizontal axis denotes the time, and the vertical axis presents the frequency of the signal. In the figure, the dark color represents the higher density of noise power. We turn on the ANC system after playing the noise several seconds. Obviously, the noise power decreases dramatically when the ANC system has turned on.

The second experiment uses multi-tonal noises with 350 Hz, 450 Hz, 550 Hz and 650 Hz as the undesired noise. The experimental result is shown in Fig 5(b) and depicts similar result as in Fig. 5(a). The noise reduction performance is very well.

Figure 5(c) illustrates the results of sweep sine signals cancellation. We generate 15 tonal noises within 200-500 Hz, and then increase the frequency to 200-700 Hz and decrease the frequency to its original frequency. After ANC has turned on, we can compare the performance as shown in Fig. 5(c). This experiment shows that the proposed system is capable to cancel the noise with changing frequency, and reliable to system change.

Figure 5(d) depicts the white noise reduction. We use the white noise with frequency 200-750 Hz as the noise source. Since the speaker is not able to play the noise below 200 Hz, meanwhile, most industrial noise has its main power at frequency below 700 Hz. Therefore, we cut the frequency lower than 200Hz and higher than 750 Hz for simplicity. This is also reasonable because human ear is not sensitive to the sound with lower frequency.

IOP Conf. Series: Materials Science and Engineering **237** (2017) 012015 doi:10.1088/1757-899X/237/1/012015

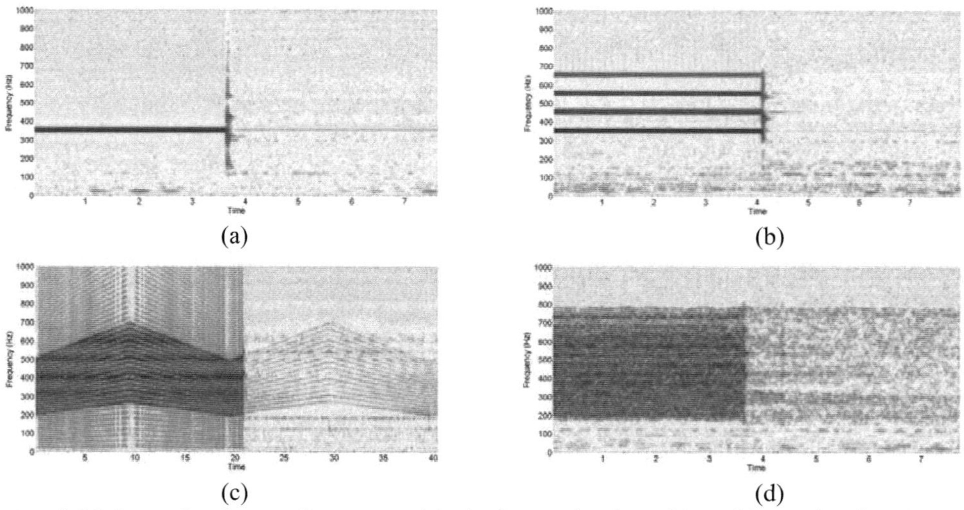

Figure 5. Noise reduction performance, (a) single tonal noise, (b) multi-tonal noise, (c) sweep sine signal and (d) white noise.

We use a sound pressure meter with A-weighting to test the noise reduction before and after ANC is turning on. The result is shown in Table 1. Obviously, the ANC system performs very well. The noise reduction can be up to 20 dB.

Table 1. Noise reduction used sound pressure meter with A-weighting.

Noise type	ANC OFF	ANC ON	Noise Reduction
Single tonal noise	86	60	26
Multi-tonal noise	80	60	20
Sweep sine signal	86	64	22
White noise	88	66	22

4. Conclusions

A single channel feed forward ANC system with feedback neutralization is proposed to cancel the duct noises. We successfully cancel the noise in a PVC duct and the length between reference microphone and secondary speaker is 135 cm. Several noises including the white noise was tested to be cancelled. Experimental results show that the causality problem can be overcomes and the noise reduction is up to 20 dB.

References

[1] Wise S, Leventhall G, 2010, *Active Noise Control As a Solution to Low Frequency Noise Problems Journal of Low Frequency Noise, Vibration and Active Control*vol 29 no.3, pp 129-137.

[2] Toochinda V, Hollot C V and Chait Y., December 2001, *On Selecting Sensor and Actuator Locations for ANC in Ducts Proceedings of the 40th IEEE Conference on Dwision and Control Orlando, Florida USA,* vol 3, pp. 2593-2598.

[3] Sawada Y, Ohsumi A, June 2010, *Active Noise Control of Sound Wave in a One-dimensional Duct Proceedings of The American Control Conference Chicago Illinois,* vol 5, pp. 3013-3017.

[4] Burian L, Fuchs P., 2005, *A Simple Active Noise Control in Acoustic DuctProceedings of the 2005 European Conference on Circuit Theory and Design, vol* 3, pp. 265-268.

[5] Burgess J C, 1981, *Active Adaptive Sound Control in a Duct: A Computer Simulation Journal of Acoustical Society of Americal,* vol.70 no. 3, pp. 715-726.

[6] Kamil Z M, Ahmed A and Tufail M, 2014, *A Duct Based Experimental Setup for Active Noise Control12th International Conference on Frontiers of Information Technology* pp 139-143.

[7] Devineni N, Panahi I, 2011, *Predictive Multi-Channel Feedback Active Noise Control for HVAC Systems, 2011 IEEE International Conferenceon Electro/Information Technology*, pp 1-5.

[8] Raghunathan K P, Kuo S M, Gan W S, 2010, *Active Noise Control for Motorcycle Helmets*, The 2010 International Conference on Green Circuits and Systems, pp. 170-174.

[9] Nazmul Huda T M, Foysol I S, 2009, *Active Noise Control for Industrial Applications, First Asian Himalayas International Conference on Internet*, pp. 1-5.

[10] Tamamura M, Shibata, 1996, *Application of Active Noise Control for Engine Related Cabin Noise JSAE Review,* vol 17 no 1, pp. 37-43.

Applying a rateless code in content delivery networks

Suherman[1,*], Muhammad Zarlis[1], Sahat Parulian Sitorus[1], Marwan Al-Akaidi[2]
[1]Universitas Sumatera Utara, Kampus USU Medan 20155, Indonesia
[2]Arab Open University, Kuwait.

*suherman@usu.ac.id

Abstract. Content delivery network (CDN) allows internet providers to locate their services, to map their coverage into networks without necessarily to own them. CDN is part of the current internet infrastructures, supporting multi server applications especially social media. Various works have been proposed to improve CDN performances. Since accesses on social media servers tend to be short but frequent, providing redundant to the transmitted packets to ensure lost packets not degrade the information integrity may improve service performances. This paper examines the implementation of rateless code in the CDN infrastructure. The NS-2 evaluations show that rateless code is able to reduce packet loss up to 50%.

1. Introduction

Content delivery network (CDN) is a collection of servers, replicating server content to other servers, to shorten the server and client proximity and to enhance service performance. CDN allows a web server (or other internet services) installed in one server, and copied to other server if necessary to make clients' accesses easier [1]. In order to do so, server selection algorithm is required. Client requests may be addressed to origin server (surrogate) and redirected to the nearest server if necessary. This requires IP redirection technique. Other solution is by using a DNS server, which forwards requests to the best server. Server selection may use polling algorithm such as round robin (RR), weighted round robin (WRR), least connection (LC) [2], geographical mapping using a domain name system (GeoDNS) [3], a web mapper [4] or other algorithms. Server update is required to maintain the integrity of data on multiple servers.

This paper examines the impact of applying rateless code [5] on CDN servers when sending short packets to client. The purpose of the implementation is to increase the service reliability for traffics that are transmitted using unreliable protocol such as user datagram protocol (UDP). Why such traffics require more reliable service? Firstly, there has been growing applications that require short packet transmissions such as whatsapp, facebook and other social media. Secondly, if those social media applications employ UDP, then it has no guarantee on data delivery [6]. Applying rateless code may improve the probability that packets are received successfully.

Rateless code adds redundancy on packet transmissions. The original m packets in sender are processed to generate m+n transmitted packets. Receiver will be able to reconstruct message if it receives at least m packets from any of m+n packets [6]. By doing this, lost on certain packet sequence could have no impact to the message integrity and the performance may increase.

Figure 1a show how the rateless code works in CDN. Originally, a social media user, M, uploads data to a social media server X (1). User N accesses data belongs to M ona social media server that is addressed to CDN DNS (2). CDN DNS selects the best (perhaps the nearest) social media server, say server Y. Server Y checks the updated version of user M by requesting update from server X (3 & 4). Server Y then sends data M to user N (5). Since the transmitted data from server X to server Y is predicted to be short, containing only update information or status, rateless code may be implemented from server X to server Y. Since some social media provides its user interface in user equipment (a gadget for instant) which store previous information, user N may only need an update data of user M Rateless code may be implemented from server Y to user N.

Server Y generates n redundant packets from m original packets (Figure 1b). User N needs to receive minimum m packets, not necessarily in order to reconstruct the transmitted information. User interface on user N decodes m packets into original message and embeds it to application data.

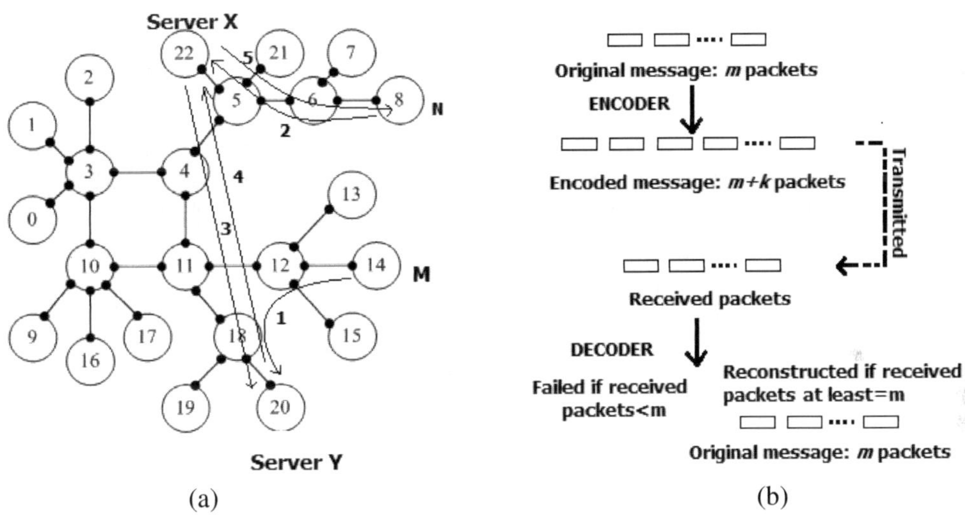

Figure 1. Content delivery network

2. Research method

In order to evaluate the rateless code performances in delivering social media application on CDN, a multi-server network is set on NS-2 simulator [7] as depicted in Figure 2. There are three servers serving 19 clients. These three servers exchange updated data among each other. 10 clients request data from servers while other 9 clients generate background traffics.

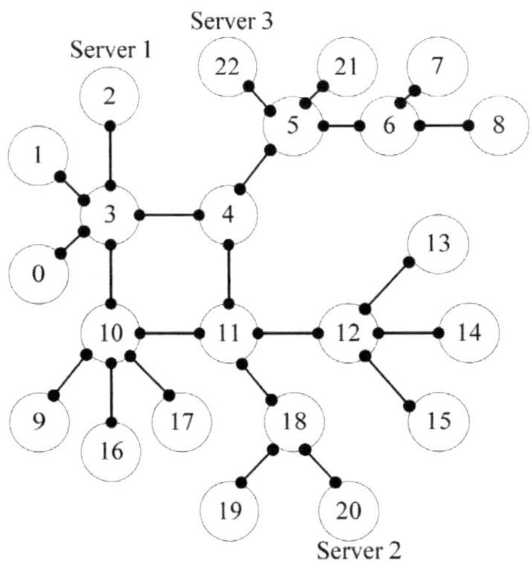

Figure 2. The evaluated network

The CDN server selection is implemented by using DNS on node 4 using round robin (RR), weighted round robin (WRR), least connection (LC) and geographically clustered service area using GeoDNS. The server, DNS, and GeoDNS coverage area are plotted in Figure 3.

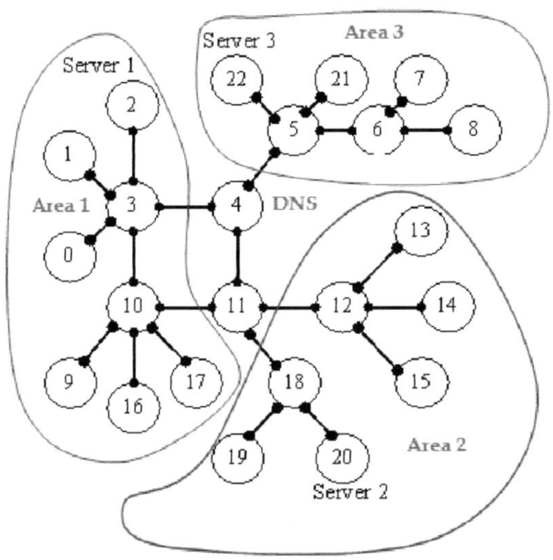

Figure 3. DNS and GeoDNS area

The rateless code encoder is implemented in server and decoder is in client. The rateless code is set to generate 50% and 100% packet redundancy. The transmitted packet is 1054 bit length with rate less than 300 kbps. The evaluated performances are in term of packet delay, packet jitters and packet loss.

3. Simulation results

3.1. Single server versus CDN

The first assessment is to compare the performances of servers when each server works as single server; each server is assigned to a certain area (GeoDNS); server selection using LC, RR and WRR algorithm. Figure 4 shows delay performance of each method. Single server experiences the worst up to 279 ms, while GeoDNS has the lowest delay down to 75 ms.

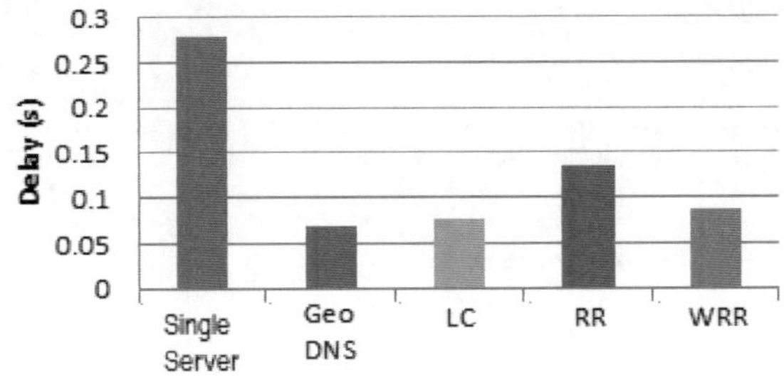

Figure 4. Delay single server vs CDN

In term of packet loss and jitter (Figure 5), least connection based server selection suffers the least. Based on these figures, assessment on rateless code impact on CDN is based on least connection server selection.

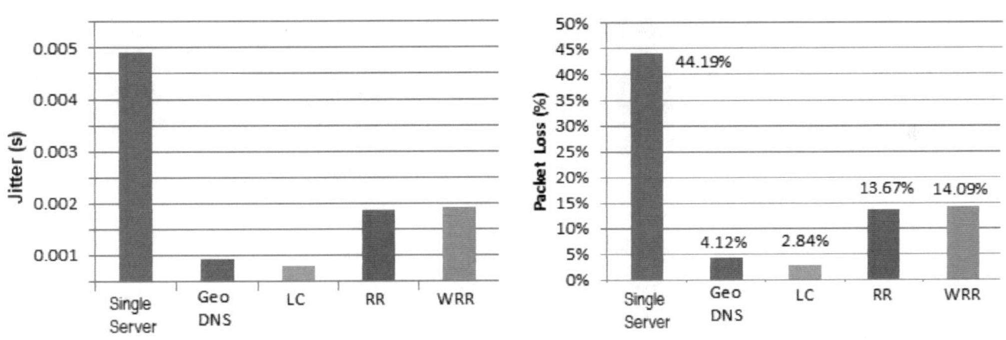

Figure 5. Jitter and packet loss single server vs CDN

3.2. Impact of Rateless Code on CDN performance

Increasing number of the transmitted packets results delay increment on packet transmission (Figure 6). Rateless code with 50% redundancy experiences 20% delay increment to about 90 ms, while 100% raises delay to 58% (119 ms). Likewise, jitter worsens by value of 20.7% and 41%. However, these values are still in acceptable range [8].

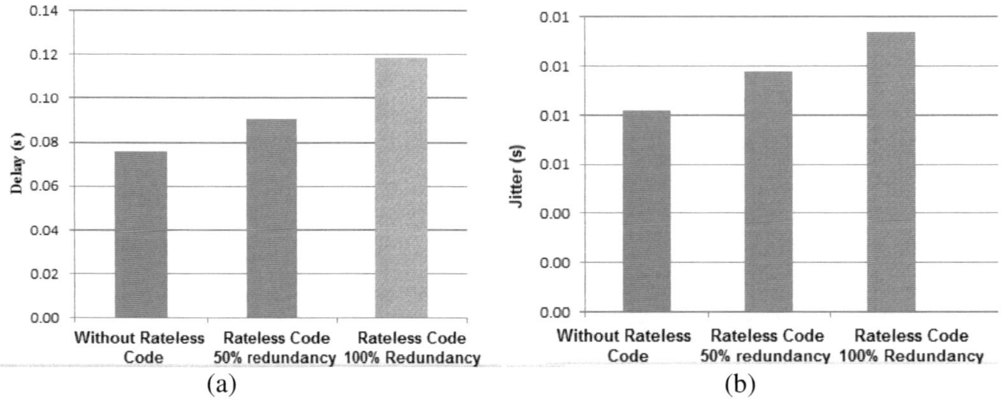

Figure 6. Delay and jitter caused by rateless code

However, number of packet loss decreases more than 50% for redundancy 50% and about 25% for redundancy 100% (Figure 7). This happens as rateless code ignores losses if the received packet reaches *m* packets, the same as the original packets.

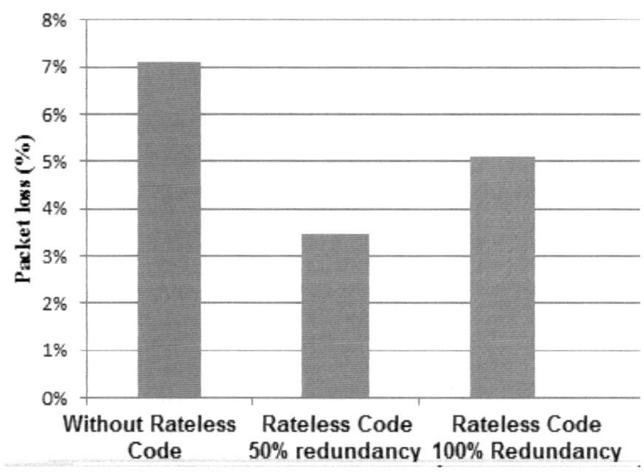

Figure 7. Packet loss caused by rateless code

On the other hand, increasing packet redundancy from 50% to 100% reduces the rateless code performance shown by the increase of packet loss. It occurs as redundant packets congested the network.

4. Conclusion

The conducted CDN assessments by using NS-2 simulations show that CDN performs much better than single server. The assessment also reveals that GeoDNS server selection generates the lowest delay as clients are served by the nearest server. However, server selection based on the least connection within the serving machine results the lowest packet loss.

Rateless code implementation on the selected CDN technique shows positive impact as number of packet loss decreases significantly. The 50% rateless code redundancy decreases lost packets up to 50%, while 100% redundancy drops the packet loss up to 25%. Even though rateless code increases delay and jitter, the values are still under the accepted level. In addition, 50% redundant performs better than 100% redundant. Therefore, redundancy rate should be carefully calculated.

References

[1]. Haßlinger, Gerhard, 2014, "Efficiency of caching and content delivery in broadband access networks." Advanced Content Delivery, Streaming, and Cloud Services: 71-89.

[2]. Soni, Gulshan, and Mala Kalra, 2014 "A novel approach for load balancing in cloud data center." In Advance Computing Conference (IACC), 2014 IEEE International, pp. 807-812. IEEE.

[3]. Khiyaita, A., H. El Bakkali, M. Zbakh, and Dafir El Kettani, 2012, "Load balancing cloud computing: State of art." In Network Security and Systems (JNS2), 2012 National Days of, pp. 106-109. IEEE.

[4]. M. Andrews, B. Shepherd, A. Srinivasan, P. Winkler and F. Zane, June 2002, "Clustering and server selection using passive monitoring." In Proc. IEEE INFOCOM, pp. 1717-1725.

[5]. Luby, Michael, Mark Watson, Tiago Gasiba, Thomas Stockhammer, and Wen Xu. 2006, "Raptor codes for reliable download delivery in wireless broadcast systems." In Consumer Communications and Networking Conference, 2006. CCNC 2006. 3rd IEEE, vol. 1, pp. 192-197. IEEE.

[6]. Suherman, Suherman, and Marwan Al-Akaidi, 2014, "Increasing uplink broadband video streaming protocol performance in WiMAX network." International Journal of Internet Protocol Technology 7, no. 3, pp. 176-185

[7]. Issariyakul, Teerawat, and Ekram Hossain, 2011, Introduction to network simulator NS2. Springer Science & Business Media.

[8]. ETSI Telecommunications and Internet Protocol Harmonization Over Networks (TIPHON), October 1998, General aspects of Quality of Service (QoS) TR 101 329 V1.2.5.

Impact of window decrement rate on TCP performance in an adhoc network

Suherman[1,*], Arief T. W. Hutasuhut[1], Khaldun Badra[1], Marwan Al-Akaidi[2]
[1]Electrical Engineering Department, Universitas Sumatera Utara, Kampus USU Medan 20155, Indonesia
[2]Arab Open University, Kuwait

* suherman@usu.ac.id

Abstract. Transmission control protocol (TCP) is a reliable transport protocol handling end to end connection in TCP/IP stack. It works well in copper or optical fibre link, but experiences increasing delay in wireless network. Further, TCP experiences multiple retransmissions due to higher collision probability within wireless network. The situation may get worsen in an ad hoc network. This paper examines the impact half window or window reduction rate to the overall TCP performances. The evaluation using NS-2 simulator shows that the smaller the window decrement rate results the smaller end to end delay. Delay is reduced to 17.05% in average when window decrement rate decreases. Average jitter also decreases 4.15%, while packet loss is not affected.

1. Introduction

Transmission control protocol (TCP) is one part of protocol stacks building computer network [1]. Its connection is reliable but sensitive to network condition. TCP adjust transmission rate by reducing or increasing its window. Window decrement rate is usually assigned by half window (0.5 x initial window). Window decrement and increment strategies are the reason TCP has many variants, such as TCP Reno and TCP Texas [2].

On the other hand, one of the networks that always give TCP difficulties is an ad hoc network [3]. The ad hoc network is a network configuration of the IEEE 802.11x wireless network which is known as Wireless Fidelity (WiFi). High demands on WiFi network make TCP performance evaluations within ad hoc network gets researcher attentions such as in [4-7]. This paper adjusts half window rates (0.5Win) on standard TCP [8] from 0.1x initial window (0.1Win) to 0.9 x initial window (0.9Win).

2. Research method

In order to evaluate window decrement rates from 0.1Win to 0.9Win, NS-2 simulator is employed. NS-2 is an event-based simulator that is flexible to make changes to some networks and protocols [9]. The simulator is employed for examining the impact of window decrement rates by directly changes the related code within 802.11 scripts. The installed NS-2 runs on Ubuntu 14.04. The evalvid from [10] is employed to evaluate network parameters such as delay, jitter and packet loss.

The evaluated 802.11 network was set by using request to send (RTS) and clear to send (CTS) multi access method, transmission power is set to cover 1000 m distance, which means whole stations are able to listen other stations; no middle temporary router is required. Network is fully in an ad hoc configuration. Modulation is fixed; 64 QAM with two-ray ground propagation model.

Number of nodes is set from 2 nodes to 20 nodes. Each station is assessed in both conditions: transmitting and receiving. The evaluated traffics are video with 300 kbps rate. TCP packet size is set to 1052 bytes.

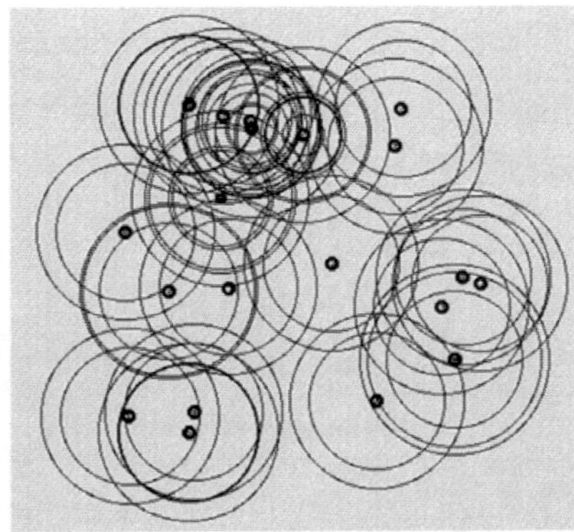

Figure 1. Snapsot of simulation configuration

Figure 1 shows the snapshot of the simulation configuration. Nodes are free moving randomly within the square area 1000x1000 m^2. Each decrement window is tested for 2 to 20 nodes. Each number of nodes is tested 20 times with duration 90 second transmissions.

3. Simulation results

3.1. Impact on end to end delay

End to end delay increases to number of nodes. By changing the value of half window from0.1to 0.9 and recording delay for each node in each simulation, the average delay is obtained as depicted in Figure 2.

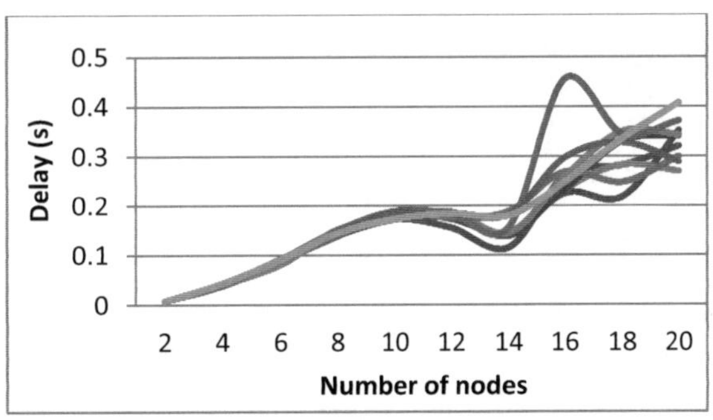

Figure 2. Delay variation for different number of nodes

Up to 10 nodes, each window decrement rate has little differences, however up to 20 nodes, delay varies. This means that window decrement rate be influential if number of the connected node is large. After the delay values for all number of nodes are averaged, the delay pattern to the decrement rate is shown in Figure 3.

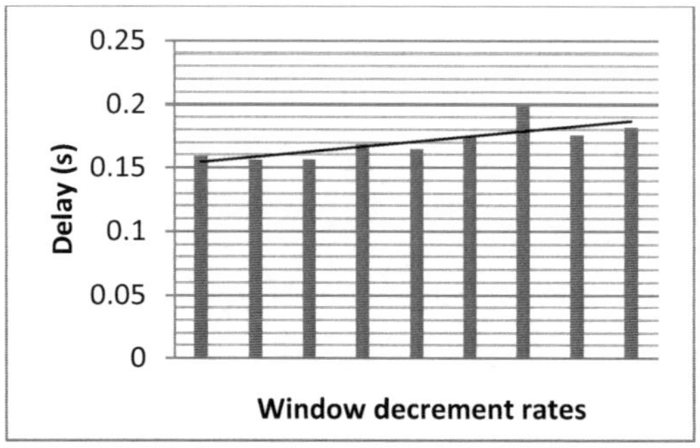

Figure 3. Average delay for each decrement rate

Figure 3 depicts the average delay of all number of nodes snapshot for increasing window decrement rate from 0.1 to 0.9. Increment causes delay increases linearly. This figure suggested that reducing window decrement rate will avoid increasing delay. Reducing decrement rate results 17.05% average delay reduction.

Suppose the duration of TCP timer waiting for the acknowledgement packet is *rto*, the flow size of packet *i*is l_i and p is the probability of dropped packet determined by the physical and medium access layers, then the expected response time for the acknowledgement is [11]:

$$r_i = \frac{l_i.rto_i}{1-p}$$

For window decrement rate k, then the average expected response time is:

$$r_i' = l_i.rto_i.2^k$$

This means, the smaller value k, the smaller ri' which causes smaller packet delay. It confirms the simulation results. Figure 4 depicts the example average response time increment to value k for packet flow 1024 bytes and *rto* 100 ms.

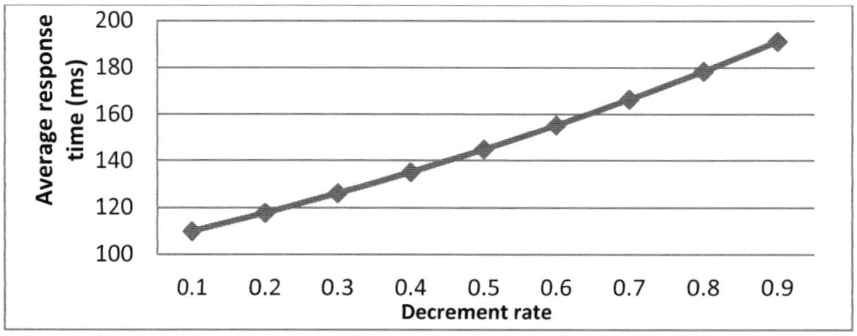

Figure 4. Average acknowledgment response time

However, since small$k$ causes significant window reduction as depicted in Figure 5, value k should be carefully chosen to avoid high overall delay. Parameter k should fit the following equation for given server rate C [11]:

$$k \geq \log_2 \frac{\sum_i^n 1/rto_i}{C}$$

IOP Conf. Series: Materials Science and Engineering **237** (2017) 012017 doi:10.1088/1757-899X/237/1/012017

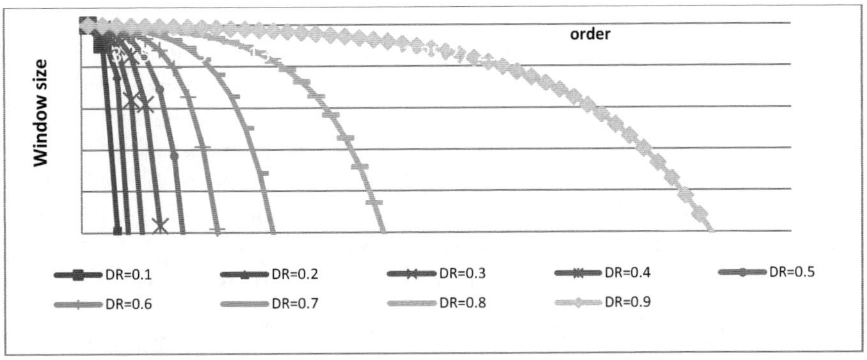

Figure 5. Decrement rate impact to window value

3.2. Impact on jitter

Impact decrement rate on jitter is the same as delay, which is influential if number of nodes large (Figure 6). Jitter pattern to decrement rate varies (Figure 7). Average jitter rises slightly if decrement rate increases. As stated in previous section, lower k results lower l_i. Smaller acknowledgement reponse time tends to reduce jitter, although in many cases, subsequent packets may experience different network conditions.

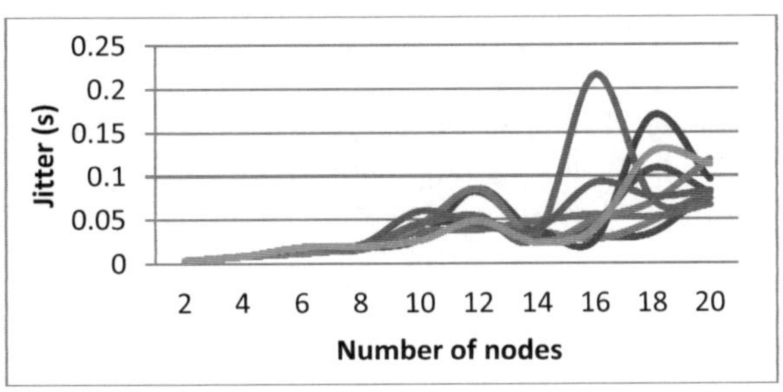

Figure 6. Jitter variation for different number of nodes

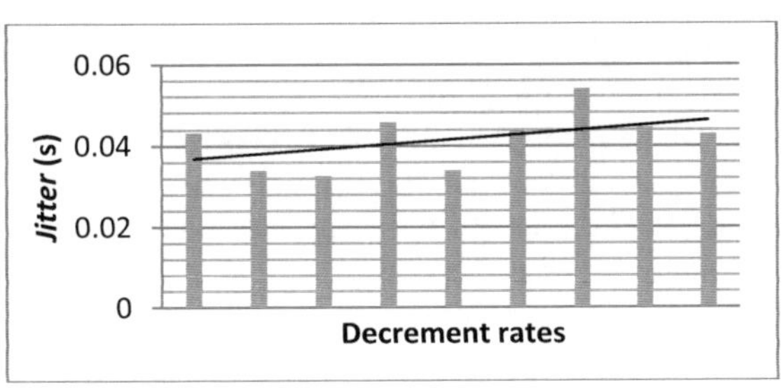

Figure 7. Jitter variation for different decrement rates

3.3. Impact on packet loss

Packet loss is not affected by the change of window as TCP keeps resending the lost packets. The average loss is zero for all conditions.

IOP Conf. Series: Materials Science and Engineering **237** (2017) 012017 doi:10.1088/1757-899X/237/1/012017

4. Conclusion

This paper has evaluated impact of changes on window decrement rate. Normally window decrement rate is 0.5 or known as half window. The assessment changes decrement from 0.1 to 0.9. As result, delay reduces in average 17.05% as window decrement rate reduced from 0.9 down to 0.1. Jitter has the same pattern, but smaller average decrement, just about 4.15%. Packet loss is not influenced by the changes as TCP keeps retransmit the lost packets.

Since half window is utilized in all TCP variants, the finding in this paper could be applied to other types of TCP. That would be of future work.

References

[1] Forouzan, Behrouz A., 2000, "TCP/IP Protocol Suite". Singapore: McGraw-Hill International Edition.

[2] Stallings W, 2007, Data and computer communications. Pearson/Prentice Hall.

[3] Suherman, Marwan Al-Akaidi, and Naemah Mubarakah, 2015, "A Transport Layer Protocol for uplink WiMAX video streaming." International Journal of Multimedia & Ubiquitous Engineering, Volume 10, Number 1.

[4] Deshpande, A. and Kaushal, A., September 2016, Feedback-based adaptive speedy transmission (FAST) control protocol to improve the performance of TCP over Ad-Hoc networks. In Advances in Computing, Communications and Informatics (ICACCI), 2016 International Conference on (pp. 2034-2041). IEEE.

[5] Deshpande, Aniket, and Ashok Kaushal, 2015, "MX-TCP and HS-TCP as Possible Options to Overcome TCP Limitations in Multi-Hop Ad-Hoc Networks." Int. J. Adv. Res. Comput. Commun. Eng. 4.12, pp. 505-508.

[6] Liaqat HB, Xia F, Yang Q, Liu L, Chen Z, Qiu T. August 2014, Poster: reliable TCP for popular data in socially-aware ad-hoc networks. InProceedings of the 15th ACM international symposium on Mobile ad hoc networking and computing pp. 429-430, ACM.

[7] Luo, Ying, et al. 2014, "An improved congestion avoidance control model for TCP Vegas based on Ad Hoc networks." Control and Decision Conference (2014 CCDC), The 26th Chinese. IEEE.

[8] Allman, Mark, Vern Paxson, and Ethan Blanton, 2009, "TCP congestion control". No. RFC 5681.

[9] Issariyakul, Terawat, and Ekram Hossain, 2011, "Introduction to network simulator NS2". Springer Science & Business Media.

[10] Klaue, Jirka, Berthold Rathke, and Adam Wolisz, 2003, "Evalvid–A framework for video transmission and quality evaluation." International Conference on Modelling Techniques and Tools for Computer Performance Evaluation. Springer Berlin Heidelberg.

[11] Mondal, Amit, and Aleksandar Kuzmanovic, 2008, "Removing exponential backoff from TCP." ACM SIGCOMM Computer Communication Review 38, no. 5, pp. 17-28.

A review on transport layer protocol performance for delivering video on an adhoc network

Suherman[1,*], Suwendri[1], Marwan Al-Akaidi[2]
[1]Electrical Engineering Department, Universitas Sumatera Utara, Kampus USU Medan, 20155, Indonesia
[2]Arab Open University, Kuwait

* suherman@usu.ac.id

Abstract. The transport layer protocol is responsible for the end to end data transmission. Transmission control protocol (TCP) provides a reliable connection and user datagram protocol (UDP) offers fast but unguaranteed data transfer. Meanwhile, the 802.11 (wireless fidelity/WiFi) networks have been widely used as internet hotspots. This paper evaluates TCP, TCP variants and UDP performances for video transmission on an adhoc network. The transport protocol - medium access cross-layer is proposed by prioritizing TCP acknowledgement to reduce delay. The NS-2 evaluations show that the average delays increase linearly for all the evaluated protocols and the average packet losses grow logarithmically. UDP produces the lowest transmission delay; 5.4% and 5.8% lower than TCP and TCP variant, but experiences the highest packet loss. Both TCP and TCP Vegas maintain packet loss as low as possible. The proposed cross-layer successfully decreases TCP and TCP Vegas delay about 0.12 % and 0.15%, although losses remain similar.

1. Introduction

According to TCP/IP stack, transport layer protocol consists of two types: reliable and unreliable protocols. The reliable means the packet transmission is guaranteed arriving at the receiver by using connection establishment, connection termination, acknowledgement, retransmission and flow control schemas [1]. In contrast, unreliable protocol means no guarantee, no confirmation and no additional services except sending the data [2].

Reliable protocol has benefits such as high reception rate, in order arriving packets, and friendly to network condition. However, some applications may experience tremendous delay. On the other hand, reliable protocol results poor packet loss and poor network congestion. However, its delay may reduce significantly. Transmission control protocol (TCP) is the representation of the reliable protocol, while user datagram protocol (UDP) is an unreliable protocol. TCP has various variants as its flow control schemas may behave differently. TCP Vegas, TCP Reno, TCP Friendly Rate Control (TFRC) are samples of the variants [3]. On the other hand, UDP could be adjusted by adding negative acknowledgement or error control code. Inter-frame Retransmission (IR) [4] and UDP-Lite are samples of the UDP improvements [5].

TCP and UDP may work properly in copper or fiber network, but deal with lots problems in wireless network. TCP retransmission may occur frequently, while UDP may experience tremendous loss.Transport layer protocols experience even more difficulties in an unstructured wireless network such as ad hoc network [6]. The challenge is even bigger when the traffic is a real time video. This paper reviews TCP, TCP Vegas, and UDP performances in delivering real time video traffics over an ad hoc network. The study has also been carried out by many researchers such as in [7, 8]. This paper also introduces the cross-layer scheme to fasten TCP acknowledgement (ACK) transmission so that TCP delay can be reduced. The fasten ACK has been proposed in previous work by using routing schema [9]. The cross layer is achieved simply by reducing the interframe space (IFS) for short acknowledgement packet.

2. Research method

In order to evaluate the protocols, NS-2 simulator is employed. NS-2 is an event-based simulator that is able to simulate network and protocol performances [10]. The simulator is running on Ubuntu 14.04 operating system. The evalvidas used in [11] is employed to record network parameters such as delay, jitter and packet loss.

The evaluated ad hoc network was set by using request to send (RTS) and clear to send (CTS) method, 1000 m coverage where whole stations are detecting other stations, modulation is fixed by using 64 QAM with two-ray ground propagation model. Number of nodes is set from 2 nodes to 20 nodes with simultaneous transmitting and receiving. The evaluated traffics are 300 kbps video. Packet size is set to 1052 bytes.Figure 1 shows the simulation configuration snapshot. Nodes are moving randomly within the square area of 1000x1000 m^2. Each number of nodes is tested 20 times with duration 90 second. In addition, TCP variant is represented by TCP Vegas.

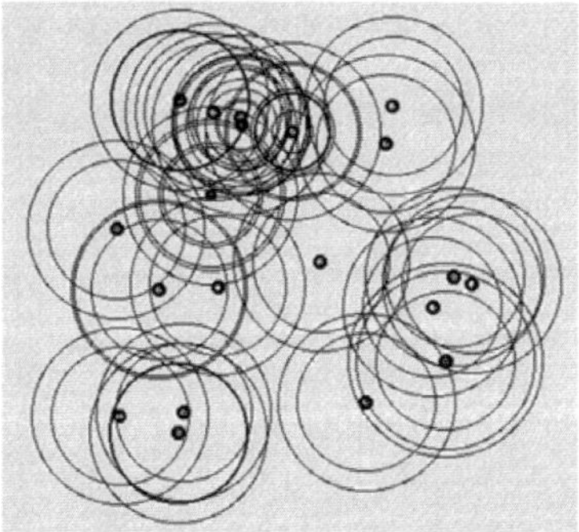

Figure 1. Snapshot of simulation configuration

3. Simulation results

3.1. Delay comparison

Figure 2a shows delay characteristics for TCP, TCP Vegas and UDP. Delay increases as number of terminal increases. TCP and TCP Vegas delays rise linearly with gradient about 1.7 and 1.6, while UDP gradient is just about 0.0002. Video packet experiences the lowest delay when transmitted using UDP, 94.6% and 94.2% lower than TCP and TCP Vegas.

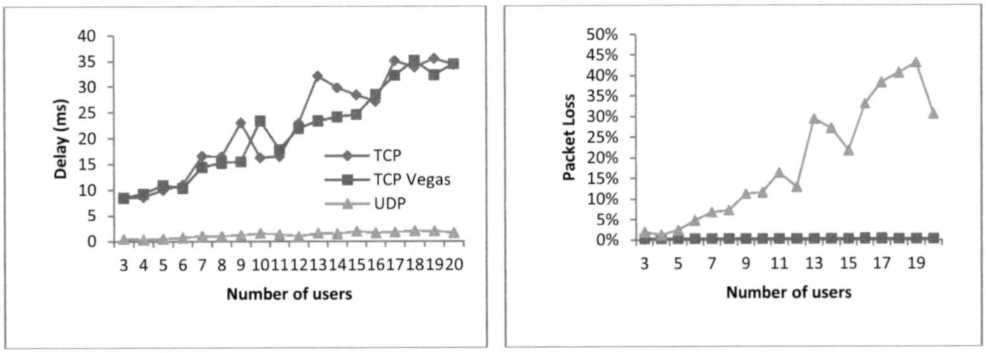

a. Delay b. Packet loss

Figure 2. Delay characteristics

3.2. Packet loss comparison

Packet loss characteristic is opposite to delay. UDP experiences tremendous loss as it does not provide acknowledgement, retransmission or even congestion control. UDP loss increase significantly to more than 40% as number of user increases (Figure 2b). TCP and TCP Vegas maintain low packet loss.

3.3. TCP and TCP Vegas comparison

TCP Vegas has more stable delay and loss than TCP. By considering r-square factor for linear trend lines, TCP Vegas has 0.054 higher r-square than TCP. The same thing occurs for packet loss, by using logarithmic trend lines, TCP Vegas r-square is 0.82 while TCP is 0.27 (Figure 3).

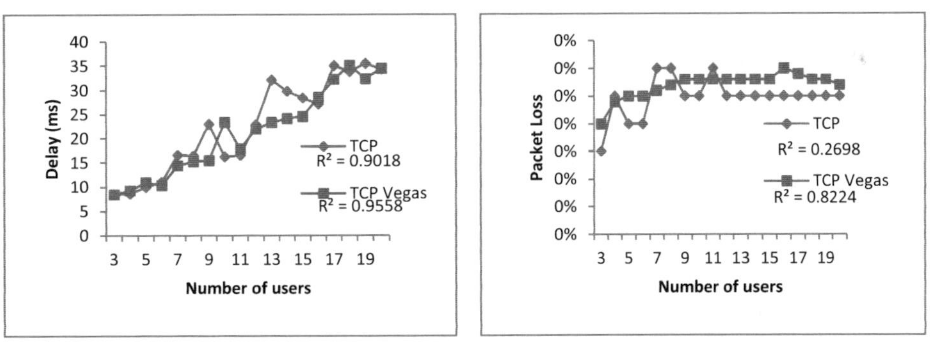

a. Delay b. Packet loss

Figure 3. TCP and TCP Vegas comparisons

3.4. Impact cross-layer on TCP Performances

Fasten ACK packet transmission effectively reduce TCP delay by 0.15 % in average (Figure 4a). Although prioritizing ACK packet may disturb data transmission, simulation shows that its impact is positive. Packet loss on the other hand remains similar.

3.5. Impact cross-layer on TCP Vegas Performances

TCP Vegas has also positive response to ACK cross-layer. Its delay reduces up to 27 % and in average decrease 0.12 % (Figure 4b). The same as in TCP, packet loss characteristic is not affected by the scheme.

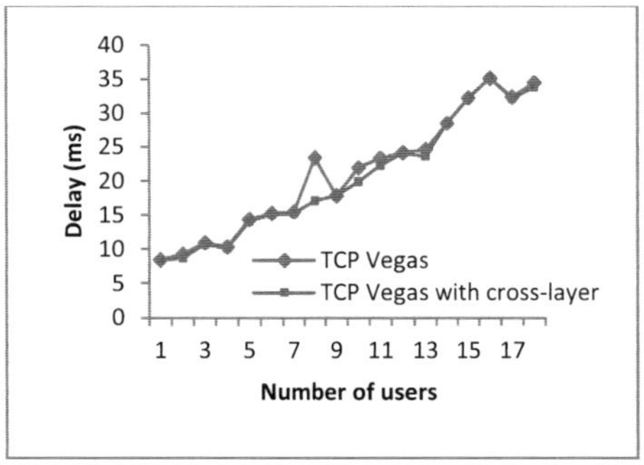

a. TCP

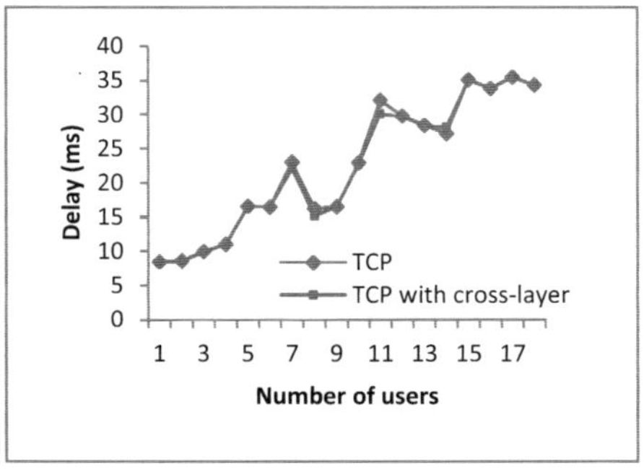

b. TCP Vegas

Figure 4. Cross-layer impacts to TCP and TCP Vegas

4. Conclusion

This paper has reviewed TCP, TCP Variant (Vegas) and UDP in ad hoc network with simulataneous transmit and receive video. UDP is excellent in term of delay, but experience enormous packet loss more than 40%. In contrast, TCP and TCP variant are excellent in maintaining packet transmission but experience higher delays. TCP variant improves TCP in term of stability.

The proposed cross-layer has positive impacts for both TCP and TCP Vegas. Delays decrease about 0.15 % and 0.12 % in averages. Since the assessment on the proposed method is achieved only by the simulations, the analytical proof is required and may be part of future work.

References

[1] Fall, K. R., and W. R. Stevens, 2011, TCP/IP illustrated, vol.1: The protocols. addison-Wesley.
[2] Forouzan, Behrouz A., 2002, TCP/IP protocol suite. McGraw-Hill, Inc.
[3] Qureshi, B., M. Othman, and N. A. W. Hamid, 2009, "Progress in various TCP variants (February 2009)." In Computer, Control and Communication, 2009. IC4 2009. 2nd International Conference on, pp. 1-6. IEEE.
[4] Suherman, Suherman, Marwan Al-Akaidi, and Naemah Mubarakah, 2015, "A Transport Layer Protocol for uplink WiMAX video streaming." International Journal of Multimedia & Ubiquitous Engineering, Volume10, Number 1.
[5] Larzon, Lars-Ake, Mikael Degermark, and Stephen Pink, 1999, UDP lite for real time multimedia applications. Hewlett-Packard Laboratories.
[6] Wu, Jiyan, Chau Yuen, Bo Cheng, Ming Wang, and Junliang Chen, 2016, "Streaming high-quality mobile video with multipath TCP in heterogeneous wireless networks." IEEE Transactions on Mobile Computing 15, no. 9, pp. 2345-2361.
[7] Saadia, Rahat, and Amir Qayum, 2014, "Performance analysis of transport layer protocols for MPEG4 video traffic in IEEE 802.11 e ad-hoc wireless networks." In Applied Sciences and Technology (IBCAST), 11th International Bhurban Conference on, pp. 504-507. IEEE.
[8] Koba, Shun, Kazuya Tsukamoto, Masato Tsuru, and Yuji Oie, 2013, "Cognitive radio-aware transport protocol for mobile ad hoc networks." In Design of Reliable Communication Networks (DRCN), 9th International Conference on the, pp. 131-138.

IEEE.
[9] Suherman, Masykur Sjani, and Evi Andriani, 2016, "Reducing multimedia transmission delay by shortening TCP acknowledgement route." In Technology of Information and Communication (ISemantic), International Seminar on Application for, pp. 114-117. IEEE.
[10] Li, Jingyu, Yunan Zhang, Jing Zhao, Yanbin Wang, Xiaomin Ma, and Wei Wu, 2017, "NS-2 Simulation of VANET for Safety Applications: Issues and Solutions." In Proceedings of the 8th International Conference on Computer Modeling and Simulation, pp. 67-72. ACM.
[11] Suherman, Suherman, and Marwan Al-Akaidi, 2013, "Increasing uplink broadband video streaming protocol performance in WiMAX network." International Journal of Internet Protocol Technology 7, no. 3, pp. 176-185.

IOP Conf. Series: Materials Science and Engineering **237** (2017) 012019 doi:10.1088/1757-899X/237/1/012019

A subjective scheduler for subjective dedicated networks

Suherman[1], Said Reza Fakhrizal[1], Marwan Al-Akaidi[2]
[1]Electrical Engineering Department, Universitas Sumatera Utara, Kampus USU Medan 20155, Indonesia
[2]Arab Open University, Kuwait

suherman@usu.ac.id

Abstract. Multiple access technique is one of important techniques within medium access layer in TCP/IP protocol stack. Each network technology implements the selected access method. Priority can be implemented in those methods to differentiate services. Some internet networks are dedicated for specific purpose. Education browsing or tutorial video accesses are preferred in a library hotspot, while entertainment and sport contents could be subjects of limitation. Current solution may use IP address filter or access list. This paper proposes subjective properties of users or applications are used for priority determination in multiple access techniques. The NS-2 simulator is employed to evaluate the method. A video surveillance network using WiMAX is chosen as the object. Subjective priority is implemented on WiMAX scheduler based on traffic properties. Three different traffic sources from monitoring video: palace, park, and market are evaluated. The proposed subjective scheduler prioritizes palace monitoring video that results better quality, xx dB than the later monitoring spots.

1. Introduction

WiMAX (IEEE 802.16) is other version of WiFi (802.11) that serving wider coverage area as well as high bit rate [1]. If WiFi employs request to send and clear to send (RTS/CTS), WiMAX uses frame allocation [2]. If WiFi allocates channel based on first come first served (random access), then WiMAX uses scheduler. The basic schedulers such as round robin (RR), weighted round robin (WRR), deficit round robin (DRR), first in first out (FIFO) can be implemented in WiMAX [3]. Cross layer scheduler based on physical (PHY) and medium access (MAC) cross layers such as the earliest deadline first (EDF) [3] and application cross layer such as frame-based [4] and non-sorting schedulers[5] are potentially improving the WiMAX performances.

This paper proposes the subjective feature of the application layer to be the basis of prioritization in WiMAX scheduler. The proposed method is an application and MAC cross layer technique as in frame-based scheduler. If frame-based scheduler enables WiMAX to prioritize I-frames of any multimedia packets, the proposed scheduler differentiates priority of packets based on its subjective feature. For instance, if a network is dedicated for football fans, then scheduler should prioritize any football-related packets. If a network is dedicated for education, scheduler prioritizes education materials. This requires application data to flag the traffic types. Although current application does not support it, future application may use it, for instance, prioritized traffics use a specific range of port numbers.

IOP Conf. Series: Materials Science and Engineering **237** (2017) 012019 doi:10.1088/1757-899X/237/1/012019

2. Research method

In order to evaluate the proposed scheduler, the NS-2 simulator was employed. The simulation scenario uses a point to multipoint system with 4 subscriber stations (SSs). The network configuration is shown in Figure 1. The transmitter power is set to provide a 1000 m coverage. The modulation scheme is 64 QAM. The Hatta propagation model is applied. The downlink to uplink ratio is set to 0.3, which means 30% of the network resources are allocated to downlink.

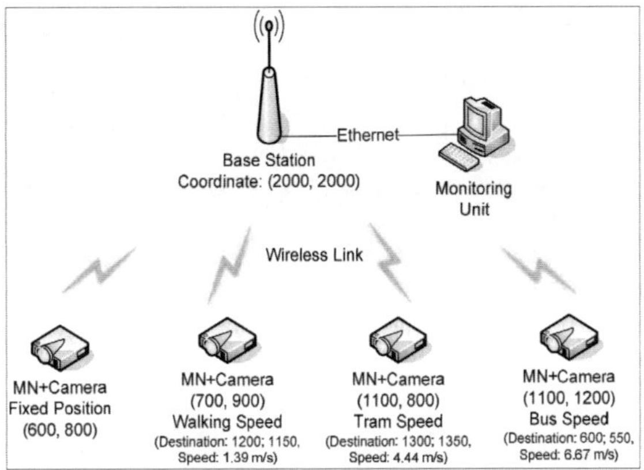

Figure 1. Network configuration

The evaluated traffics are uplink video streaming flowing from SS to a server connected to base station. Video streams are monitoring videos: palace, park and market. The priority is set to be the palace video. Each video source assigns flag to its packet header; 1 for palace, 2 and 3 for park and market. The evaluated performance metric is image quality, peak to noise ratio (PSNR). The objective features are frame type and GoP; and the subjective feature is flag 1.

3. Simulation results

By comparing the proposed scheduler to frame-based and non-sorting scheduler, the average PSNR values for prioritized and non-prioritized video are shown in Figure 2.

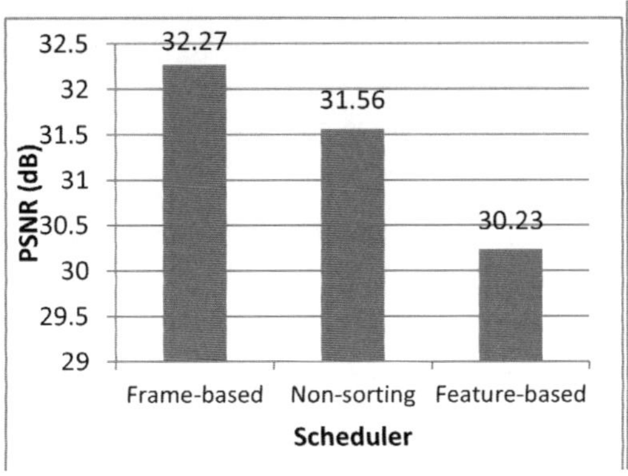

a. Non-place image

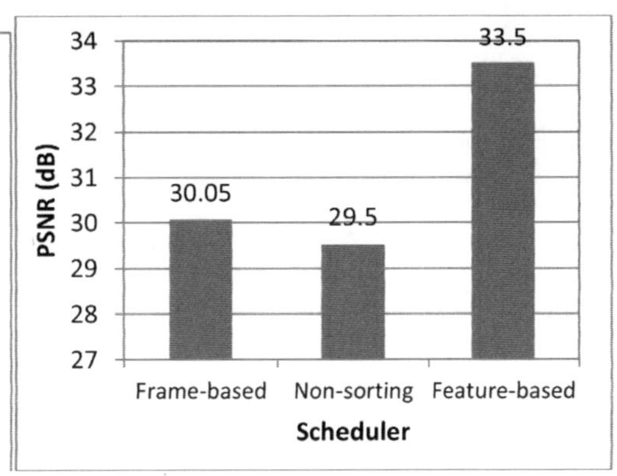

b. Palace image (prioritized)

Figure 2. Subjective feature performance

The non-prioritized videos: park and market produce higher image quality for frame-based and non-sorting schedulers. As both schedulers do not care what video flows in network. In opposite, the proposed scheduler is able to determine the prioritized video. As result, palace image has the clearest and highest quality than other images within the proposed and the compared schedulers. Palace image is 33.5 dB, 1.23 dB higher the best image produced by other scheduler. It shows that the proposed scheduler successfully achieves its goal.

4. Conclusion

The NS-2 evaluation of the proposed scheduler, which added video type property to frame type and GoP, is able to increase image quality of the prioritized video. The scheduler achieves the highest PSNR, 33.5 dB, compared to frame-based and non-sorting schedulers. Implementation of subjective scheduler to other type of networks could be of future work.

References

[1]. H. L. Vu, S. Chan, L. Andrew, 2010, "Performance Analysis of Best-Effort Service in Saturated IEEE 802.16 Networks", IEEE Transactions on Vehicular Technology, vol. 59, no. 1, pp. 460-472.

[2]. T. Koon Hoo, Zhifeng Tao, and Jinyun Zhang, 2007, "The mobile broadband WiMAX standard [standards in a nutshell]." IEEE Signal Processing Magazine 24, no. 5, pp. 144-148.

[3]. P. Dhrona, N. Ali, H. Hassanein, October 2008, "A performance study of scheduling algorithms in Point-to-Multipoint WiMAX networks", in Proc. Local Computer Networks Conference, (Montreal, Que.), pp. 843-850.

[4]. S. H. Kang, A. Zakhor, April 2002, "Packet scheduling algorithm for wireless video streaming", in Proc. International Packet Video Workshop, Pittsburgh, PY.

[5]. S. Suherman, M. Al-Akaidi, 2013, "Adjusting WiMAX for a dedicated surveillance network", International Journal of Electrical and Computer Engineering, vol 3, no 4.

Assessing the ERP-SAP implementation strategy from cultural perspectives

Gunawan Wang, Bakhri Syaiful, Sfenrianto, Fajar Ahmad Nurul
Master in Information Systems Management, Bina Nusantara University, Jakarta, Indonesia.

gwang@binus.edu; sfenrianto@binus.edu; afajar@binus.edu

Abstract. Implementing ERP-SAP projects in Indonesian large enterprises frequently create headaches for the consultants, since there are always be a large gap between the outcomes of the SAP with the expected results. Indonesian enterprises have experience with a huge amount of investments and ended up with minor benefits. Despite its unprecedented benefits, the SAP strategy is still considered as a mandatory enterprise system for every enterprise to compete in the marketplaces. The article examines the SAP implementation from cultural perspectives to present new horizon that commonly ignored by major Indonesian enterprises. The article applies the multiple case studies with three large Indonesia enterprises, such as KS, the largest steel producer; GEM, a subsidiary of conglomerate enterprise operates in the mining industry, and HS, a subsidiary of the largest retailer in Asia with more than 700 stores in Indonesia. The outcome of the article is expected to provide a comprehensive analysis from cultural perspectives regarding to common problems faced by SAP consultants.

1. Introduction

Enterprise resource planning (ERP) system is an integrated information system that enables to influence internal business process and external firms. It also enables to blend global standard with local situation (1). An ERP system allows an organisation to establish a convergent and integrated view of the enterprise information through the use of centralised databases and integrated business processes across different divisions and departments (2). Successful ERP implementation is believed may improve organisational capabilities such as: agility, flexibility, internal communication and cross-functional coordination, product innovation, process improvement, and information access that help the firms to respond quickly to the customer needs (3).

Due to its advantages, the ERP implementations in many firms are considered as an investment wastes rather than promising real benefits (4)(5). The similar cases also can be found with the ERP-SAP (SAP) implementation in major large enterprises in Indonesia. The SAP implementation is commonly perceived as complex tasks and surrounding with cultural barriers, with high failure rates that can reach more 80% (6)(7)(8)(9). The risk of SAP implementation increases along with the size of the enterprise. With large amount of money invested in SAP system, less than 75% of the SAP system's effectiveness was utilized (10). Scholars argue that ERP as an outcome of human actions and interactions (1)(11), whenever prepared and applied effectively, the ERP enabled to deliver benefits like faster transaction processing, error reduction, and improved productivity for the adopting organisations (12)(13).

In many firms, the SAP strategy is dominated by more for marketing and political purposes rather on the real needs. Whenever the SAP systems do not fit, the adoption would ultimately deliver badly impact to the post-adoption performance, and their underperformance would act as snow ball effects to create major problems for the firm (14). Scholars addressed the organisations with good IT governance and organisation culture to have more effective ERP investments compared to those who do not have. The firms with good enterprise system design enables to improve their innovation through supporting internal idea generation and implementation, and it delivers impact to product or process improvement (15). The article examines the key success factors of the SAP implementation strategy in the leading enterprises in Indonesia. The article takes case study of three leading enterprises, such as: KS (the largest steel producer and a member of national strategic enterprises), GEM (subsidiary of national conglomerates in energy producer), and HS (a subsidiary of the largest retailer in Asia region). Those are the leading firms in their fields.

2. Literature Review

In the complex competition environment, the use of ERP system enables to provide the firms with a competitive advantage by improving the organizational value stream (16)(17)(18). However, applying ERP system is not an easy task. Davenport argues the ERP projects often acting against the prevailing firm culture (12)(9). Olie (1995) summarised the concept of culture as: (1) culture is a collection of individuals who share common values, beliefs, idea, etc.; (2) culture is learned; (3) culture has a historical dimension; (4) culture has different layers, such as: symbols, heroes, rituals, and values. Boersma and Kingma (2005) examines the complexity of ERP as a condition of organisations where it constitutes the intersections of segments of autonomous systems of goals. They emphasize that cultural processes play important roles in the concept of sense making to address the gap between the expectations and evaluations of ERP projects.

ERP has important implications for the organizations such as (17): (1) integration, ERP promises seamless integration of all the information flowing through a company that includes financial and accounting information, human resource information, supply chain information and customer information; (2) packages, ERP are commercial packages that are designed to support generic business processes; (3) some assembly required, there is no such single enterprise system enables to meet all the information-processing needs of the majority of organizations, integration challenges are needed to cope with this. The ERP project outcomes can be measured on that specific knowledge (20) such as: (1) software specific knowledge; (2) business process specific knowledge; and (3) organisation specific knowledge;

The ERP system faces pressure in narrowing the gap between pre-packaged information system and perceived functionality required by the firm (21). Some gaps are resolved in business process reengineering (BPR), or software customisation through knowledge management framework (22)(23), while other gaps will never be resolved (21). Knowledge management efforts includes to (22): (1) capture knowledge; (2) convert personal knowledge to group-available knowledge; (3) connect people to people, people to knowledge, knowledge to people, and knowledge to knowledge; and (4) measure that knowledge to facilitate management of resources and help understand its evolution.Transferring explicit knowledge to the adopting organisation is part of the standard ERP implementation strategy.

3. Methodology

3.1. Research Method

The article examines the multiple-case studies for SAP implementation in the leading enterprises, that enables to explore the differences between cases (24). The case study method is selected due to its advantages in studying phenomena in their natural setting and learn about the state of the art (25). It enables deeper exploration into selected respondents. The objective of the study is expected to replicate the findings across similar cases with the objectives to predict the similar results among the enterprises that have implemented SAP in Indonesia(25). The article has the proposition to examine: how the cultural issues affect the SAP implementation strategy in large Indonesian enterprises.

3.2. Enterprise Overview
There are three enterprises selected as a case study in the article such as: (1) KS enterprise, is the largest steel producer in Indonesia, has produced more than 3 million tonnes of steel each year. Currently, KS has more than 8 subsidiaries to serve local demands and international markets. Since 1995, KS has installed major SAP R/3 modules such as: FICO (Financial Controlling), SD (Sales and Distribution), HR (Human Resources), PSIM (Project System & Investment Management), MM (Material Management), QM (Quality Management), PM (Plant Maintenance), PP (Production Planning), and XI (Xchange Interface); (2) GEM enterprise, is a subsidiary of national conglomerate in energy industry. GEM has a major player in the trading sector in mining products/services. Since 2010, GEM has installed major SAP modules such as: FICO (Financial Controlling), MM (Material Management), SD (Sales and Distribution), and PP (Production Planning). GEM with its subsidiaries has operated large mining fields in Jambi (Sumatra) and Kalimantan (Borneo); (3) HS enterprise, is a subsidiary of the largest retailer in Asia region with more than 700 stores in Indonesia. HS has numerous subsidiaries in retailers, health and beauty stores, convenience stores, and mini markets. Since 2010, HS has implemented SAP modules such as: FI (Financial), ME (Merchandise), ME (Merchandising) and RL (retail).

3.3. *Data gathering method*
The article applies the interviews, focus group discussion and field observations to the selected 118 respondents in those three enterprises. They are summarised as follow:

Table 1. Distribution of the respondents.

No.	Respondents	KS	GEM	HS	Total	percen-tage (%)
1	**Age:**					
	< 30 years	48	19	5	72	61.02%
	31 – 40 yrs.	9	15	5	29	24.57%
	41 – 50 yrs.	8	1	6	15	12.71%
	> 50 yrs.	2	0	0	2	1.70%
2	**Education:**					
	High School	4	0	1	5	4.24%
	Vocational	6	1	4	11	9.32%
	Bachelor	52	28	8	88	74.58%
	Post Graduate	5	6	3	14	11.86%

No.	Respondents	KS	GEM	HS	Total	percen-tage (%)
3	**Working Exp:**					
	< 3 yrs.	32	20	6	58	49.15%
	3 – 5 yrs.	21	12	4	37	31.36%
	6 – 10 yrs.	6	3	4	13	11.02%
	> 10 yrs.	8	0	2	10	8.47%
4	**Title:**					
	Staff (*Analyst*)	37	33	15	85	72.03%
	Consultant	26	0	0	26	22.03%
	ERP PM	2	1	1	4	3.39%
	Head of Department	2	1	0	3	2.54%

4. Case Study

4.1. Findings and Analysis: Characteristics of the Enterprises.
Interviews and observations have been done to examine the characteristics of those three enterprises (see Table 2 below). The findings in table 2 infers the organisational issues are highly influenced by the culture andit affects the SAP system implementation (7): (1) mismatch with local culture and SAP cultural readiness, SAP strategy requires cultural ready and systematic thinking which are lacking in almost all Indonesian enterprises. Lack of these

factors affects the SAP strategy design, preparation and implementation; (2) lack of ownership culture, there is no share-ownership culture of all departments to SAP strategy. The case study above addresses the SAP decision is a top-down decision, and ownership culture is perceived belongs to top management issues; (3) management and communication culture, the SAP adoption in most Indonesian enterprises are led more on the political and marketing purposes rather than based on real needs. There is also lack of communication strategy to articulate the interest of top management; (4) cultural change, SAP strategy states the importance of revising the current organisation culture. The table 4 shows that top management puts minimum efforts on the adopting the cultural change to support the SAP strategy. It can be seen in: top management support and commitment, clear vision, goals and objectives of SAP system, end-user involvement, organisation fit for SAP, business process reengineering (BPR) initiatives, and process management, etc. (26); (5) information flow, SAP requires the role of KM strategy to manage the information flow of tacit/explicit knowledge amongst the stakeholders, such as: users across departments/divisions, consultants, and management. However, the appropriate KM strategy hardly can found in the case study; (6) sectoral differences, due to the vast of organisations, places and distribution of people, the sectoral ego dominates the Indonesian large enterprises. It creates political and psychological barriers to SAP adoption. Understanding the sectoral differences should be considered as an essential component of SAP implementation strategy.

Table 2. Comparison of the Characteristics of the Enterprises.

No.	**Business-IT Issues**	**KS** (State-owned Enterprise)	**GEM** (Energy enterprise)	**HS** (Retailer enterprise)
1	Top Management Supports	Lack of appreciation for the staffs to encourage motivation to use IT-based applications.	Internal conflict intra/inter department and lack of support to SAP implementation.	Lack of supports to SAP implementation team.
2	Organisation Culture	External intervention (government bodies) to the KS strategy and operation. Intervention can go deep into ERP strategy and SAP implementation.	Leadership is controlled by business family (Conglomerate). External intervention from other subsidiaries/holdings, although is not frequent, may impact the ERP strategy and implementation.	External intervention especially from holding company. All SAP operation is managed centralised and may slow down the decision process in the subsidiaries.
3	Challenges of Project Manager (PM)	There are always unexpected things required by the users or top management (beyond initial plan).	Multiple roles required for PM. GEM requires the PM should be ready for all projects complemented with SAP applications such as legacy applications.	PM should develop adequate knowledge related to retail business. Current PM has a high dependency on the knowledge of key users in each department.
4	Business-IT Plan	Focus on extending SAP with legacy applications;	Focus on extending SAP with legacy applications;	DFG (Holdings) controls all SAP operation in all subsidiaries including HS;

Note: PM: project manager;

4.2. Findings and Analysis: Technical Perspectives

Technical aspects are also the major cultural issues need to be examined in SAP implementation. Findings address the SAP consultant needs to understand: (1)SAP specific knowledge and interconnection with legacy applications (27)(28). ERP strategy should deal with integrating legacy and SAP culture; (2) business process specific knowledge, relates to managing cultural change or resistances in BPR initiatives, applying KM culture relates to tacit/explicit knowledge transfer, and project management framework dealing with technical process management (29); (3) organisation process specific knowledge, relates to political decision making and marketing expectation of SAP strategy. SAP implementation strategy also should deal with narrowing the gap between expected SAP results and SAP outcomes.

5. Lesson Learned

Scholars have addressed the importance of culture preparation to smoothing the SAP implementation strategy. The SAP software has embedded its success culture based on the best practices in the developed countries, while the Indonesian enterprises are not familiar with this culture. It creates a gap between expected SAP results with the outcomes. The SAP implementation in major enterprises is surrounded with the political environment with a top-down approach, while other related to marketing purposes. Very rare, the SAP adoption reflects the real needs of the firm. Case study above infers the cultural issues related to SAP implementation strategy can be summarised into three specifics knowledge. They are (20): (1) software specific knowledge, in the article refers to expertise of SAP vendors and best practice associated with legacy applications; (2) business process specific knowledge, it relates to preparing BPR initiatives, KM (30) and project management framework; and (3) organisation specific knowledge, it relates with understanding the decision process, leadership style and managing change culture (31)(32). The effective consultant should be able to anticipate that knowledge into SAP implementation strategy.

6. Conclusion

SAPproject is a complex task with high failure rates in implementation, especially when applying in Indonesian enterprises. The article examines these issues from cultural perspectives to provide holistic analysis to the existing condition. Literature studies have shown that cultural issues often ignored in preparing SAPimplementation strategy, and it delivers significant impacts to SAPoutcomes. SAP strategy requires changing in organisation culture comply with the best practices practised by well developed firms. Indonesian enterprises are still relatively new and may not be exposed to this kind of culture, and they need special attention on it. The article examines the cultural issues that commonly ignored by the Indonesian enterprises whenever they implemented SAP system such as: SAP specific knowledge (best practices) and its integration with legacy applications, business process specific knowledge (relates to BPR initiatives, KM and project management framework), and organization specific knowledge (relates to understanding decision making, leadership style and changing culture).

References

[1]. Boersma K, Kingma S. Developing a cultural perspective on ERP. 2005;11(2):123–36.
[2]. Musa U, Usman Z. Knowledge Management in Success of ERP Systems. 2012;3(1):21–8.
[3]. Hwang W. The Drivers of ERP Implementation and Its Impact on Organizational

Capabilities and Performance and Customer Value. 2011;(June):1–301.
[4]. Esteves de Souza JM. Definition and analysis of Critical Success Factors for ERP implementation projects. Thesis. 2004;1–313.
[5]. Ehie IC, Madsen M. Identifying critical issues in enterprise resource planning (ERP) implementation. Comput Ind. 2005;56(6):545–57.
[6]. Hidayanto AN, Hasibuan MA, Handayani PW, Sucahyo YG. Framework for measuring ERP implementation readiness in small and medium enterprise (SME): A case study in software developer company. J Comput. 2013;8(7):1777–82.
[7]. Zaglago L, Apulu I, Chapman C, Shah H. The Impact of Culture in Enterprise Resource Planning System Implementation. Proc World Congr Eng. 2013;1:3–8.
[8]. Rabaai A. the Impact of Organisational Culture on Erp Systems Implementation : Lessons From Jordan the Impact of Organisational Culture on Erp. 2009;
[9]. Skok W, Doringer H. Potential impact of cultural differences on enterprise resource planning (ERP) projects. Electron J Inf Syst Dev Ctries [Internet]. 2002;7:1–8. Available from: http://www.ejisdc.org
[10]. Bernroider EWN. IT governance for enterprise resource planning supported by the DeLone-McLean model of information systems success. Inf Manag. 2008;45(5):257–69.
[11]. Williams R, Edge D. The social shaping of technology. Res Policy. 1996;25(6):865–99.
[12]. Davenport TH, Harris JG, Cantrell S, Davenport TH, Harris JG, Cantrell S. Enterprise systems and ongoing process change. 2004;
[13]. Akram K. Role of Knowledge Management to Bring Innovation : An Integrated Approach. Int Bus Res. 2011;11(11):121–34.
[14]. Ugrin JC, Morris JJ, Ott RL. A study of how underperforming firms follow industry leaders when adopting ERP systems and the economic effects of their adoption decisions. Commun Assoc Inf Syst. 2016;39(1):52–72.
[15]. Bhattacharya PJ, Seddon PB, Scheepers R. Enterprise systems for innovation in products and processes: Beyond operational efficiency. In: ACIS 2012: Location, location, location: Proceedings of the 23rd Australasian Conference on Information Systems 2012. 2012. p. 1–11.
[16]. Hayes,David C; Hunton, James E.; Reck JL. Market Reaction to Auditor Changes.pdf. Journal of Information Systems, Vol.15, No.1, pp 3-18; 2001.
[17]. Markus ML, Tanis C. The Enterprise System Experience — From Adoption to Success. Fram Domains IT Manag Proj Futur Through Past. 2000;173–207.
[18]. Nah FF, Delgado S. Critical Success Factors for Enterprise Resource Planning Implementation and Upgrade. J Comput Inf Syst. 2006;47(Special Issue):99–113.
[19]. Olie R. The'culture'factor in personnel and organization policies. Int Hum Resour Manag London Sage. 1995;
[20]. Sedera D, Gable GG. Knowledge management competence for enterprise system success. J Strateg Inf Syst. 2010;19(4):296–306.
[21]. Fryling M. ERP Customization vs. Business Process Reengineering: Technical and Functional Perceptions. 2014;1–15. Available from: www.aitp-edsig.org
[22]. O'Leary D. Knowledge management across the enterprise resource planning systems life cycle. Int J Account Inf [Internet]. 2002;3:99–110. Available from: http://www.sciencedirect.com/science/article/pii/S1467089502000386
[23]. Leary DEO. Knowledge Management in Enterprise Resource Planning Systems : Support for Financial Transactions. 2000;

[24]. Baxter P, Jack S. Qualitative Case Study Methodology : Study Design and Implementation for Novice Researchers Qualitative Case Study Methodology : Study Design and Implementation. 2008;13(4):544–59.
[25]. Yin RK. Case study research: Design and methods. Sage publications; 2013.
[26]. Tarhini A, Ammar H, Tarhini T, Masa'deh R. Analysis of the Critical Success Factors for Enterprise Resource Planning Implementation from Stakeholders' Perspective: A Systematic Review. Int Bus Res. 2015;8(4):25–40.
[27]. Warriss G. Data Migration from various Legacy Systems to SAP R/3. 2000.
[28]. Keller P. SAP Test Data Migration Server: Creating High-Quality Test Systems with a Reduced Set of Data. 2009.
[29]. Bosilj-Vukšić V, Spremić M. ERP System Implementation and Business Process Change: Case Study of a Pharmaceutical Company. J Comput Inf Technol. 2005;13(1):11–24.
[30]. Smith MH. Knowledge management competence for ERP implementation success Introduction Knowledge management competence for ERP implementation success Conceptual framework on KM competence-ERP success Case study outline Key contributions Conclusion. 2012;(April).
[31]. Shad SA, Chen E, Malik FAF. Performance Enhancement Factors of ERP Projects in a Telecom Public Sector Organization of Pakistan : An Exploratory Study. Interdiscip J Contemp Res Bus [Internet]. 2011;2:95–109. Available from: http://www.scribd.com/doc/97916899/Performance-Enhancement-Factors-of-ERP-Projects-in-a-Telecom-Public-Sector-Organization-of-Pakistan-An-Exploratory-Study%5Cnhttp://arxiv.org/abs/1207.2862
[32]. Kähkönen T, Kähkönen T, Maglyas A, Smolander K. What Are the Factors Affecting ERP System Integration? - Observations from a Large Manufacturing Enterprise. Proc 16th Int Conf Enterp Inf Syst [Internet]. 2014;(January):5–17. Available from: http://www.scitepress.org/DigitalLibrary/Link.aspx?doi=10.5220/0004866000050017

IOP Conf. Series: Materials Science and Engineering **237** (2017) 012021 doi:10.1088/1757-899X/237/1/012021

Design of motor induction 3-Phase from waste industry to generator for microhydro at isolated village

Rimbawati [1,*], Abdul Azis Hutasuhut [1], Faisal Irsan Pasaribu [1], Cholish [1], Muharnif [2]

[1]Electrical Enginering Department, [2]Mechanical Enginering Department, University of Muhammadiyah Sumatera Utara, Jln.Kapt Mukhtar Basri No. 03 Medan, Indonesia

* rimbawati.umsu@yahoo.co.id

Abstract. There is an electric machine that can operate as a generator either single-phase or three-phase in almost every household and industry today. This electric engine cannot be labeled as a generator but can be functioned as a generator. The machine that is mentioned is "squirrel cage motors" or it is well-known as induction motor that can be found in water pumps, washing machines, fans, blowers and other industrial machines. The induction motor can be functioned as a generator when the rotational speed of the rotor is made larger than the speed of the rotary field. In this regard, this study aims to modify the remains of 3-phase induction motor to be a permanent generator. Data of research based conducted on the river flow of Rumah Sumbul Village, STM Hulu district of Deli Serdang. The method of this research is by changing rotor and stator winding on a 3 phase induction motor, so it can produce a generator with rotation speed of 500 rpm. Based on the research, it can be concluded that the output voltage generator has occurred a voltage drop 10% between before and after loading for Star circuit and 2% for Delta circuit.

1. INTRODUCTION

1.1. Background

Induction motor is one of the driving forces that are most often used in indusrial applications. Besides its function is as a driving force, an induction motor can also be used as a generator, either large or small generator power capacity. The construction of an induction motor is generally equal to an induction generator, when the slip is made negative, or in other words the rotational speed of the rotor (nr) is made larger than the synchronous speed (ns) then the machine will function as a generator and the output voltage will be returned to the voltage flowing. Therefore, the induction motor can be operated as an induction generator [1].

An induction motor as a generator is widely used in Micro Hydro Power (MHP). [2] and [3] conducted a study to meet the needs of electrical energy in a rural community near the source of irrigation channels and the flow of rivers in the highlands, to try to make Power Micro Hydro low power < 3 KW by utilizing a single-phase induction motor as an induction generator.

Due to the number of the remains of 3-phase induction motor from water pump, drying machine, blower, and industrial machines, it opens the chance to make modification to utilize the motors to be a *hydroelectric generator* that can be operated in Micro Hydro Power in all areas that have small water discharge every year.

Rumah Sumbul village, STM Hulu District of Deli Serdang with a population of ± 25 families, has the potential water resources to be utilized as Micro Hydro Power with a capacity of 14,15 kVA.

During this time, the resident's lighting is only an oil lamp with the expensive cost and it is sometimes not available. Because the geographical location of the village is far from the capital district, it causes that village is more declining than other villages. Based on statistical data on population, the degree of education in that village is very low, so to actualize MISG that has a capacity of 10 kVA is very required.

2. REVIEW OF RELATED LITERATURE

2.1. Induction motor as a generator

Basically, an induction motor can be used as a generator by adding capacitors at the output generator. The capacitor installed in the output of the generator will supply reactive power which will raise the output voltage on the generator. Capacitor value installed in the output generator is various because it is based on increasing voltages in the generator. To get the maximal output, it is needed a control device that can measure the value of capacitors which can be used appropriately. The control device is *Induction Generator Control (IGC)*.

The use of an induction motor as a generator has been applied widely in Micro Hydro Power even though it has shortcoming in terms of efficiency and voltage regulation. Induction generator is widely used because it is easy to be obtained, cheap, has simple construction and easy to be maintained.

[4] states that to operate an induction motor as a generator, it takes mechanical power as prime driving force that will rotate the rotor more than the speed of the synchronization – in other words, slip induction is always negative in the generator. It can be seen in this formula:

$$S = \frac{n_s - n_r}{n_s} x100\% \qquad \text{...........} \quad (1)$$

where S = Slip
Ns = the speed of stator
Nr = the speed of rotor

To generate a voltage at the output terminals of induction generator, reactive power is required in accordance with the need of reactive power of the induction machine. Reactive-power requirement can be met by installing a unit of capacitor at the output terminals of induction generator. There is a residual flux or magnetic field in the stator coil at the induction generator. Voltage generation process will not occur if there is no residual flux.

2.2. Advantages and Disadvantages for Induction Motor as Generator

Induction motor as a generator has advantage and disadvantage. The advantages of induction motor as a generator are:

1. It is cheap prize and easy to find in the market.
2. It is available from low power to high power, single-phase to 3-phase.
3. It does not need the maintenance because it does not have carbon brush and slip ring.

Besides the advantages, the followings are some disadvantages of Induction Motor as Generator:

1. It has low power (< 0,8).
2. It has large voltage drop (>10%).

IOP Conf. Series: Materials Science and Engineering **237** (2017) 012021 doi:10.1088/1757-899X/237/1/012021

3. The stability voltage is not maintained at a constant value.
4. It requires a special controller which is quite expensive to increase the power factor.

3. Research Method

The following is the image of test series that will be conducted to determine the parameters of the design generator which will be used as the MHP:

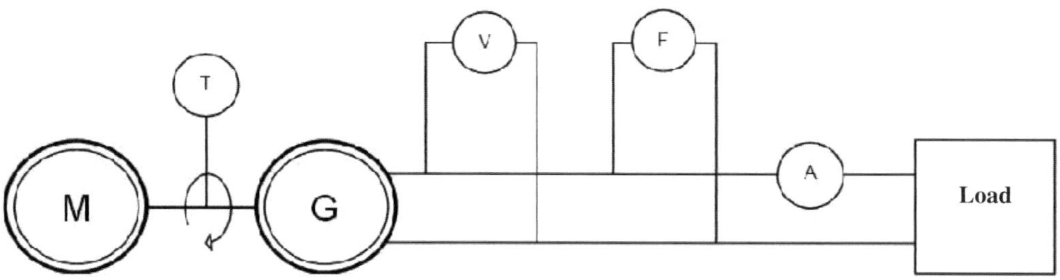

Figure1. Testing-Tool Scheme of Generator Design That Will Be Modified

The explanation of the picture:

M : Motor as the first driving force
G : Tested generator
T : Tachometer
V : Voltmeter
A : Ampere meter
F : Frequency Counter

Before the generator is installed in Micro Hydro Power, the generator is tested first. Installation testing tools are as in Figure1. Since the generator must be rotated at a constant speed of 500 rpm, the induction motor is coupled with a generator by using V-Belt Pulley with a ratio *Pulley* 3: 2 for the appropriate rotation.

The next step by estimating the water discharge, then first step that needs to be done is to calculate speed of water and calculate the area cross section. The area cross section area is the total number of five squares, with depth as measured at points D1, D2, D3, D4 and D5. The width of each box is one-fifth of the channel / river width. The calculation data is done based on the calculation data conducted on the river flow of Rumah Sumbul Village, STM Hulu district of Deli Serdang with average depth (D) 0,188m width of river (L) = 1.8 m and mean water velocity 1, 9 m / s so obtained the calculation of water debit as follows;

A c s

$$A = D \times L$$
$$A = 0{,}188 \times 1{,}8$$
$$A = 0{,}3384\ m^2$$

W d ha

$$Q = A \times V$$
$$Q = 0{,}3384\ m^2 \times 1{,}9 \text{m/s}$$
$$Q = 0{,}642 m^3/s$$

From the results of debit and head analysis, an electrical power that can be generated can be calculated as follows;

$$P = g \times Q \times H \times \eta$$

$$P = 9{,}8\ x\ 0{,}642\ x\ 3\ x\ 75\%$$
$$P = 14{,}15\ k$$

Before doing a modification, it is required to know the construction and the equivalent circuit of the motor to simplify the design modification. To get more clearly about equivalent circuit of the induction motor and the generator, it can be seen in the following picture.

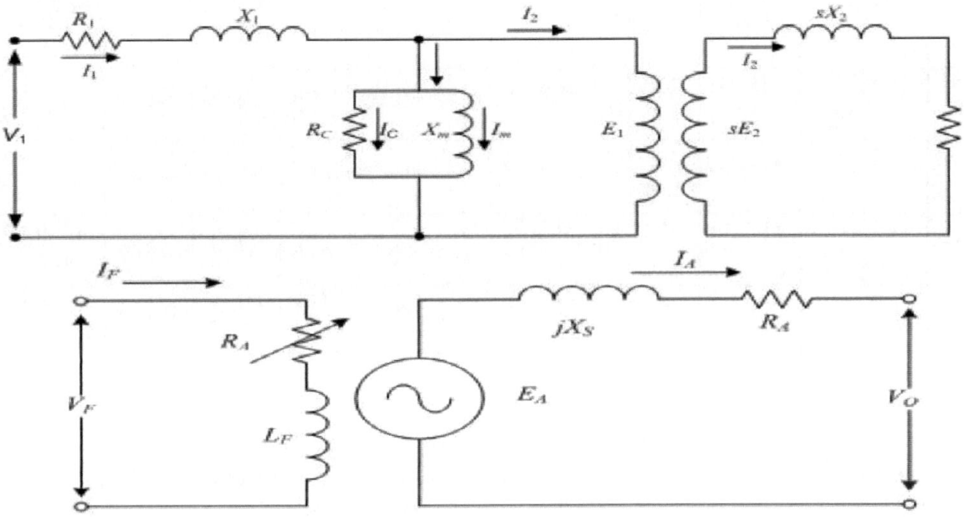

Figure 2. Equivalent Circuit of Induction Motor that will be modified

The above induction motor circuit will be converted into an equivalent circuit generator as in the following picture:

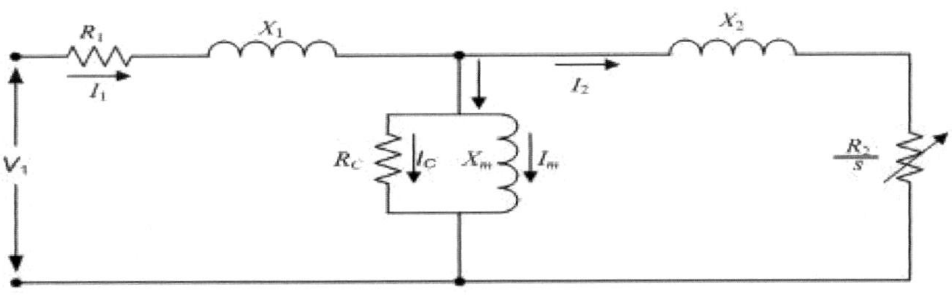

Figure 3. Generator Equivalent circuit after modification

4. Analysis and Discussion

4.1. Load Voltage Test

A. The Test of Load 3-Phase Star-Connection Generator.

The test was conducted to the 3-phase star-connection generator. The load that was given was 3 lamps with 60 watt/220 volt. The data with the variation in generator rotation were shown in table 4.1.

Table 4.1 Data of load 3-phase Star connection Generator Test

No	Rotation (Rpm)	Output voltage(V)			Current (A)			Cos ϕ		
		Phasa R-S	Phasa S - T	Phasa R- T	R	S	T	R	S	T
1	350	201,90	202,07	202,39	1,23	1,23	1,23	0,859	0,861	0, 892
2	450	259,59	260,13	260,22	0,96	0,96	0,96	0,857	0,891	0,933
3	500	288,43	288,61	288,79	0,86	0,86	0,86	0,862	0,894	0, 952
4	550	317,27	317,31	317,36	0,78	0,78	0,78	0,864	0,898	0,988
5	600	342	342,51	343,41	0,72	0,72	0,72	0,903	0,923	0,992

Based on the table 4.1, it can be seen that output voltage has risen significantly with the changes of the modification generator speed. Despite a high enough voltage drops after the generator weighed is ± 10%, but that rate is still high to calculate working voltage of a generator.

B. The Test of Load 3-Phase Delta-Connection Generator

The test was conducted to the 3-phase delta-connection generator. The load that was given was 3 lamps with 60 watt/220 volt. The data with the variation in generator rotation were shown in table 4.2

Table 4.2 Data of Test of Load 3-Phase Delta-Connection Generator.

No	Rotation (Rpm)	Output voltage(V)			Current (A)			Cos ϕ		
		Phasa R-S	Phasa S - T	Phasa R- T	R	S	T	R	S	T
1	350	107,96	110,01	110,19	1,23	1,23	1,23	0,859	0,861	0, 892
2	450	143,29	144,56	143,63	0,96	0,96	0,96	0,857	0,891	0,933
3	500	161,93	161,05	162,13	0,86	0,86	0,86	0,862	0,894	0, 952
4	550	175,67	175,69	174,74	0,78	0,78	0,78	0,864	0,898	0,988
5	600	191,1	191,38	191,87	0,72	0,72	0,72	0,903	0,923	0,992

Result of testing on the output voltage generator connected by delta indicates that the output voltage produced by the generator is lower than the star circuit. This is suitable to the basic principles of relations with the load generator.

5. Conclusion

After having been analyzed, it can be concluded that:

1. The output voltage generator occurred voltage drop 10% between before and after being loaded for star circuit and by 2% for circuit Delta.
2. The test will still be conducted to the perfection of hydroelectric generators produced.

References

[1] Zuhal, 1988 "Dasar Teknik Tenaga Listrik dan Elektronika Daya", Gramedia Jakarta

[2] B.S, M. Isnaeni, 2005, "Motor Induksi Sebagai Generator", *Seminar Nasional Ketenagalistrikan*, Teknik Elektro Fak Teknik Universitas Diponegoro.

[3] Ion, Catalin Petra. Serban Ioan. 2006: *Single-phase Dump Load For Stand Alone Generating Units with Induction Generator,* Annals of the University of Craiova.

[4] Effendy M, 2009, "Rancang Bangun Motor Induksi Sebagai Generator (MISG) pada Pembangkit Listrik Tenaga Mikrohidro", TRANMISI, Jurnal Teknik Elektro, Volume 11, Nomor 2, Juni 72 2009, hlm. 71-76. UMH Malang.

[5] ESDM, 2003, *"Kebijakan Pengembangan Energi Terbarukan dan Konservasi Energi (Energi Hijau)"*, Departemen ESDM, Jakarta, Indonesia.

[7] Sekeroney F, 2009, "Penggunaan Motor Induksi Sebagai Generator Arus Bolak-balik", *Jurnal Teknologi , Volume 6 Nomor 2, 2009; 697-702*, Politeknik Negeri Ambon.

[8] Surya W, (2009), "Analisa Karakteristik Motor Induksi Sebagai Generator (MISG) pada PLTMH", Teknik Elektro UPI, Bandung.

[9] Yahya S, 2008, "Motor Induksi Split Phase Sebagi Generator Induksi Satu Fasa" Prosiding Seminar Nasional Teknoin 2008 Bidang Teknik Elektro, Politeknik Negeri Bandung.

[10] Yulianus S, 2009, "Perencanaan Motor Induksi Sebagai Generator Induksi", Jurnal: Adiwidia Edisi Desember 2009, No. 2, Teknik Elektro UKI Paulus Makasar.

[11] Chapallaz, 2002, J.M., J. Dos Ghali, P. Eichenberger, G. Fisher, "Manual on Induction Motors Used as Generators", GTZ Eschborn.

[12] Ekanayake Dr, 2002, *"Small Hydro Schems",* Power Engineering Journal; April 2002

Real time control "es-dawet" mixer using dasboard based on PLC and WSN

Pandapotan Siagian[1,*], Sindak Hutauruk[2], Kisno[1]
[1]Institut Teknologi Del, Faculty of Informatics and Electrical Engineering, Computer Network Management Study Program, Jl. Sisingamangaraja, Sitoluama, Toba Samosir, Sumatera Utara, 22381, Indonesia
[2] Universitas HKBP Nommensen, Electrical Engineering Department, Jl. Sutomo No.1, Medan, North Sumatera, 22381, Indonesia

* siagian.p@gmail.com

Abstract. The aim of this study is to monitor and acquire the remote parameters like Speed control a DC Motor, IR Sensor, Temperature of pasteurize mix of ice cream, and send these real values over wireless network. A proposed system is dashboard monitoring system for PLC based system wirelessly using ZigBee protocol. To implement this a ZigBee model is connected to a programmed digital signal controller which would transmit the data to Zigbee coordinator which is connected to a PC through RS232 serial communication. Person can need only to send the reply about the process that is to be carried out and PLC will check the status of the web base sent by person and take the action according to it where wired communication is either more expensive or impossible due to physical conditions. A low cost system for measured the parameters of motor such as IR Sensor, Speed control a DC Motor by PWM and temperature with Zigbee protocol connectivity. A database is built to execute monitoring and to save the motor parameters received by radio frequency (RF) data acquisition system. Experimental results show that the proposed system is less costly, provides higher accuracy as well as safe and gives visual environment.

1. Introduction

An automation system, the communication needs are traditionally served by wired technologies. These communication systems have been specifically designed to meet the stringent real time and reliability requirements found in many industrial applications. A low cost communication is wireless technologies should provide the same type and quality of services to industrial users as the traditional, wired technologies do. A lot of devices and machine can be controlled and data can be received and sent a data by wireless technology. A PLC based fault protection system has been introduced in which massages the type of faults on the display, but data can't be display and stored as well as the system becomes costly. In some cases, motor parameters have been used to display the electrical and mechanical performance of the motor through wires on a PC. The aim of this paper is to design dashboard web for monitoring, measuring a acquire the parameters of motor such as IR Sensor, Speed control a DC Motor by PWM and temperature with ZigBee protocol connectivity. A ZigBee model is connected to a programmed digital signal controller which would transmit the data to ZigBee coordinator which is connected to a PC through RS232 serial communication. Sensors are controlled and monitored through dashboard web correspondent its set value points. For RF module is to communicate their remote location areas. Depends on that corrected values from set points it can be functioned. Each process of sensors values can be transferred for dedicated system.

2. Related Works

2.1 ZIGBEE IEEE 802.15.4

Zigbee devices are capable of peer-to-peer, point-to-multipoint and mesh communications. They offer convenient low power wireless solutions for embedded systems where power consumption is a critical factor. A Zigbee network consist of three different types of ZigBee devices a coordinator, router, end-device. Each network has a 16 bit PAN ID. All devices in a Zigbee network is assigned a single PAN ID Protocol, Zigbee is a low power spin off of WiFi [6], [7], [10].

2.2 DC Motor Control and Sign/ Magnitude PWM Control

This motor is used for this project for control speed and position [4]. Controlling a DC motor with a sign magnitude PWM control signal requires one PWM control input, one directional control input, and the voltage source. Sign magnitude PWM requires more pins than locked anti-phase PWM control but offers twice the precision control of the speed [1],[3].

2.3 Features Of Mitsubishi Melsec PLC Controller

The name itself suggests the type of PLC and I/O it provides like LQ-F is a series name, 10 stands for (6+4) 6 inputs and 4 outputs, FX series. The delta PLC is cheaper than other brands of the PLC and easy to configure with any application of industry as well [8],[11]. The programmable controller PLC used in this study is Mitsubishi Melsec PLC LQ-F Series FX5.

2.4 Temperature Sensor LM35

The LM 35 series are precision integrated-circuit temperature sensors, whose output voltage is linearly proportional to the Celsius temperature [13]. The LM35 does not require any external calibration or trimming to provide typical accuracies of ±1⁄4°Cat room temperature. LM35 is rated to operate over a −55° to +150°C temperature range. In can be used with single power supplies or with plus and minus supplies. As it draws only 60 μA from its supply, it has very low self-heating, less than 0.1°C in still air.

2.5 Es-Dawet

Es-Dawet or ice cream from Sumatra in Indonesia is made of the combination of coconut, milk, maezena flour, sugar, milk, and avocado. Es-Dawet making process takes 50 rpm for 3-4 hours [12]. This time is much shorter than the time used when screening is done manually by hand.

3. Design and implementation

3.1 Proposed Model

Architecture contains the pair of ZIGBEE USB modules, Mitsubishi Melsec PLC, SCADA software installed in computer. PLC is a central unit to control the process and sensors and other inputs are connected at input terminal of the PLC, Raspberry Pi 3, Temperature of pasteurize mix ice cream and ZIGBEE USB is connected with PLC via communication port RS485 (RS485 to RS232) converter is PC, that is shown in Figure 1.

Figure 2 shown Es-Dawet making processed which takes 50 rpm for 3-4 hours [12]. This time is much shorter than the time used when screening is done manually (by hand). The plan is to design mechanical a pasteurize mix that is show in Figure 3 and the architecture plan, that is shown in Figure 4 needs a device a LM35 Sensor, IR Sensor, Zigbee, RS232 converter, Raspberry pi3, regulator supply 24 V, 5 V and Mitsubishi Melsec PLC Controller.

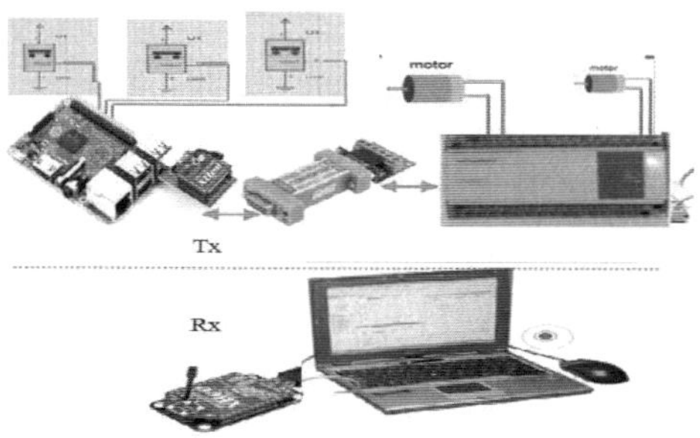

Figure 1. System Design

Figure 2. Traditional pasteurize mix *by hand*

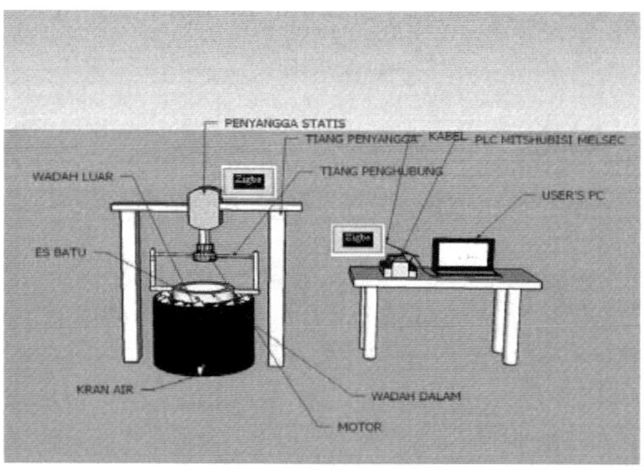

Figure 3. Mechanical design a *pasteurize mix "Es-Dawet"*

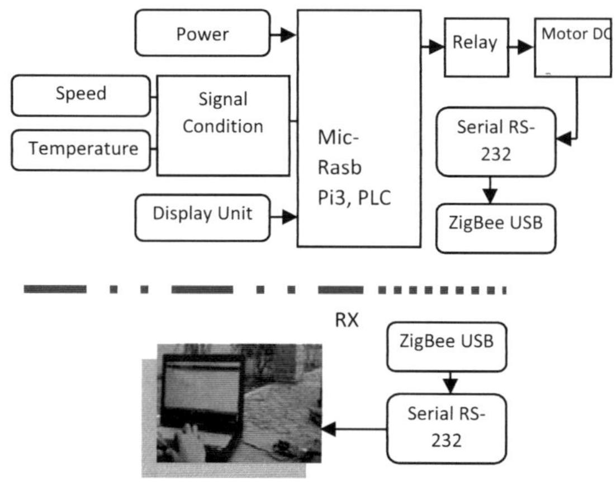

Figure 4. Block System

3.2 Regulated power supply

A regulated + 5, + 24 VDC power supply is designed to power up the raspberry microcontroller. A 220 V to 6-0-6 volt, and 25-0-25 volt, 500mA center tapped transformer is used to step down the AC Voltage. Bridge rectifier made of UF4007 is used for AC to pulsating DC conversion; capacitors are used to filter ripples. The voltage regulator IC 7805 is used to produce regulated +5V and IC 7905 is used to produce regulated -5V dc supply and voltage regulator IC 7825 is used to produce regulated +25V.

3.3 Mechanical a pasteurize mix and DC Motor Control

A DC Motor are specifications for this motor: continuous stall torque 0.402 Nm, peak stall torque 2.825 Nm, no load max. speed 6000 RPM, terminal voltage 24-60 Vdc, motor inertia 0.00006286 Kg-m^2, motor weight: 1.588Kg 2.4.3. A mechanical a pasteurize mix "Es Dawet" need median rpm is 50 rpm. The DC motor is ideal motors for applications using 6000 RPM. so, pasteurize mix "Es Dawet ' needed 50 RPM. To doing a round tubes in a centrifuge to 50 RPM by selecting rotation ratio gear reducer (gear box) is 1:90. If speed DC 6000 RPM motor, so that output rotation reducer becomes 6000/86 = 69.788 = 70 RPM. Reducer pulley diameter 5 "and pulley shaft tube 7 ", so that the speed of rotation tube = 5/7 X70 = 50 RPM.

3.4 Real Time Control "Es-Dawet"Mixer using PLC Control and Feedback Sensors

The objective is to control the speed of two motors i.e. DC Motor, a motor 2 control a body of pasteurize Mix and motor 2 is pasteurize mixer in the Es-Dawet. The speed of the motor is sensed by wireless temperature sensor on the wireless sensor output the corresponding PWM are triggered to rotate the motor. PLC accepts 24 Volts signal, regulated power supply is input for device others and the sensor output is in the form of 5 Volts. A DC motor performance will be controlled by ladder set in PLC, procedure coder and set ladder that is shown in Table 1 and Figure 5.

Table 1. PLC Ladder Diagram

Input	Output
X_0 = System Start	Y0 = Temperature ON
X_1 = System Stop	Y1 = Shutter motor forward direction
X_2 = Sensor 1	Y2 = Torque motor forward direction
X_3 = Temperature Sensor	Y3 = Shutter motor reverse direction
X_4 = Sensor 2	Y4 = Torque motor reverse direction
X_5 = Sensor 3	Y5 = LED

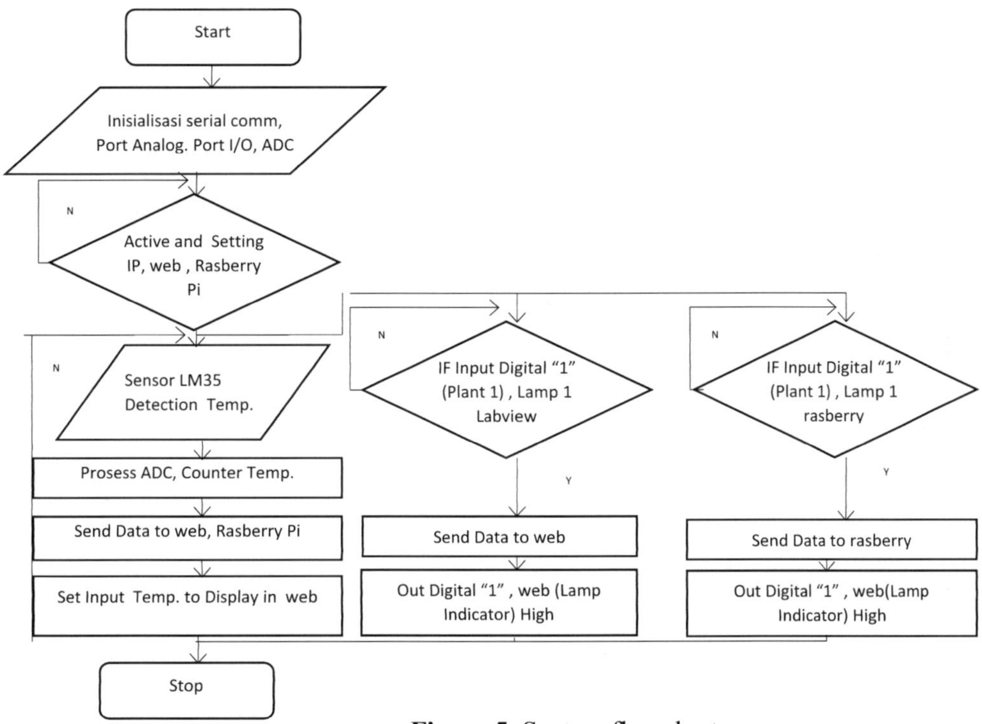

Figure 5. System flowchart

The dashboard web design for remote wireless temperature monitoring system places multiple nodes in different regions. Each node collected measured the temperature of the object information. Then it transfers data to the coordinator node via wireless multi-hop routing. Dashboard web interface for communication with Zigbee protocol, the first step a system proses is will initialize the flash drive and display all files on the flash drive to the COM port on the computer. It will then setup a TCP socket where the controller is the server, and print out the IP address of the dashboard web. This only works if you are connected to Ethernet, otherwise an error message will display. Once this socket is set up, it will wait for the web client to establish a connection to it. On the web server it will display "Wait for connection..." until a proper connection is established. Once this happens, the web client will display both the IP of the board and the IP of the computer running the dashboard web application. For this system used a default port of 40 but any port could be used. When this is done, it will wait for the run time and time interval from the dashboard web application and, once received, proceed to initialize all of the system, the real time task (setting the period to the time interval received), and start the timer.

4. Result and Discussion

Dashboard web interface for communication with Zigbee protocol, the first step a system proses is initializing the flash drive and display all files on the flash drive to the COM port on the computer. It will then setup a TCP socket where the controller is the server, and print out the IP address of the dashboard web. This only works if you are connected to Ethernet, otherwise an error message will display.

4.1 Set Zigbee parameters to PC for receiving data

After the start of the receiving node (the coordinator node) network, receiving temperature data from a remote node, and serial data transfer to the PC. The PC software displays real time information. Setting set a communication zigbee to PC that is shown in Figure 6 and serial communication between zigbee and reception of data on X-CTU that is shown in Fig. 7.

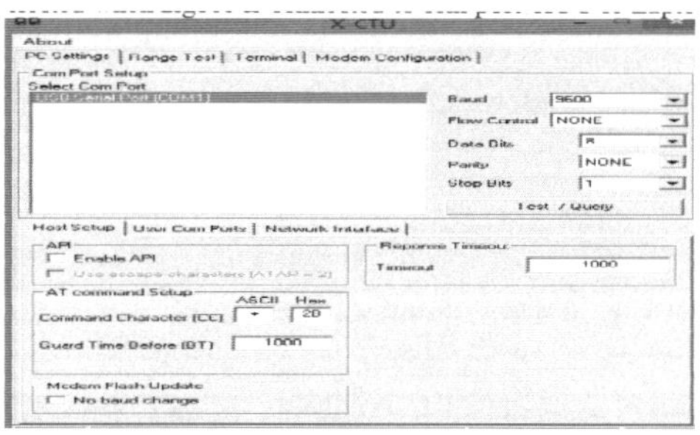

Figure 6. Zigbee COM port setting

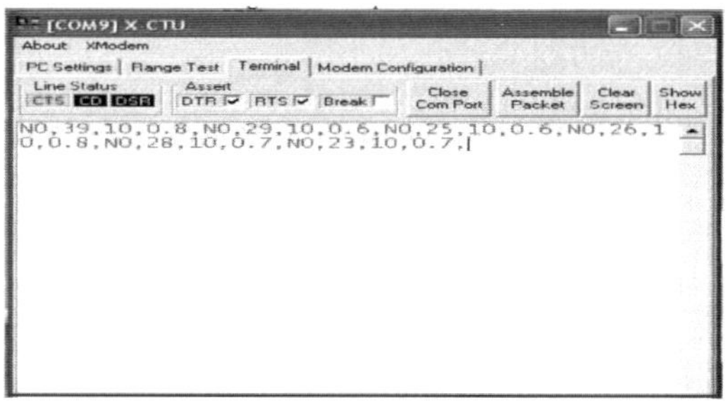

Figure 7. Communication zigbee and reception of data on X-CTU

4.2 Temperature Sensor

The Temperature Read Thread is the real-time task which periodically reads from the temperature sensor. The data is set up in a Temperature structure which is defined as follows: Every time the real-time task executes, it will allocate memory for a piece of mail of type Temperature, which is very similar to a FIFO (essentially an allocatable queue), Once this is done, it will pass the mail and shared buffer into the function calcTemp() to calculate the data received from the temperature sensor and store it on the mail and buffer. It will also signal the semaphore for the LED and speaker thread. It then puts this mail onto the Mailbox. Also, you must remember to free this memory after you put it onto the mailbox, otherwise you will likely run out of memory. The USB Write Thread will have already called mail_boxes.get() which will block until the Temperature Read Thread has put this data onto the Mailbox. Once this has happened, it will write the data to the text file "Temperature_Loger.txt" on the flash drive, and also print the data through the COM serial port.

```
Typedef struct {
float     temperatureC;
float     temperatureF;
float     timeStamp;
float     avgC;//averrage change in celcius per second
float     avgF;//averrage change in Farenheit per second
}temperature;
```

PC interface displays dashboard web real-time temperature information after the coordinator node transfers data to a PC through the RS232. The specific requirements of the system, using C # host computer interface dashboard development the design of the PC software data flow: 1) serial data is received; 2) to determine whether there is complete data head (the software is simple "PAN ID :"); 3) extracted four node PAN ID value (PAN ID value is mainly used to distinguish between different devices); 4) Extraction temperature data; 5) the extracted data is graphed.

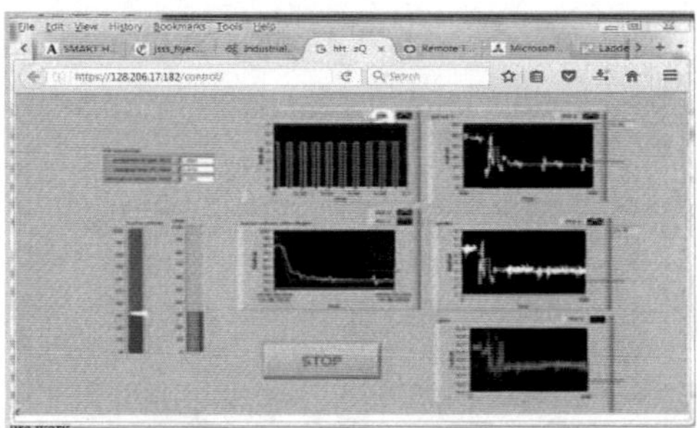

Figure 8. Dashboard Web control application a DC motor

Figure 9. Prompt for IP Address of of Server display a sensor graph

The design of dashboard real time wireless temperature information about pasteurize mix, places multiple nodes in different regions. Each node collected measured the temperature of the object information. Then it transfer data to the coordinator node via wireless multi-hop routing. A testing, the serial transmission baud rate is set to 9600 -38400, after the system is running, real-time temperature information is displayed by the host computer that is shown in Figure 8 and Figure 9. Through the COM port, the following message appeared: Temperature_Log.txt is where all the information will be stored to 128.206.17.181/graph. It then proceeds to display the IP Address in which this case is 128.206.17.181/graph. It then proceeds to wait for a connection from the web application. When a connection is received it will, take the runtime and interval from the application and start the real time task that is shown in fig.8, wait to display graph of a sensors, one item was sent through the socket, and the USB file was opened successfully that is shown in Figure 10.

A set PWM control motor can do it without entering a run time and interval, and then press the start button. After this, both system will start running and displaying the data. The minimum interval used was 2 second. The data being sent through the socket, each time a system send the current time in seconds and temperature to the web application. Since it is rapidly being sent, you will notice that multiple of the same pieces of data are being sent. A code is mess with wait statements to decrease the amount of same results being sent, but found it to occasionally make the application lag slightly so decided to leave them out and allow it to continue to send multiples, see the data was successfully written to the text file twice in Figure 11.

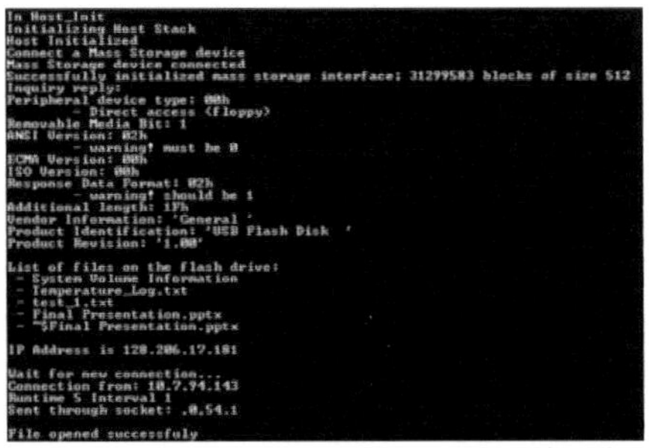

Figure 10. A connection of received data

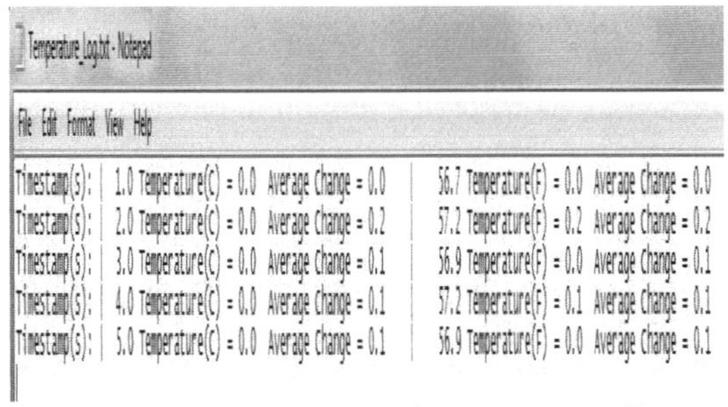

Figure 11. Data Written to Text File

5. Conclusion

This system has been done to monitor and control the DC motor parameters which is widely used in industries. Use of Zigbee communication protocol helps in reducing the wires and hardware significantly. That the dashboard real time task and multithreading involved in the software implementation performed as desired. As the correct data was displayed on all dashboard web application, and USB Text File.

References

[1] Ahmad Ismail, Man (2008) DC Motor Speed And Position Control Using Analog PID Controller.

[2] Charles L.Phillips, H. Troy Nagle, (1995), Digital Control System Analysis And Design, 3rd edition, Prentice Hall,Inc. A Person Education Company.

[3] Data sheet, LM35 Precision Centigrade Temperature Sensors, National Semiconductor,Nov., 2000 3.

[4] Mahendra P. Bodkhe K.N.Pawar, (2008)," Parameter monitoring Using Zigbee Protocol for Three Phase Induction Motor" International Journal of Emerging Technology and Advanced EngineeringWebsite: www.ijetae.com (ISSN 2250-2459,ISO 9001:2008 Certified Journal,Volume 4, Issue 1, January 2015.

[5] Mohamed Farid Bin Mohamed Faruq,(2008), PID Controller Design for Controller DC Motor Speed Using MATLAB Application, Universiti Malaysia Pahang.

[6] Okan Bingol, Kubilay Tasdelen,(2014), Web-based Smart Home Automation: PLC-controlled Implementation, Vol. 11, No. 3, 2014.

[7] T. Krishnan, B. Ramaswamy, (1974), "A fast response dc motor speed control system", IEEE Transactions on Industrial App, vol., IA-10, No. 5, Sept./Oct. 1974, pp. 643-651.

[8] T.Kalaivani, Aallirani, P.Priya, (2011), "A Survey on Zigbee Based Wireless Sensor Networks in Agriculture": Department of ECE, 2011 IEEE.

[9] X. XingMing, (2009), "PLC control system reliability design", Automation and Instrumentation.

[10] ZigBee Alliance, (2005), ZigBee Specification[z]. Version 1.0, http://www.ZigBee.org, 2005-06-27.

[11] Z. Wan-zhen,(2004), "PLC analysis and design applications", Electronic Industry Press.

[12] ____, Dewan Standardisasi Nasional (DSN). 1993. Standar Nasional Indonesia (SNI).

[13] http://www.ti.com/lit/ds/symlink/lm35.pdf.

Behavioral analysis of use personal service e-balance Indonesian social security

Gunawan Wang, Novi Fitriani, Ahmad Nurul Fajar
Master in Information Systems Management, Bina Nusantara University, Kebun Jeruk Street, Jakarta, Indonesia

* gwang@binus.edu

Abstract. Indonesian Social Security is the one of the government agencies that is trusted to organize social security. With help of Information technology that growing these day, Indonesian Social Security is also developing E-Balance application, where previously all activities for checking balance is done by giving their slip details through the nearest branch to be distributed to each company. So far there is no research that reviewing e-Balance. Hence, the authors is interested to do research related factors that influence the behavior of the use of E-Balance Indonesian Social Security in the Jakarta area and model that can describe those factors Authors distributing questioners to 193 respondents and perform data processing. The result of this study is to know the factors that influence the behavior of use Personal Service E-Balance Indonesian Social Security and model that can describe those factors. The result shows that UTAUT 2 model is not match with this research and need to be enhanced. After enhancement, there are 3 factors that being significant. Such as Behavioral Intention, Effort Expectancy and Social Influence while the others are not supported and need to be customize.

1. Introduction

Indonesian social security has implemented in Indonesia. Based on the data a few years ago, Indonesian Social Security program has given the good impact to the members because they can save money by deduct from their payroll every month for their old age. But, using paper every year to inform their balance is not effective and they cannot see it real time. It became difficult when the member move to another company, they need to check with the nearest branch for the information. Implementation refers into all organization activity for adoption, management and routine innovation. Successful implementation need some factors, such as : 1)Support from Management. 2) Behavioral Intention from users. 3) Supporting team to give information about system and influence users. From the implementation of this application, the registered workers do not need to get a slip details for the balance every year, but they can perform real-time checking of the balance that's automatically updated if there is a balance replenishment. This system was developed by the outsourced vendor and further developed by the Department of Information Technology Development Employment BPJS itself. E-Balance has been implemented into all branches since 2015 until today. So far the process of checking the balance of JHT and JK can be done by E-Balance. The decision to adopt a system of information technology in the hands of the manager, but the success of these technologies depends on the user and the reception and use of each individual wearer (Hartono, 2007).

UTAUT 2 models shows that the intention to behave (behavioral intention) and behavior to use a technology (use behavior) affect to expectation of performance (performance expectancy), hopes of effort (effort expectancy), social influence (social influence), supporting conditions (facilitating conditions), pleasure and comfort (hedonic motivation), the benefits of the cost (price value), and actions performed repeatedly (habit). The seventh factor is moderated by a factor of sex , gender , age and experience (Vankatesh et al, 2012). The objective in this study is to analyze the behavioral the system which is called E-Balance Indonesian Social Security.

2. Literature Review

Behavior is basically oriented on the goal. In other words, our behavior is generally motivated by a desire to achieve a particular goal. The specific objectives are not always known consciously by the individual concerned (Winardi, 2004). Definition of behavior as a result of the construction theory and research, as follows: 1) Behavior is something that is caused by something. 2) Behavior shown toward the specific goal. 3) Behavior that can be observed also can be measured and 4) Behavior cannot observed directly (example: think and perception) is also important in order to achieve objectives. Self-Service Technology (SST) is defined as an interface technology that allows consumers to generate a service independently of direct employee involvement (Meuter et.al, 2000). Self-Service Technology is one of the Technology Infusion Strategy or strategies to measure the extent to which technology can penetrate an organization that is trying to use technology to improve and satisfying experience of service encounters customers, both in terms of customization and flexibility, improved service recovery, as well as the provision of spontaneous delight. Implementation of self-service technology enables the company to provide additional services with more value from core business which is running. The advantages of self-service technology can be viewed from two sides, such as service providers and customers. From services provider enterprise technology, it could be an opportunity to accelerate the delivery of services, precision, customization, reduce costs and improve productivity. For customer self-service technology also provides several advantages such as saving time and costs, higher control on the services, shortening the waiting time, convenience of location, the pleasure in the process of production and the ease of use of. Unified Theory of Acceptance and Use of Technology 2 is the development of models that's stated UTAUT by Venkatesh et.al (2003).UTAUT has four main variables that plays a fundamental role, such as performance expectancy, effort expectancy, social influence and facilitating condition that affects behavioral intention and use behavior. In addition, there are four moderators: gender, age, experience and voluntariness of use, which is positioned to support the impact of constructs - constructs on behavioral intention and use behavior. Later on UTAUT is developed by Venkatesh, Thong and Xu. According to Venkatesh et.al. (2012: 157), developed of models UTAUT to be UTAUT 2 is to study the acceptance and use of technology in the context of the consumer. Venkatesh et.al. added three new constructs into UTAUT models, such as Hedonic Motivation, Price Value, and Habit and include three moderator variables, ie Age, Gender, and Experience.

3. Research Method

This study includes data applied research (Applied Research) which is used to improve existing practices and increase effectiveness. The study was conducted using a quantitative approach that emphasizes the analysis of numerical data were processed with statistical methods. This study employs a structural equation modeling (SEM) approach to develop a model that represents the relationships among the seven variables and the dependent variable in this study: PE, EE, SI, FC, HM, PV, HB and BIU, UB (dependent variable). Data were collected using a survey questionnaire which comprises questions on participants' characteristics and multiple items for each variable in the study. Participants were 193 of workers' registered on Indonesian Social Security in Jakarta area. The mean of the age of participant under productive age on 20 – 25 years old (51%) and 63.7% were Male and their experience of work between 1 – 5 years. This research is using UTAUT 2 Model, it is a model which is can shows the user behavioral and acceptance of the system that has implemented in organization.

4. Results and Discussion

The stages for behavioral analysis : (1). Create Path Diagram, (2). Model Identification, and (3). Testing feasibility model. The Diagram Initial Model described in figure 1 below:

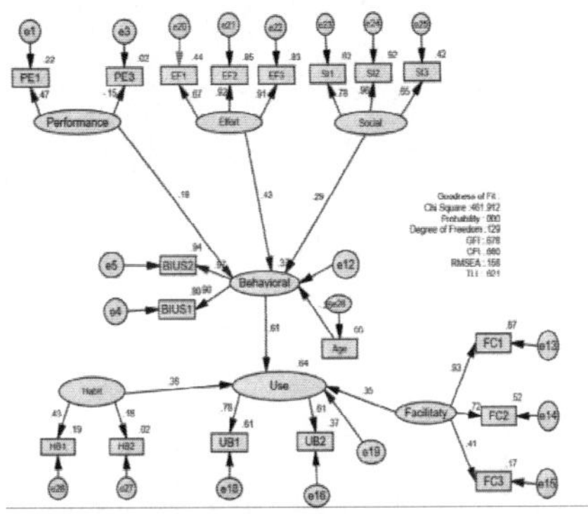

Figure 1. Output Path Diagram Initial Model

AMOS output results shows that the indicators of age, PE1, PE3, FC3, HB1 and HB2 factor loading values <0.50. Furthermore, the indicator eighth deleted. Therefore Performance Expectancy indicators and Habit has no indicator then this variable deleted from the model. Then, after removal of all the indicators and variables, the GOF results are not much better than the previous test results. Figure 2 is a diagram of the resulting new path after the abolition of indicators and variables.

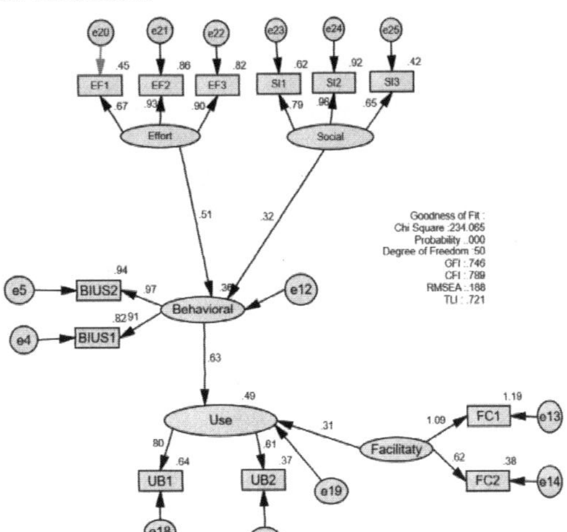

Figure 2. Output Path Diagram After Indicators and Variables Elimination

Because the research of the model still does not meet the recommended value GOF, the study concluded that the model does not fit with the data obtained. Hence, hypothesis testing / structural model cannot be done.

Conducting Interpretation and Modifying Model

Hence, the modification of the model is done for subsequent research, to know the factors that influence the behavior of the level of usage of the application of E-BalanceIndonesian Social Security.

Modifications model performed in this study based on the theory described by Arbuckle in AMOS 16.0 User's Guide (2007) which discusses how to modify the model to see Modification Indices produced by AMOS. Arbuckle explains that the Modification Indices provide some additional recommendations dashed lines / connection to minimize the chi-square value so as to make the model becomes more fit. In addition based on the theory of Arbuckle, the determination of where the connection lines are added is also based on some other theory. According to James L. Arbuckle (2007) at the AMOS 16 User's Guide, the numbers in the table Modification Indices that are below, indicating substantial minimum chi-square value would go down if the corresponding variable is connected. For example, if the variable Effort Expectancy associated with Social Influence, then the chi-square value will drop a minimum of 24.045.Figure 3 is the output path diagram generated after linking variables - variables.

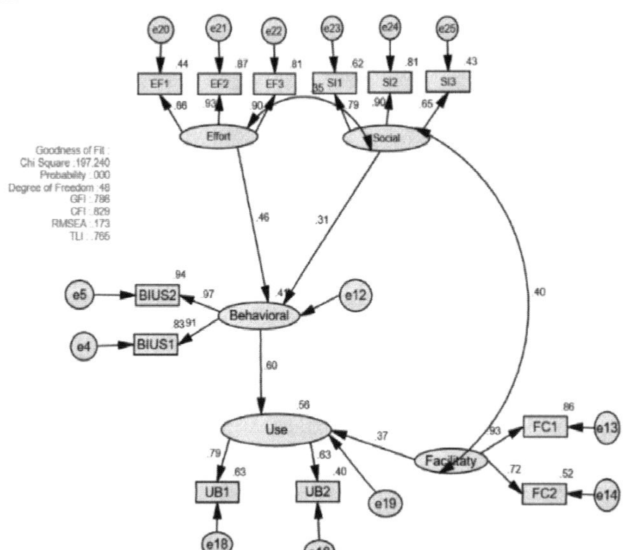

Figure 3 Output Path Diagram Initial Modification

Based on test results GOF, the model results of this initial modification is still not fit with existing data. Hence, the addition of a line connection needs todone again, each time you make modification, the results table for Modification Indices AMOS output will vary. Extra connection error between variables is performed continuously until finally declared fit models. After conducting several additional correlation error variables, finally obtained a model which is declared fit with existing data. Figure 64 is a model last modified after the addition of the correlation connection.

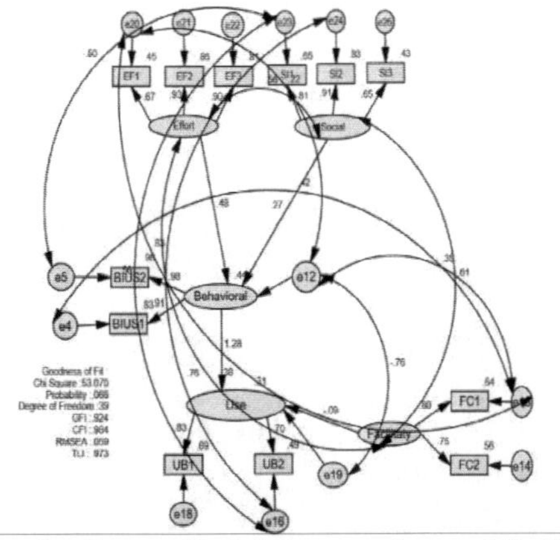

Figure 4. Output Path Diagram Latest Modification

Hypothesis testing is done by comparing the value C.R at 4:20 table with critical values that Identic with t value is 1.65 at the 5% significance level. If the value C.R greater than the critical value at significance level of p <0.05, then the hypothesis is accepted. However, if the value of C.R not reach the critical value at significance level of p> 0.05, then the hypothesis is rejected.Based on the table it can be seen that from the model that most contribute to and influence on the use of e-Balance is user intent to use (Behavioral Intention) e-balance by 126%, and ease of use (Effort Expectancy) by 53%, and the influence of the surrounding environment (Social influence) by 25%.So it can be assessed that the intention of the user to use E-Balance is a very influential factor in the behavior of the user. With the intention to access the application will be the start / beginning of influencing factors - other factors. In this case, it is expected that Indonesian Social Security can increase the sale and continue to invite the participants to use E-Balance.Factor in the ease of use of E-Balance (Effort Expectancy) are also factors that have an influence on the behavior of the user. In this case the user would find it easy to use e-applications will often use this balance and user behavior, the better. The user-friendly application is certainly a key success so that users want to use the application.The last factor that influence the effect of the environment around the user and encouragement from the people around (Social Influence) which is very important in influencing the behavior of the use of E-Balance.

5. Conclusions

This study shown that UTAUT 2 Model is not match as a behavior modelof use Personal Service e-Balance Indonesian Social Security in the Jakarta area. It is because, Indonesian citizen and environment has characteristic and specificity. In order to solve the problem, we should enhanced and modified the modelin advance. Then, it can then be testing the hypothesis. Our Modifications were done is remove the variables age, performance expectancy and habit. While the variable price value and hedonic motivation has been removed first for only the variables which are not allowed in AMOS then some variables that are not correlated deleted. Deletion of the age variable in the model, which indicates that the age factor cannot be used as a measurement of the level of application usage behavior E-Balance Indonesian Social Security, due to the age of the workforce who were respondents in this study are in the age range 20-25 years which is equivalent , the inappropriateness of this model to study the behavior of use among participants Indonesian Social Security in the Jakarta area caused by the data distribution is not normal and also the indicator / gauge does not measure variables well.

References

[1] Arbuckle, J. L. (2007). AMOS 16.0 User's Guide. PA: Amos Development Corporation.
[2] Arikunto, S. (2010). Prosedur Penelitian Suatu Pendekatan Praktek. Jakarta: Penerbit Rineka Cipta.
[3] Chang, A. (2012). UTAUT AND UTAUT 2: A Review and Agenda For Future Research. Journal The Winners, 106-114.
[4] Gioliano Putra, M. A. (2013). Pengaruh Faktor-Faktor Dalam Modified Unified Theory or Acceptance and Use of Technology 2 (UTAUT 2) TerhadapNiat Prospective Users UntukMengadopsi Home Digital Services PT. Telkom di Surabaya. JurnalManajemen Indonesia, 59-76.
[5] Hair, J. F., Black, W. C., Babin, B. J., Anderson, R. P., &Tathan, R. L. (2006). Multivariate Data Analysis (6th ed). New Jersey: Prentice-Hall,Inc.
[6] Handayani, T., &Sudiana. (2015). Analisis Penerapan Model UTAUT (Unified Theory of Acceptance and Use of Technology) Terhadap Perilaku Pengguna Sistem Informasi (Studi Kasus : Sistem Infroamsi Akademik pada STTNAD Yogyakarta). Jurnal Angkasa, 165-180.
[7] Hartono, J. (2007). Sistem Informasi Keprilakuan. Yogyakarta: Andi.
[8] HeliawatyHamrul, B. S. (2013). Analisis Perbandingan Metode TAM dan UTAUT Dalam Mengukur Kesuksesan Penerapan Sistem Informasi Akademik (Studi Kasus Penerapan Sistem Informasi STMIK Dipanegara Makassar). Seminar Nasional Informatika, 140-146.
[9] House, E. R. (1980). Evaluating with Validity. California: Sage Publication.
[10] Ir. Listyo Dwi Harsono, M., &Suryana, L. A. (2014). Factors Affecting the Use Behavior of Social Media Using UTAUT 2 Model. Proceedings of the First Asia-Pacific Conference on Global Business, Economics, Finance and Social Sciences (AP14Singapore Conference) , 1-14.
[11] Jogiyanto. (2005). Analisis dan Desain Sistem Informasi. Yogyakarta: Andi.
[12] Lawrence, G. (1980). Health Education Planning A Diagnostic Approach. Baltimore: The John Hopkins University, Mayfield Publishing Co.
[13] Marhaeni, G. A. (2014). Analisis Perilaku Penggunaan Aplikasi Pesan Instan Dengan Menggunakan Model Unified Theory of Acceptance and Use of Technology 2 di Kota Bandung. Universitas Telkom, 1-15.
[14] Mei-Ying Wu, P.-Y. Y.-C. (2012). A Study on User Behavior for I Pass by UTAUT: Using Taiwan's MRT as an Example. Asia Pacific Management Review, 91-111.
[15] Meuter, L. (2000). Self Service Technology : Understand Customer Satisfation with Technology Based Service Encounter. Journal of Marketing, 64.
[16] Morissan. (2012). Metode Penelitian Survei. Jakarta: Prenadamedia Group.
[17] Notoatmodjo, S. (2003). Pengantar Pendidikan Kesehatan dan Perilaku Kesehatan. Jakarta: PT RinekaCipta.
[18] Sihombing, S. O. (2004, 05 29). Hubungan antara Sikap dan Perilaku Memilih Satu Merek: Komparasi Antara Theory of Planned Behavior dan Theory of Trying. Retrieved

05 29, 2009, from Scribd: http://www.scribd.com/doc/12910783/Disertasi-Komparasi-TPB-%20dan-TT-

[19] Sihotang, M. E., &Nugroho, W. S. (2015). Investigasi Faktor - Faktor yang Mempengaruhi Partisipasi Diskusi Online: Studi Kasus Student - Centered E-Learning Environment Magister Teknologi Informasi Universitas Indonesia. Journal of Information Systems, 1-10.

[20] Sofyan, Y. (2009). SPSS Complete Teknik Analisis Statistik Terlengkap dengan Software SPSS. Jakarta: Salemba Infotek.

[21] Sugiyono. (2010). Metode Penelitian Kuantitatif Kualitatif dan R&D. Bandung: Alfabeta.

[22] Tan, P. J. (2013). Applying the UTAUT to Understand Factors Affecting the Use of English E-Learning Websites in Taiwan. SAGE Open, 1-12.

[23] Tan, P. J. (2013). Applying the UTAUT to Understand Factors Affecting the Use of English E-Learning Websites in Taiwan. SAGE Open, 1-12.

[24] Touré, B. K. (2014). Understanding Information and Communication Technology Behavioral Intention to Use: Applying the UTAUT Model to Social Networking Site Adoption by Young People in a Least Developed Country. Journal of The Association for Information Science and Technology, 1662–1674.

[25] Venkatesh, V. ,. (2003). User acceptance of Information Technology: Toward a Unified View. MIS Quarterly, 27(3), 425–478.

[26] Venkatesh, V., Thong, J. Y., & Xu, X. (2012). Consumer Acceptance and Use of Information Technology : Extending The Unified Theory of Acceptance and Use of Technology. MIS Quarterly, 157-178.

[27] Widiyanti, S., Winarno, W. W., & Sunyoto, A. (2015). Analisis Penerimaan Pengguna Sistem Informasi Akademik Berbasis Web di UNISRI Surakarta. Jurnal Sainstech Politeknik Indonusa Surakarta, 1-8.

[28] Winardi. (2004). Manajemen Perilaku Organisasi. Jakarta: Cetakan kedua. Kencana Prenada.

[29] Yogesh K. Dwivedi, N. P. (2011). A Meta-analysis of the Unified Theory of Acceptance and Use of Technology (UTAUT). IFIP International Federation for Information Processing 2011, 155-170.

[30] Yukari, Y. K. (2005, 02 01). Use of SEM Programs to Precisely Measure Scale Reliability. Retrieved from Sigmath: http://www.sigmath.es.osaka-u.ac.jp/~kano/research/paper/dvi/kano

Real time chromametry measurement for food quality detection using mobile device

Gunawan Witjaksono[1*], Nur Haziqah Farah Binti Mohamad Hussin[1], Almur Abdelkreem Saeed Rabih[1], Sagir Alfa[2]
[1]Department of Electrical and Electronic Engineering, Universiti Teknologi PETRONAS, Seri Iskandar, 32610, Perak Darul Ridzuan, Malaysia
[2]Mechanical Engineering Department, Mercu Buana University
Jl. Meruya Selatan no.1 Jakarta Barat, Indonesia

[*]gunawan.witjaksono@utp.edu.my

Abstract. Freshness of the food is the main factor in determining the quality and safety of the consumed food and hence consumers satisfaction. Current technologies for food quality determination depend on colour changing labels to indicate the freshness level, which is subjective to human eyes. The goal of this paper is to design and develop chromatic algorithm based on RGB colour reading and correlation with pH values for real time determination of freshness level of shrimp. The results show that the developed algorithm is able to measure, analyse and display the freshness level of food directly on the screen of a mobile app technology. The mobile app is developed on Android platform and is tested in the shrimp freshness range by stating whether it is "fresh, good or spoiled".

1. Introduction

Quality of food is considered to be the main parameter that determines its acceptability by the consumers [1]. Food inspection for high quality and safety assurance has been in high demand due to many factors such as strict rules of the regulatory agencies, costumers' demand for high quality food and increasing food borne illness outbreaks [2]. Richard J. Schonberger [3]stated that "quality is like art, everybody praises it, everybody recognizes it but each one has its own understanding of what it is". Freshness of is one of the main contributors to high quality of food. Many methods have been proposed to detect and ensure the freshness of food; including subjective (sensory), objective (non-sensory) methods, and smart packaging. Sensory assessment which is the most commonly used method is fast and non-destructive; however it is subjective to human senses, which might differ from person to another [4].For instance, the attribute for quality assessment of raw fish is to check the eyes, skin gills and the texture of the fish itself [5].On the other hand, non-sensory food assessment involves instrumental usage to measure physiochemical properties of food [6]. Some of the non-sensory attributes include total volatile base-nitrogen (TVB-N), trimethylamine (TMA), adenosine 5'-triphosphate (ATP) breakdown compounds, K and related values, biogenic amines, texture and total viable count (TVC) [7]. Non-sensory methods are not subjective to human senses, yet they uses only one aspect of food, hence it cannot determine the overall consumer acceptance unless correlated to the sensory assessment methods [8]. Instruments such as gas-chromatography, mass spectroscopy, colorimetry, infrared spectroscopy [9], ultrasound, x-ray, magnetic resonance imaging [2] have been used extensively to evaluate the quality of food.

However, these instruments are costly, bulky, bench top instruments, and they require qualified operators. Smart packaging devices are defined as small labels or tags that are attached onto the packages of the food to indicate the food quality and safety [10]. There are two types of smart packaging devices. The first type is called data carriers like barcode labels and radio frequency identification (RFID) tags. The second type of smart packaging devices is called package indicators, which are either used to monitor environmental changes such as time-temperature indicators and gas indicators. Fadable ink for time-temperature control of food freshness has been widely used in printing labels on food packages for the freshness indication. The working principle is that, as time passes, the ink will disappear and indicating that the food is no longer fresh[11]. The label ink will react to the oxygen which will make the colour to fade with time or temperature change. The rate of color change is varied by the chemical composition used to make the ink. In a study [12] pH indicator was used to determine the freshness of shrimp, where the pH values were found to vary from pH 6.24 at fresh to pH 9.04 for the spoiled shrimp after 7 days of the storage. The aim of this paper is to design and develop chromatic algorithm based on relating pH values of spoiled shrimp to RGB values of its color change, and to implement this algorithm in mobile app technology for real time determination of the freshness level of the shrimp.

2. Materials and methods

2.1 RGB data

A set of different colors were generated by computer from their corresponding RGB numbers. Then these colors were captured using smart phones, and the RGB values were re-generated and compared with their original values.

2.2 pH testing

Theoretically, the different pH values will result in different colours, which are dark purple for pH 14, green for pH 7 and red for pH1 as shown by **Figure. 1**.

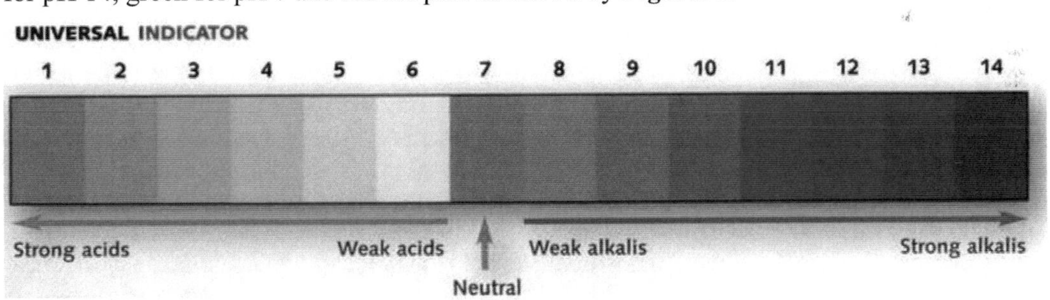

Figure 1. Universal pH indicator[13]

2.3 Chromatic parameterand correlation to RGB values

The chromatic number is the main parameter used to develop the algorithm. To calculate the chromatic parameter associated with each pH value and RGB value the formula given by Eq. 1 was formulated and proposed.

$$X = \sqrt{(R-r)^2 + (G-g)^2 + (B-b)^2} \quad (1)$$

where: R, G and B are RGB values obtained from the experimented pH paper, while r, g, and b are the baseline RGB values of particular pH paper. In this case the values of r, g and

b were set to be 196, 159 and 56, respectively.

3. Results

3.1 RGB data collection results

Figure 2 shows the computer generated (**Original code**) results of the 7 colors with their RGB values and the captured counterparts (**Colour produced**) with the RGB values after captured 3 times using smartphone. Variables such as distance between camera and the captured color and lighting levels were kept constant to avoid any inconsistencies between color captures. The colors were captured 3 times to minimize the errors which might occur due to the different lighting or quality of the camera.

Colours / RGB	Original Code (R G B)	1st (R G B)	2nd (R G B)	3rd (R G B)	Average (R G B)	Colour Produced
	255 155 155	187 122 132	190 124 138	186 116 126	187 120 132	
	255 097 41	191 070 036	196 070 032	196 071 035	194 070 034	
	255 255 102	190 182 051	190 183 057	185 178 053	188 181 054	
	000 128 000	008 104 007	008 104 006	006 104 007	007 104 007	
	000 000 255	009 059 197	009 060 201	010 060 202	009 060 200	
	084 000 084	124 007 073	128 007 074	127 008 075	126 007 074	
	065 005 038	042 002 013	041 000 008	043 000 08	042 001 010	

Figure 2. Original colors and smartphone captured colors

It can be seen that the original colors show a slight difference compared to the captured colors.

3.2 pH testing results

The studies have been conducted using acidic and alkaline solutions with different pH levels ranges from pH 5.5 to pH 8.5. The test was performed using pH paper to observe the reaction of the pH paper with the solutions. The pH values of the different solutions were accurately measured using pH meter 700.The experiment was conducted using two types (yellow- and 4-coloured) of pH papers as shown in **Figure. 3** to find their reactions with the different pH solutions. From the experiments the yellow-colored pH paper gave better sensitivity for pH testing compared to the 4-coloured pH paper. The pH papers were immersed inside 6 different solutions with pH of 5.5,6, 6.5, 7, 8 and 8.5. Only the yellow-colored pH paper was used for the further studies. **Figure. 4** (a to f) shows the color change of the yellow-colored pH paper for the different solutions with pH sensitivity differs with 0.5.

Above studies were conducted for pH ranging between pH 5.5 to pH 8, which is acid-neutral-alkaline, to observe the sensitivity of the yellow-color pH papers. To cover wider range of acid-neutral-alkaline, different solutions with pH ranging between the pH 13 of alkaline to pH 2 of acids were investigated and the corresponding RGB values were found.

3.3 Results of chromatic parameter calculation

From the 12 samples of pH paper shown in **Figure. 5**, the colors are measured and analyzed to obtain their RGB values and the chromatic parameters. The pictures for each of the 12 samples of pH paper were captured 3 times (Take 1, Take 2 and Take 3) to reduce the error that might occur due to different in lighting or distance between the sample and the camera. It is clear that the colour of the pH paper changes from red for the strong acidic solution of pH 2 to purple for the strong alkaline solution with the pH 13.

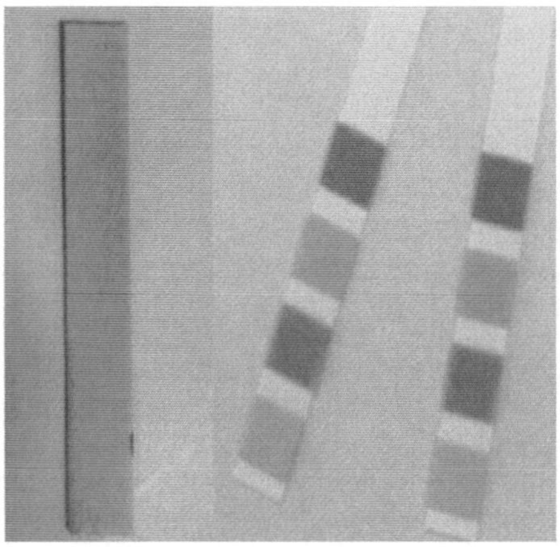

Figure 3. Two types of pH papers; (a) yellow-colored, and (b) 4-coloured paper

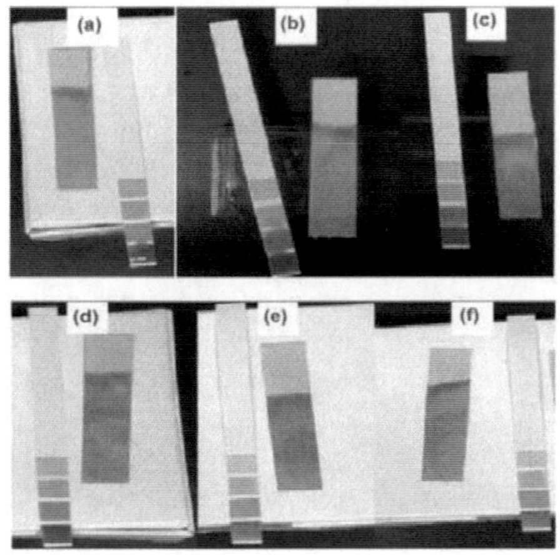

Figure 4. Colors change of pH papers for different solutions with (a) pH 5.5, (b) pH 6, (c) pH 6.5, (d) pH 7, (e) pH 8 and (f) pH 8.5

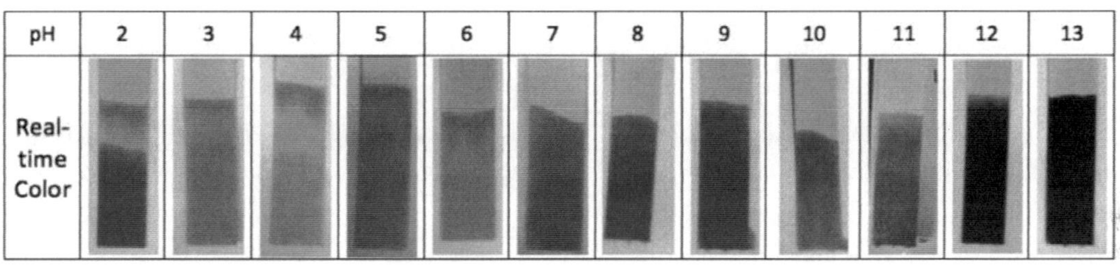

Figure 5. Reaction of pH paper with different pH values solutions

The RGB values of these different samples were found and given in **Table 1,** while the chromatic numbers are shown in **Table 2**.

Table. 1. RGB values for different pH solutions.

pH	Take 1			Take 2			Take 3		
	R	G	B	R	G	B	R	G	B
2	134	35	36	134	37	37	131	36	36
3	184	76	38	184	79	40	187	79	40
4	168	116	23	167	115	22	170	118	23
5	154	106	18	154	103	17	154	103	17
6	151	109	12	145	105	13	144	104	13
7	105	94	21	101	92	22	104	95	22
8	77	84	17	67	77	18	68	77	20
9	77	86	24	79	86	24	77	84	26
10	94	10	24	95	10	26	83	23	31
11	43	62	32	46	64	35	47	67	32
12	28	27	27	29	27	28	33	29	29
13	42	38	42	40	37	40	42	37	41

Table. 2. Chromatic values for 3 experiments.

pH	X(take 1)	X(take 2)	X(take 3)
2	139	138	140
3	85	82	81
4	58	60	56
5	75	78	78
6	78	83	85
7	116	120	116
8	145	156	155
9	142	141	143
10	183	182	178
11	182	178	176
12	215	214	210
13	196	198	197

4. Development of the chromatic algorithm and discussion

It was mentioned earlier that the fresh shrimp has a pH 6.24 while for the spoiled shrimp the pH increases to pH 9.04. The data from **Table 2** suggests that based on the chromatic parameter, the pH value could be correlated to the freshness of the shrimp. It can be seen that pH 6, pH 7 and pH 8 have different numbers of chromatic parameter X ; fall under range of 78 – 85, 116 – 120, and 145 – 155, respectively. From above values an expressions that indicates the freshness of shrimp based on the chromatic parameter could be formulated as below:

pH 6: $78 < X < 85$ = The shrimp is fresh
pH 7: $116 < X < 120$ = The shrimp is good
pH 8: $145 < X < 155$ = The shrimp is spoiled

Above expressions were used to design a mobile applications to assist in measuring, analyzing and displaying the correct freshness of the shrimp based on reading the RGB of the litmus paper.

Two types of Android base development have been tested for the implementation of the algorithm. Firstly, Android studio was used. Apps such as simple mathematics and camera calling were performed. Secondly, user uses MIT inventor for developing the Apps. Some simple Apps from Android Studio were generated such as camera call function and calculator. Camera calling and calculations play a significant role in developing the required App. Android Studio has been used for further developing the Apps. After a few trials of try and error, the App "real-time chromametry measurement" has been developed. The generated Apps were tested with 3 pH papers of pH 6 (fresh shrimp), pH7 (still good shrimp), and pH 8 (spoiled shrimp) to confirm that the Apps show the freshness of food based on the color of the pH paper. **Figure 6** shows the results of the App from the screen of the mobile phone. From the developed algorithm and mobile App shown in the figure, it is clear that the developed algorithm is able to predict the freshness level of the shrimp. When the pH paper reads pH 6, the implemented App displays that the food is fresh, while for pH 7 the App displays that the food is still good and it displays spoiled in pH 8, though the pH paper colors of pH 7 and pH 8 looks almost the same.

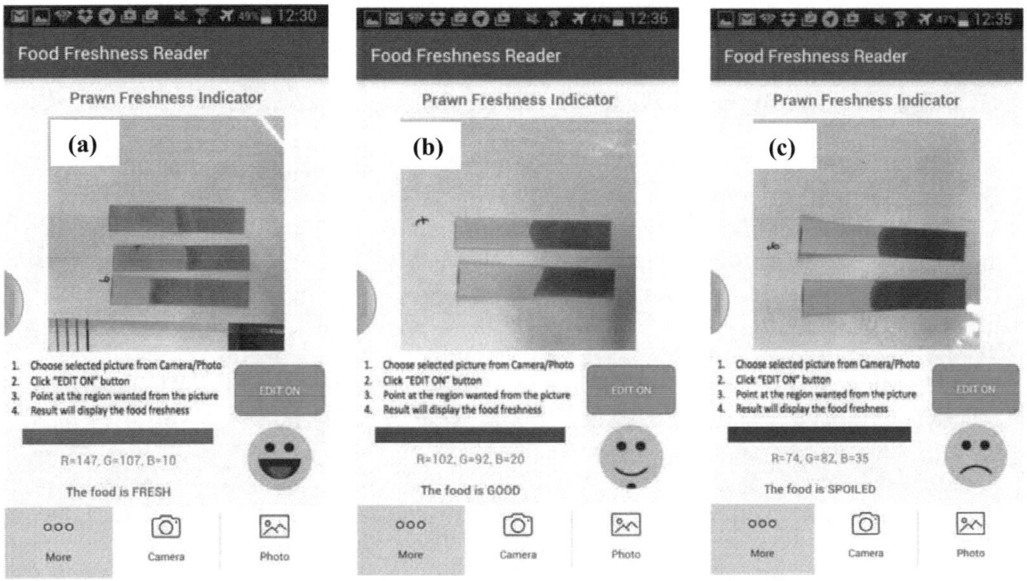

Figure. 6. Testing of Apps on pH papers of (a) pH 6, (b) pH 7 and (c) pH 8

5. Conclusion

This paper introduced chromatic algorithm for food freshness indication. The algorithm is based on chromatic parameters assigned to RGB values correlated with different pH levels, which are associated with the freshness level of the food. The algorithm was implemented in mobile App to indicate the freshness level of the food even when there is slight change on the color of the pH paper which is hard for the human eyes to see. The mobile app has been tested on yellow-colored pH papers in the shrimp freshness range by stating whether it is “fresh, good or spoiled”.

Acknowledgment

The authors of this paper acknowledge the financial support by UTP under STIRF-UTP, fund/ project code:.0153AA-F71.

References

[1] İ. Y. Genç, E. Esteves, and A. Diler, *Handbook of Seafood Quality and Safety Maintenance and Applications*: Nova Science Publishers, 2016.
[2] J. Qin, M. S. Kim, K. Chao, W. F. Schmidt, S. Dhakal, B.-K. Cho, *et al.*, "Subsurface inspection of food safety and quality using line-scan spatially offset Raman spectroscopy technique," *Food Control,* vol. 75, pp. 246-254, 5// 2017.
[3] R. J. Schonberger, *World Class Manufacturing: The lessons of simplicity applied*: The Free Press, 1987.
[4] B. A. C. Chebet Lillian, "Alternative Methods for Fish Grading," The United Nations University2007.

[5] H. Massaquoi, "Quality Changes of Fish from Catch to Processing and During Storage with Focus on Cooling Practices and Practical Application of Sierra Leone," 2011.
[6] D. Kilcast, *Instrumental Assessment of Food Sensory Quality: A Practical Guide: A Practical Guide*: Woodhead Publishing, 2013.
[7] C. Alasalvar and T. Taylor, *Seafoods-Quality, Technology and Nutraceutical Applications*: Springer, 2002.
[8] S. P, B. P, and Y. BK, "Importance of Objective and Subjective Measurement of Food Quality and their Inter-relationship," *J Food Process Technol,* vol. 6, 2015.
[9] M. G. O'Sullivan and J. P. Kerry, "12 - Instrumental assessment of the sensory quality of meat, poultry and fish A2 - Kilcast, David," in *Instrumental Assessment of Food Sensory Quality*, ed: Woodhead Publishing, 2013, pp. 355-373.
[10]K. L. Yam, P. T. Takhistov, and J. Miltz, "Intelligent Packaging: Concepts and Applications," *Journal of food science,* vol. 70, 2005.
[11]Y. Galagan and W. F. Su, "Fadable ink for time–temperature control of food freshness: Novel new time–temperature indicator," *Food Research International,* vol. 41, pp. 653-657, 7// 2008.
[12]B. Kuswandi, Jayus, T. S. Larasati, A. Abdullah, and L. Y. Heng, "Real-Time Monitoring of Shrimp Spoilage Using On-Package Sticker Sensor Based on Natural Dye of Curcumin," *Food Analytical Methods,* vol. 5, pp. 881-889, 2012// 2012.
[13]D. Walker, *Acids and Alkalis*: Evans Brothers, 2006.

IOP Conf. Series: Materials Science and Engineering **237** (2017) 012025 doi:10.1088/1757-899X/237/1/012025

Traffic control at crossroad that cut railroad based on friendly environment

Case : Crossroads at Prof. M. Yamin road with Gaharu road in Medan

Sindak Hutauruk
Department of Electrical Engineering, Nommensen HKBP University, Jln. Sutomo No. 4a Medan 20234, Indonesia.

Sindak45@gmail.com

Abstract. The frequent congestion causes accumulation of vehicles at the intersection of roads cutting off railroads, reducing the functionality of highway services, the loss of vehicle functions as a means of transportation, waste of fuel energy, and CO2 increase in intersections. The occurrence of congestion caused by the absence of integration of traffic light controls by opening and closing the railway barrier, therefore need to be done integrated control especially during and after a while train crossing the highway. The case raised in this study is the crossroads Prof. H.M. Jalan Yamin with Gaharu road in Medan. Researchers tried to solve the problem by designing and creating a tool that would coordinate and regulate traffic lights with railroad doors, thus reducing congestion for 23 minutes, reducing fuel by as much as 60,340 liters per day, thereby reducing CO2 exhaust gases in the area. This tool uses microcontroller as its control center. This design is done with a miniature simulation, meaning it is not done in the actual situation, but the field conditions can be represented in the simulation in the form of miniature.

1. Introduction

Traffic Light plays a very important role in maintaining the smoothness of traffic, especially at crossroads that are prone to congestion. Sometimes even though there is traffic light on the road, there are still frequent congestions. One example, this occurs at a crossroads close to the railway track crossing. Each train passes on a road close to the intersection of the road, than there will be a traffic jam on the intersection of four. This traffic jam also lasts for several minutes after the train crossed the road, due to the large number of vehicles that have accumulated at the intersection because of the lack of coordination between the traffic light control at the intersection with the open lid railway open. The example case is at Gaharu Street and Prof. Muhamad Yamin Street intersection in Medan, as in Figure 1. Traffic congestion often occurs at the intersection of Prof. H.M. Yamin Street with Gaharu Street in Medan, especially when the train passes or leaves the railway station which is ± 300 meters from the railway passageway, and worsens due to frequent train passing at the track, while the Traffic light at the intersection Gaharu Street and Prof. M. Yamin Street runs as usual so as to lead to the accumulation of vehicles in front of the railway door.

The accumulation of vehicles from 3 directions towards one direction of the railway track causing the occurrence of congestion causing losses, including: the reduced function of road services, the loss of function of the vehicle as a means of transportation, wasted fuel energy wasted, and increased levels of CO , CO2 at the crossroad areas.

In congestion, about 20% of machine work time is spent in 0 km / h (stop). Stopping with engine turned on is the same as disposing fuel, because motor vehicles are losing their function as a means of transportation. As it is known that motor vehicles dispose harmful substances that can cause negative impacts, both on human health and the environment [5]. Air pollution resulting from motor vehicles that emit harmful gases greatly contributed to the occurrence of air pollution and one consequence is the presence of global warming [2]. The exhaust gas emissions are the residual result of a fuel burning process inside the engine. Composition of exhaust emissions in the form of water (H2O), toxic carbon monoxide (CO), carbon dioxide (CO2), which is a greenhouse gas, sulfur (SOx), nitrogen oxide (NOx) compounds, hydrocarbon (HC) and particulate compounds Dust including lead (PB). In this case its only limited to gases in the form of CO, and CO2.

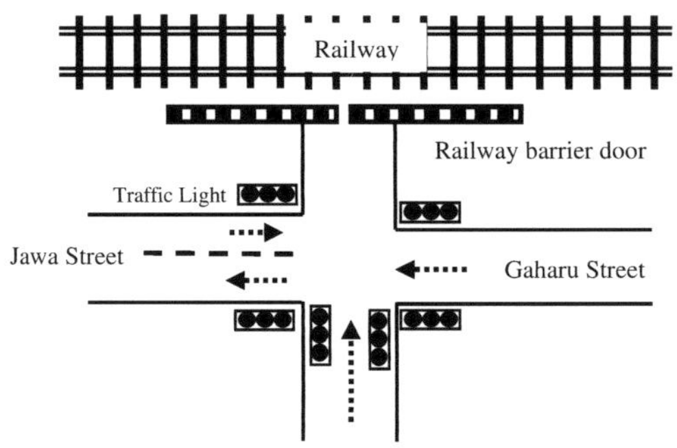

Figure 1. Map of crossroad Gaharu Street with Prof. M. Yamin Street

When the train passes on the track, the railway track is closed, and while traffic light lights up as usual, there is a buildup of motor vehicles in front of the railway track to traffic light. The duration of the congestion at the intersection of four depends on the length of the train carriage Lokomotip brought and the speed of the train. On the track, the train always runs at a slow pace because it cuts off the highway and close to the railway station, this will cause a lot of motor vehicle to be in a stop position, resulting in wasteful energy consumption, increased vehicle exhaust gas, and rise the temperature in the area that lasts during a jam. To parse this bottleneck it takes about 2-3 minutes because even when the traffic light turns green, there is still congestion in front of it.

Coordination of open-lid control of railway latches with traffic light is required so that the congestion quickly decomposes so that it can reduce the motor vehicle exhaust pollutants in the area which will automatically save the vehicle fuel as well, for that needed a controler device that indicates traffic light system with open lid railway doors, which in this case designed a microcontroller-based control tool in the form of miniature simulations that works to take over the work of traffic light automatically when the train will or is crossing the door of the train track.

2. Research Methodology

This research was conducted by taking the case of congestion that occurred in the intersection of Prof. M. Yamin streets with Gaharu streets in Medan which occurs whenever train crosses, observation time and data retrieval is done at 8.30 am. until 05.00 pm at the area in front of the railway track, the time when heavy traffic occurs. The data taken is the number of trains crossing the road on a regular basis, the number of occurrences on the road, the time length to close the railway barrier door of the train track on a regular cross, the time length to close the railway barrier door of the train trajectory at the time, the number of vehicles located in the area in front railway tracks, and length of congestion breaking down as the railway railway barrier door opened.

The data obtained is used as material for designing a controller that can synchronize the open lid of the railway barrier door with the traffic light system at the intersection, so that when the train passes the road the controller will take control of the taffic light and return it after the railway barrier door the track is open 30 seconds, so in that way can reduce the duration of the congestion on the railway barrier door area.

The reduction in the duration of the congestion on the railway track area will provide benefits such as improving the function of the road service, maximizing the function of the vehicle as a means of transportation, reducing fuel consumption, reducing the exhaust gas of CO and CO2 in the area. The design of the control device is done in the Telecommunication Laboratory of the Faculty of Engineering, Nommensen HKBP University, this controller uses a microcontroller equipped with supporting components to form a system that can be applied or implemented to control equipment with many control parameters. Or it can be said that this microcontroller can replace the function of a computer for a particular application. The microcontroller used is Atmel AT89C2051 which has memory system, timer, serial port and 32 bit I / O so it is possible to form a system consisting of single chip only, thereby reducing the design space [1,3]. This microcontroller serves as a data processing center that receives input from railway arrives sensors and outputs to ligth traffic and to the railway barrier door controller. Overall, the circuit block diagram is designed as in Figure 2.

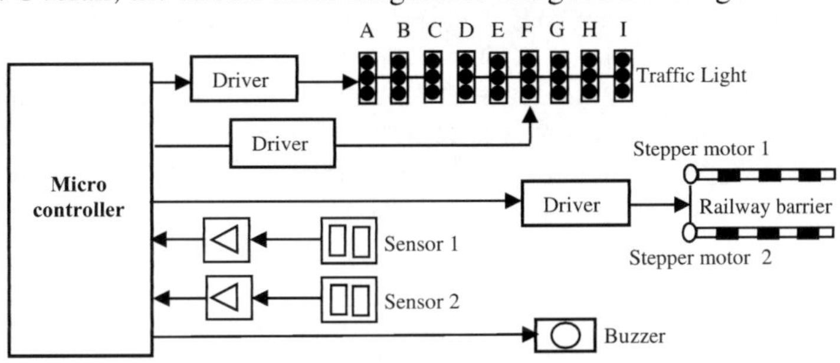

Figure 2. Block diagram controller

3. Results and Discussion

Based on observations and data we get from train officers who work as Officers that open the lid of the railway barrier door when the train cross on the railway barrier door of the railway track road Prof. M. Yamin is that in a day the train that crosses the line between 8.30 am to 05.00 pm is 16 times and the railroad that cross the road averages 7 times as can be seen in Table 1.

The railway barrier door attendant will close the railway barrier door as the train is about 500 meters away from the railway entrance door, and the average railway latch's time is 2 minutes, and when the train is switcher the crossing door is closed during 3 minutes,while the average time needed to break down the traffic jam that occurred in front of the door crossing to return to normal is 1 minute.

Table 1. Length of time close crossing barrier door of railway

Direction of train (round trip)	Frequency (times)	Length of time close the barrier (minute)	Length of time decomposes (minute)	Total of time (minute)
Medan-Binjai	12	12 x 2 = 24	12 x 1 = 12	36
Medan-Belawan	4	4 x 2 = 8	4 x 1 = 4	12
Switcher (everage)	7	7 x 3 = 21	7 x 1 = 7	28
Total	23	53	23	76

Based on the observation, the number of motor vehicles accumulated in front of the railway barrier door to the traffic light average of 50 vehicles four wheels and 100 vehicles of two and three wheeled vehicles. The amount of exhaust volume depends on the amount of fuel combustion cylinder space, if it is assumed that the exhaust volume of four-wheeled vehicles is equal to 4 times the volume of exhaust gases of two or three wheeled vehicles then we can assume that the number of 4 wheel vehicles accumulated as much as 50 + (100 / 4) equal to 75 cars of four wheels. In one day (8.30 am to 05.00 pm) there were 23 closures of the railway crossbar with the total number of vehicles 75 multiply by 23 equal to 1.725 four-wheeled vehicles with a total time of congestion for 76 minutes. The four-wheeled vehicle that stops (in a stuck situation, the live engine) has a rpm (revolution per minute) smaller than 500 rpm. The engine rotation of the vehicle in a state of stopping (traffic jams) is below 500 rpm (revolution per minute), according to research conducted by Sastra Negara et al. That the exhaust gas generated by four-wheeled vehicles in rotations below 500 rpm fueled 88 octane premium is 5.5% (CO), and 11.626% (CO2). The number of vehicle exhaust gases at each railway door closing is 247 (CO), and 532 (CO2) which lasts for an average of 3.3 minutes. Based on the Pacific Consultant International (PCI) equation, the fuel spent by four-wheeled vehicles in standstill is 0.583 ml / sec [8], resulting in the amount of fuel spent during traffic congestion (8.30am. - 17.00 pm.) Is 0.583 ml / sec multiply by 75 vehicles multiply by (76 times 60) seconds equal to 199.386 ml equal to 199,386 liters. Description of the road junction and position of the lamp, stepper motor, sensors and railway gate in the minitaur designed as in Figure 3. If the train passes through sensor 1 or sensor 2, then sensor 1 or sensor 2 will give input to microcontroller, then microcontroller will Activate the stepper motor driver so that the stepper motor will rotate, consequently the latch will be closed following the stepper motion stepper motor as far as 90 degrees. At the same time the microcontroller will also activate the lamp driver so that all lamps at the crossroads are in the red phase. Furthermore, once the train has passed the sensor 2 or sensor 1, it will give the input back to the microcontroller to return to the original state where the stepper motor will rotate in the opposite direction with the first rotation direction as far as 90 degrees, so that the latch will reopen and the condition of the lamp on The crossroads will return to the first phase. These conditions can be seen in Table 2, where the train is assumed to pass through sensor 1 to sensor 2.

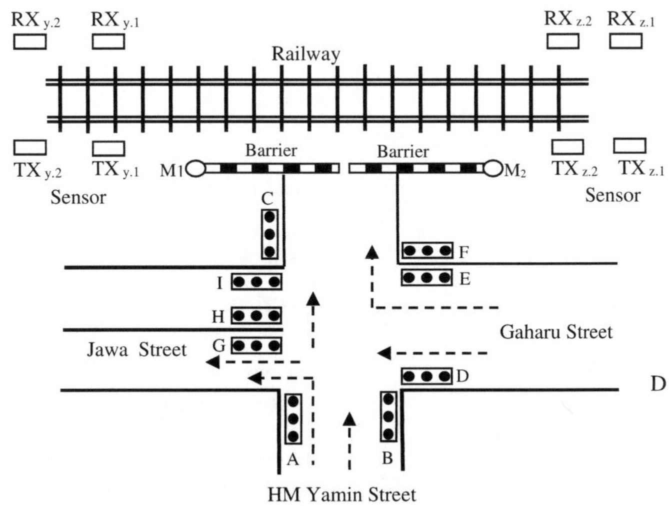

Figure 3. Crossroads, sensor position, stepper motor, and railway barrier door

Table 2. Condition of sensor, stepper motor, and traffic light on system instant before and after the train crosses sensor 1 towards sensor 2

No	State	Sensor 1		Sensor 2		Motor stepper		Traffic Light	
		$Rx_{z.1}$	$Rx_{z.2}$	$Rx_{y.1}$	$Rx_{y.2}$	M_1	M_2	A=B=C	D=E=F=G=H=I
1	Normal (t_1)	OFF	OFF	OFF	OFF	OFF	OFF	Green	Red
2	Normal (t_2)	OFF	OFF	OFF	OFF	OFF	OFF	Yellow	Red
3	Normal (t_3)	OFF	OFF	OFF	OFF	OFF	OFF	Red	Green
4	Normal (t_4)	OFF	OFF	OFF	OFF	OFF	OFF	Red	Yellow
5	Cross train (t_1)	ON	OFF	OFF	OFF	OFF	OFF	X	X
6	Cross train (t_2)	ON	ON	OFF	OFF	ON	ON	Red	Red
7	Cross train (t_3)	ON	ON	ON	OFF	ON	ON	Red	Red
8	Cross train (t_4)	ON	ON	ON	ON	ON	ON	Red	Red
9	Cross train (t_5)	OFF	ON	ON	ON	ON	ON	Red	Red
10	Cross train (t_6)	OFF	OFF	ON	ON	ON	ON	Red	Red
11	Cross train (t_7)	OFF	OFF	OFF	ON	OFF	OFF	Red	Red
12	Normal (t_1)	OFF	OFF	OFF	OFF	OFF	OFF	Green	Red

Table 3. Phases of lamps at the crossroads in normal state

Phase	HM.Yamin Street			Gaharu Street			Jawa Street			Long Phase (second)
	M_1	K_1	H_1	M_2	K_2	H_2	M_3	K_3	H_3	
1	1	0	0	0	0	1	0	0	1	50
2	1	0	0	0	1	0	0	1	0	3
3	0	0	1	1	0	0	1	0	0	50
4	0	1	0	1	0	0	1	0	0	3

The overall toolwork process is illustrated by the system flowchart as in Figure 4.

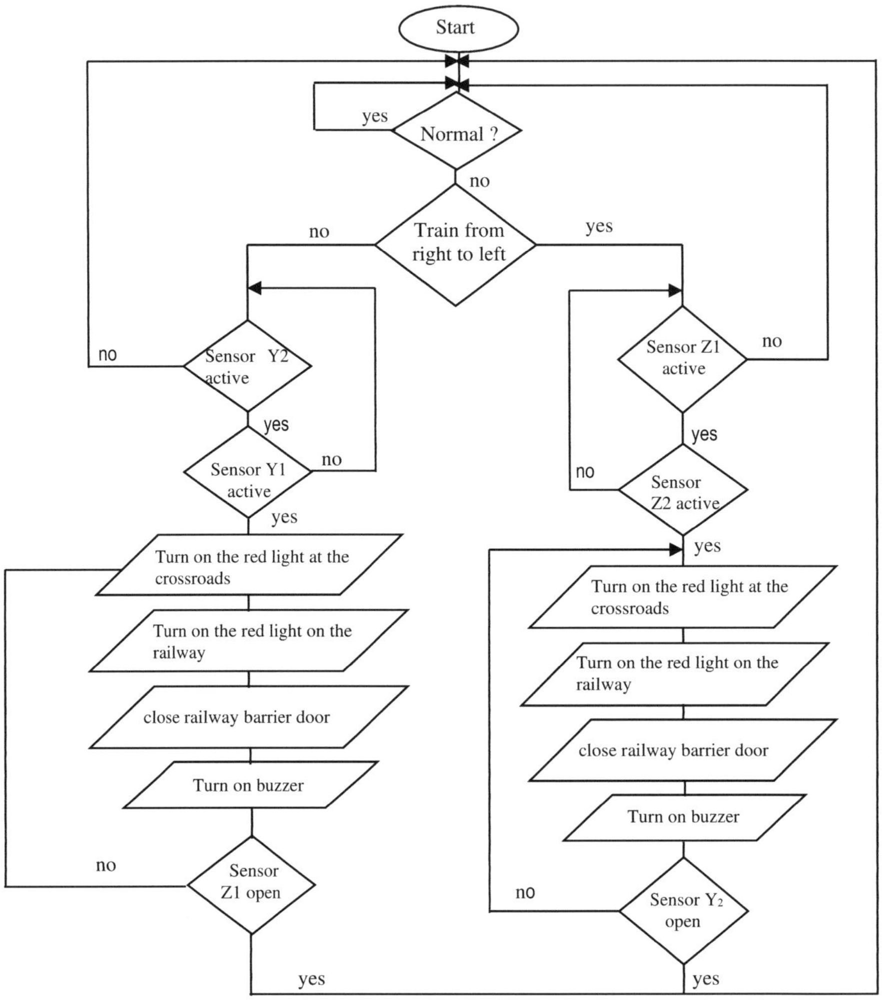

Figure 4. Flow chart Overall System

With the result of designing this control tool, it can be eliminated the time required to unravel the congestion that occurs in the railway door area when the railway barrier door is closed. This is because when the railway barrier door is closed (the train is 500 meters away from the track door) all the traffic light is red at all intersections so that no vehicle accumulates and precedes each other when the railway barrier door opens again, which means 1 minute congestion time to unravel the congestion can be eliminated so that the congestion time is reduced to just 2 minutes by the time a regular train passes and reduced to 3 minutes when the train switcher.

The total reduced congestion time in one day is 23 minutes so the total one-day bottleneck time becomes 53 minutes. Reduced fuel loss in one day is 23 minutes multiply by 60 seconds multiply by 0.583 ml / second multiply by 75 vehicles equal to 60.340 ml equal to 60.340

liters. The exhaust time of CO and CO2 is reduced for 23 minutes with 1,725 vehicles in one day so as to reduce exhaust pollution in the four-point area.

4. Conclusion

1. The design of traffic light control system can reduce congestion on the crossroad of M. Yamin street with Gaharu street for 23 minutes.
2. The design of this traffic light control system can reduce vehicle fuel waste by 60,340 liters per day and can reduce CO2 and CO2 emissions from the car for 23 minutes in one day in the area.
3. The design of this traffic light control system improves the function of road and vehicle functions as a means of transportation in the area.

5. References

[1] Agfianto E P 2002 *Learning AT89C51 / 52/55 Microcontroller (theory and application)* (Yogyakarta; Gajah Mada)
[2] Arifin Z 2009 *Vehicle Pollution Control* (Bandung; Alfabeta)
[3] Danny C, and Kris P 2004 *Basic Guide of MCS51 Family Microcontroller* (Surabaya; Innovative Electronics)
[4] Sastra N, Budiarsa S, Suarna 2012 *The Effect Of Fuel Octane Value And Engine Round On Motor Vehicles To The Characteristic Of Exhaust Emissions. Proc. Ecotrophic* 4(2) 106-111
[5] Kuswara 2006 *Emission Inventory of pollution CO2, NOx, HC and SPM in Bandung Regency J. Teknik Lingkungan* **3 (1)** 215-224.
[6] Government Regulation of the Republic of Indonesia Number 41 1999 *Air Pollution Control*
[7] Rann Y, Fanny R A, Ferina F 2011 *Analysis of the amount of Air Pollution Based on Motor Vehicle exhaust gas level in the Parking Area Gunadarma University Kalimang* (Jakarta: PKM University of Gunadarma)
[8] Yudha W 2009 *Analisis Speed of Vehicles On the Road Section Brigjen Sudiarto (Majapahit) Semarang City and Its Influence on Fuel Consumption* (Semarang : Thesis Magister of Civil Engineering University Dipanegoro)

IOP Conf. Series: Materials Science and Engineering **237** (2017) 012026 doi:10.1088/1757-899X/237/1/012026

A review: search visualization with Knuth Morris Pratt algorithm

Robbi Rahim[1*], Iskandar Zulkarnain[2] and Hendra Jaya[3]
[1]Postgraduate Ph.D. Student Universiti Malaysia Perlis, Universiti Malaysia Perlis, Malaysia
[1]Department of Health Information, Akademi Perekam Medik dan Infokes Imelda, Jl. Bilal Ujung Medan 20116, Indonesia
[2]Department of ComputerSystem, STMIK Triguna Dharma, Jl. Jenderal Abdul Haris Nasution No.73, Medan 20219, Indonesia
[3]Department of Computer Engineering, STMIK Triguna Dharma, Jl. Jenderal Abdul Haris Nasution No.73, Medan 20219, Indonesia

*usurobbi85@zoho.com

Abstract. In this research modeled a search process of the Knuth-Morris-Pratt algorithm in the form of easy-to-understand visualization, Knuth-Morris-Pratt algorithm selection because this algorithm is easy to learn and easy to implement into many programming languages.

1. Introduction
Searching is a work that is often performed in everyday life. There are times when we look for something with the purpose of just knowing whether the data is in a set of data or not, while at other times we may want the position of the data sought [1].

A significant difference exists in the string matching algorithm, if in the search process it is not necessary to apply a string matching algorithm to test the data accordingly or not enough using IF logic[1], whereas in the string matching algorithm by default use the search concept to make the process better in terms of string matching[2].

The Knuth-Morris-Pratt (KMP) algorithm is one of the string matching algorithms (pattern) with the examination process from left to right[2][3][4][5], in this research the Knuth-Morris-Pratt algorithm is described in the form of visualization to facilitate the string matching process, Visualization of the Knuth-Morris-Pratt process in this study makes it easy for other researchers to find out how the string matching and searching process works. It is expected that the Knuth-Morris-Pratt algorithm works to facilitate its application to various search processes such as search engines, plagiarism detection and so on.

2. Theory
The Knuth-Morris-Pratt algorithm was developed by D. E. Knuth, along with J. H. Morris and V. R. Pratt[1][3][6]. The Knuth-Morris-Pratt algorithm is the development of the previous string search algorithm, the Brute Force algorithm. Brute-Force algorithm is the simplest basic algorithm in solving string matching problem that examines every position in the text between 0 and n-m, where n is the length of text/number of filenames stored on the computer and m is the character length of a pattern (the word to search)[7].

KMP algorithm (Knuth Morris Pratt) is an algorithm used to perform string matching process. This algorithm is a type of Exact String Matching Algorithm which is a precise string matching with the arrangement of characters in a matched string having the number and sequence of characters in the same string. Example: the algorithmic word will show a matching only with the algorithmic word [2][3].

In the KMP algorithm, the information used to do a further shift, not just one character like the Brute Force algorithm. This algorithm performs matching from left to right[5].

3. Result and Discussion

Here is the process of applying the Knuth Morris Pratt algorithm to search for a word in a sentence sequence, as for the process as follows:

a. Given the S string variable with an array of letters as follows:

D	A	E	L	Y	M	M	A	K	R	I	N	A	A	M	R	S	W

b. Given pattern, variable P is the word to be searched in variable S

R	I	N	A

c. The first step
Compare pattern P [1] with string S [1], here is the result

D	A	E	L	Y	M	M	A	K	R	I	N	A	A	M	R	S	W
R	I	N	A														

Pattern [1] does not match sequence [1] then the pattern will shift one position to the right.

d. Step Two
Compare pattern P [1] with string S [2], here is the result

D	A	E	L	Y	M	M	A	K	R	I	N	A	A	M	R	S	W
	R	I	N	A													

Pattern [1] does not match sequence [2] then the pattern will move one position to the right.

e. Step Three
Compare pattern P [1] with string S [3], here is the result

D	A	E	L	Y	M	M	A	K	R	I	N	A	A	M	R	S	W
		R	I	N	A												

Pattern [1] does not match sequence [3] then the pattern will move one position to the right.

f. Step Four
Compare pattern P [1] with string S [4], here is the result

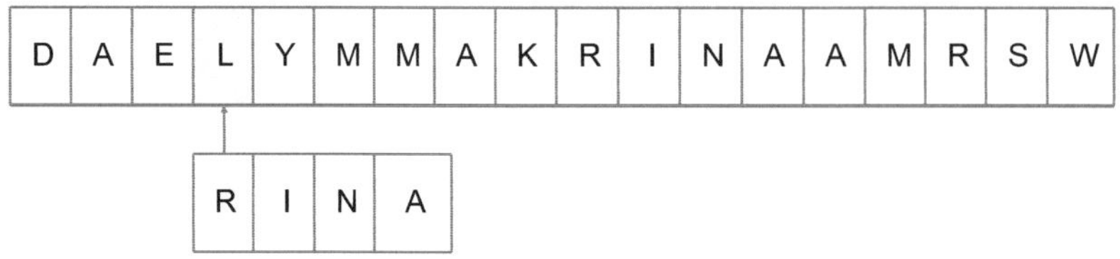

Pattern [1] does not match string [4] then the pattern will shift one position to the right.

g. Step Five

Compare pattern P [1] with string S [5], here is the result

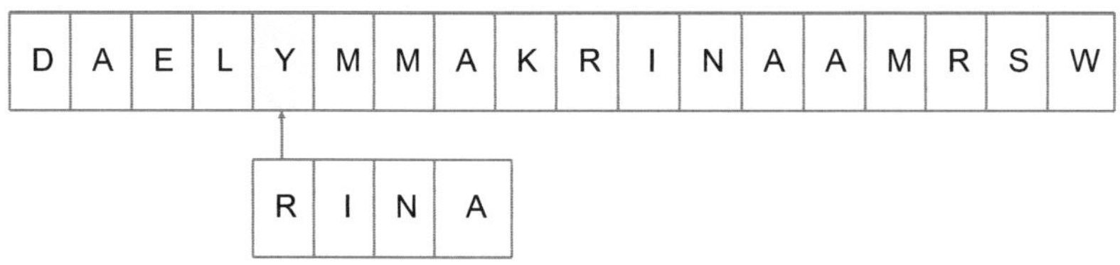

Pattern [1] does not match sequence [5] then the pattern will shift one position to the right.

h. Step Six

Compare pattern P [1] with string S [6], here is the result

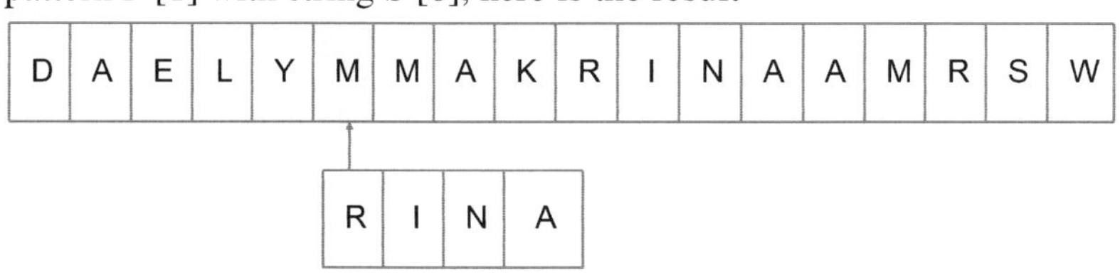

Pattern [1] does not match sequence [6] then pattern will shift one position to right

i. Step Seven

Compare pattern P [1] with string S [7], here is the result

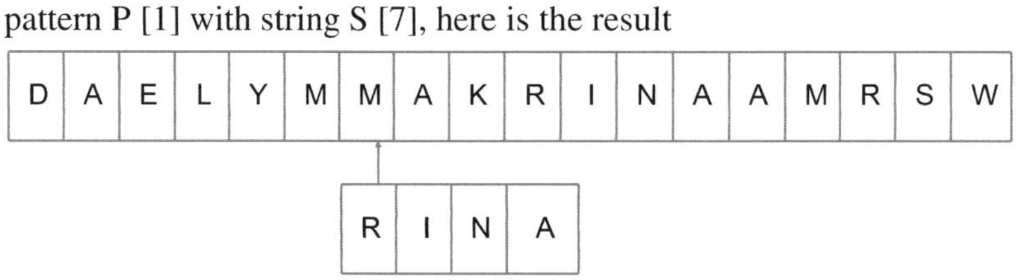

Pattern [1] does not match string [7] then pattern will move one position to right

j. Step Eight

Compare pattern P [1] with string S [8], here is the result

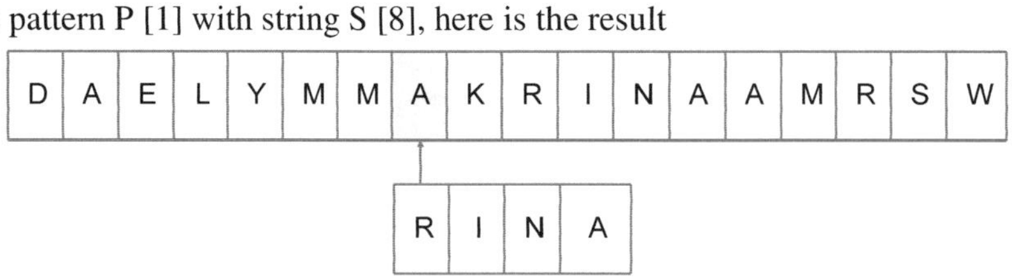

Pattern [1] does not match sequence [8] then pattern will shift one position to right

k. Step Nine

Compare pattern P [1] with string S [9], here is the result

IOP Conf. Series: Materials Science and Engineering **237** (2017) 012026 doi:10.1088/1757-899X/237/1/012026

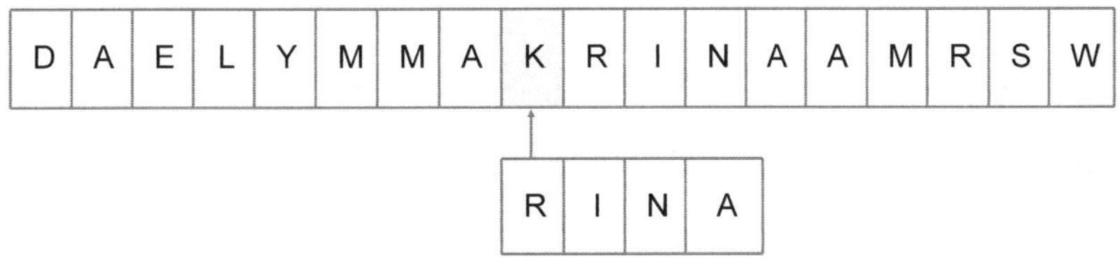

Pattern [1] does not match string [9] then pattern will move one position to right

1. Step Ten

   Compare pattern P [1] with string S [10], here is the result

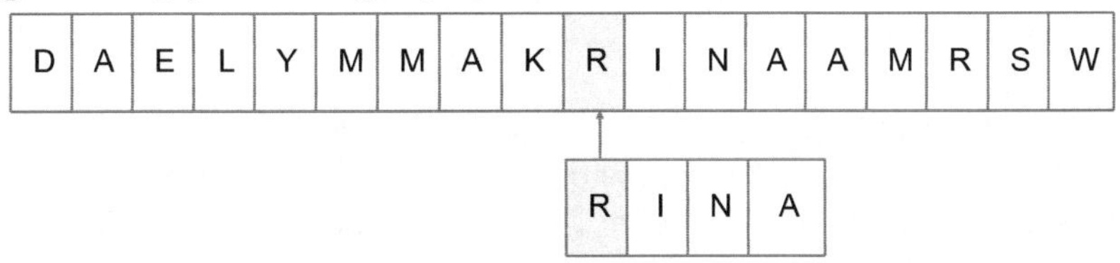

Pattern [1] matches the string [10]. Because there is a match, the Knuth Morris Pratt algorithm will store this information, and the pattern will not shift and continue matching pattern [2] with string [11].

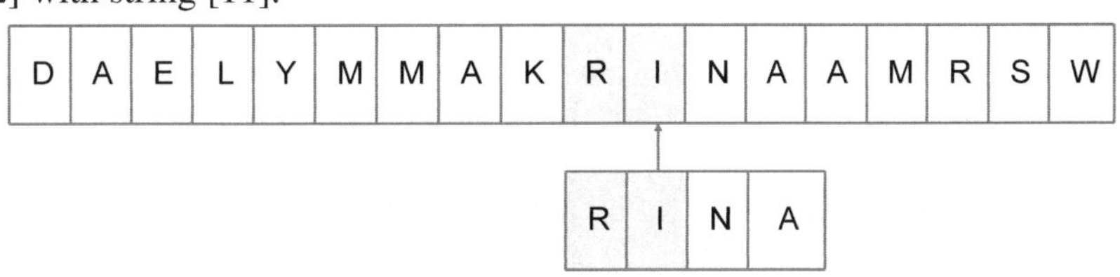

Pattern [1, 2] matches the string [10, 11]. Because there is a match, the Knuth Morris Pratt algorithm will store this information, and the pattern will not shift and continue matching pattern [3] with string [12].

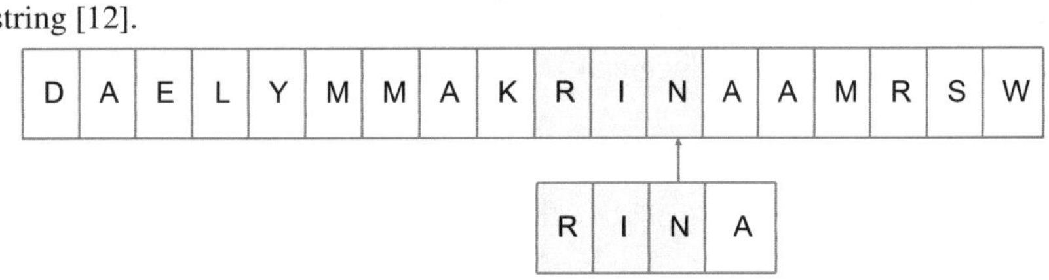

Pattern [1, 2, and 3] matches the string [10, 11, and 12]. Because there is a match, the Knuth Morris Pratt algorithm will store this information, and the pattern will not shift and continue matching pattern [4] with string [13].

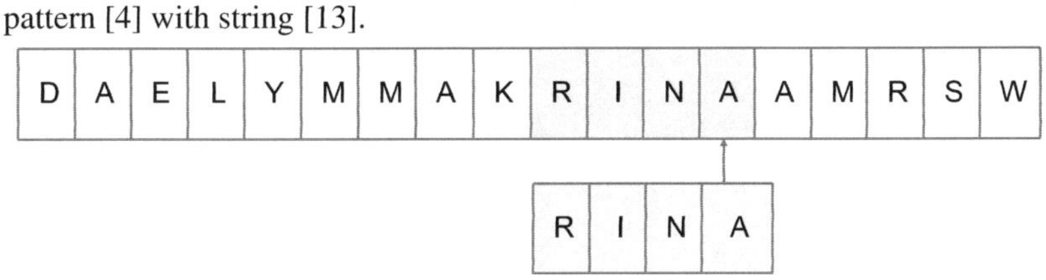

Pattern [4] matches the string [13]. Because there is a match, the Knuth Morris Pratt algorithm will store this information, and the pattern will not shift and continue matching pattern [5] with string [14]. However, since the number of patterns is only four letters, the search will be stopped and the result is that the P pattern matches S-string by 100 percent.

4. Conclusion

The process visualization of the Knuth-Morris-Pratt (KMP) algorithm allows researchers or scholars to learn how the KMP algorithm works, and in the application development it will be easier to create a function for word search and can be implemented into many search processes.

References

[1] N. R. Dalal and P. Jadhav, 2015, "A Composite Algorithm for String Matching," *International Journal of Modern Trends in Engineering and Research (IJMTER),* vol. 2, no. 7, pp. 68-73.

[2] R. Y. Tsarev, A. S. Chernigovskiy, E. A. Tsareva, V. V. Brezitskay, A. Y. Nikiforov and N. A. Smirnov, 2015, "Combined string searching algorithm based on Knuth-morris-pratt and Boyer-Moore algorithms," in *XIX International Scientific Conference Reshetnev Readings*, Rusia.

[3] I. M. Abu-Zaid and E. K. El-Rayyes, 2012, "Parallel Search Using KMP Algorithm in Arabic String," *International Journal of Science and Technology,* vol. 2, no. 7, pp. 427-431.

[4] R. Janani and S. Vijayarani, 2016, "An Efficient Text Pattern Matching Algorithm for Retrieving Information from Desktop," *Indian Journal of Science and Technology,* vol. 9, no. 43, pp. 1-11.

[5] C. S. Rao, K. B. Raju, and S. V. Raju, 2013, "Parallel String Matching with Multi-Core Processors - A Comparative Study for Gene Sequences," *Global Journal of Computer Science and TechnologyHardware & Computation,* vol. 13, no. 1, pp. 26-41.

[6] G. Pandey and G. L. Prajapati, 2016, "Applying Bi-Directional Search Strategy in Selected String Matching Algorithms," *International Journal of Computer Applications,* vol. 143, no. 11, pp. 40-43.

[7] D. Shapira and A. Daptardar, 2006, "Adapting the Knuth–Morris–Pratt algorithm for pattern matching in Huffman encoded texts," *Elsevier: Information Processing and Management,* vol. 42, pp. 429-439.

IOP Conf. Series: Materials Science and Engineering **237** (2017) 012027 doi:10.1088/1757-899X/237/1/012027

Double hashing technique in closed hashing search process

Robbi Rahim[1,2,*], Iskandar Zulkarnain[3] and Hendra Jaya[4]
[1]Postgraduate Ph.D. Student Universiti Malaysia Perlis, Universiti Malaysia Perlis, Malaysia
[2]Department of Health Information, Akademi Perekam Medik dan Infokes Imelda, Jl. Bilal Ujung Medan 20116, Indonesia
[3]Department of Computer System, STMIK Triguna Dharma, Indonesia
[4]Department of Computer Engineering, STMIK Triguna Dharma, Jl. Jenderal Abdul Haris Nasution No.73, Medan 20219, Indonesia

*usurobbi85@zoho.com

Abstract. The search process is used in various activities performed both online and offline, many algorithms that can be used to perform the search process one of which is a hash search algorithm, search process with hash search algorithm used in this study using double hashing technique where the data will be formed into the table with same length and then search, the results of this study indicate that the search process with double hashing technique allows faster searching than the usual search techniques, this research allows to search the solution by dividing the value into the main table and overflow table so that the search process is expected faster than the data stacked in the form of one table and collision data could avoided.

1. Introduction

Searching is a job that we often do in everyday life. In a text editor, we often do a word search, or search for a word and replace it with another word, or search for a particular word and calculate the frequency of occurrence of the word in the document[1][2].

The search methods are various and each method has advantages and disadvantages, one of the search methods that has better place efficiency is Hash Search[3][4][5]. Hash Search uses a certain formula to perform the process of placement and search data, Hash Search is divided into 2 kinds, namely Closed Hashing and Open Hashing. In Closed Hash there is the possibility of more than one data having the same function value resulting in a collision. One way to overcome this collision can be done with Double Hashing technique [6][7].

This study aims to show how the work of placement and search data with hash search algorithm with double hashing techniques and strategies to overcome the collision of data that may occur,

2. Theory

Hash Search method is almost similar to the method of direct search that is by using a certain formula both at the time of placement and search data[8]. Hash Search has better place space efficiency than direct search [9]. The function (I - 1) used by the direct search method has poor place use efficiency, so the Hash Search method improves it by using the modulo operation function [1][8]. Modulo operation function is often referred to as the Hash Function and the data shelter is called the Hash Table. The Hash function is not a one-to-one function such as the sharing function of the direct search method so it is possible that some data have the same function results [4][5]. This results in collisions occurring at the time of placement of data into tables so that a strategy is needed to overcome this collision. Strategies to deal with these collisions vary and each has their respective advantages and disadvantages. This strategy will be explained in the next section [6][7].

In general, Open Hashing means the use of table size is limited, but each index of the table utilizes a linked list to place data so that the amount of data is not limited [3], while Close Hash is a function with limited table size so that the amount of data that can be placed is also limited [3][7].This Hash function causes the data stored in the Hash Table to have 2 types of address (address) that is:

a. Home Address, is the location obtained by using the Hash Function.
b. Real (Physical) address is the location where data is stored in table.

This double hashing strategy overcomes / collisions by dividing the table into 2 pieces, the Main Table and the table overflow with the provisions of the size of the table as large[5]. The way of placement of data in the Main Table is done by using the Hash function. In the event of a collision, the data is placed on the overflow table. How to place data in the overflow table can be done in a sequential way or using the new Hash function. If the table is full, then the data entered into the overflow table will be rejected [3][4].

3. Proposed Method

Closed Hash data on the placement is done by using a hash function. Placement of data in Closed Hash is represented by using tables. Because the hash function allows multiple data to produce the same value, then the data placement process on Closed Hash can cause a collision. The problem can be solved by using the Double Hashing strategy.

Suppose the double hash table size is 10 then the hash function h (x) = x mod 10 and the overflow function g (x) = x mod 10. The table numbering starts from 0, for the numbers of the hashing tables in the contents as follows: 43,12,76,32,11,66,39,28,21,59,5,91,9,33, for the placement process for each table as follows:

43	**12**	**76**	**32**	**11**	**66**	**39**	**28**	**21**	**59**	**5**	**91**	**9**	**33**
1	2	3	4	5	6	7	8	9	10	11	12	13	14

Figure 1. Arrangement of numbers before the hashing process

a. Data '43'
Main Table -> h (x) = 43 mod 10 = 3, location 3 is empty so that '43' data is placed at location 3
b. Data '12'
Main Table -> h (x) = 12 mod 10 = 2, location 2 is empty so data '12' is placed on location 2
c. Data '76'
Main Table -> h (x) = 76 mod 10 = 6, location 6 is empty so that '76' data is placed at location 6
d. Data '32'
Main Table -> h (x) = 32 mod 10 = 2 -> collision occurs with data '12' the search continues to Overflow Table.
Table Overflow -> g (x) = 32 mod 10 = 2, location 2 is empty so '32' data is placed on location 2

e. Data '11'
Main Table -> h (x) = 11 mod 10 = 1, location 1 is empty so the data '11' is placed at location 1

f. Data '66'
Main Table -> h (x) = 66 mod 10 = 6 -> collision with data '76' the search continues to Overflow Table
Table Overflow -> g (x) = 66 mod 10 = 6, location 6 is empty so '66' data is placed on location 6

g. Data '39'
Main Table -> h (x) = 39 mod 10 = 9, location 9 is empty so '39' data is placed on location 9

h. Data '28'
Main Table -> h (x) = 28 mod 10 = 8, location 8 is empty so that '28' data is placed at location 8

i. Data '21'
Main Table -> h (x) = 21 mod 10 = 1 -> a collision with data '11' the search continues to Overflow Table.
Table Overflow -> g (x) = 21 mod 10 = 1, location 1 is empty so data '21' is placed at location 1

j. Data '59'
Main Table -> h (x) = 59 mod 10 = 9 -> collision with data '39' the search is continued into Overflow Table.
Table Overflow -> g (x) = 59 mod 10 = 9, location 9 is empty so '59' data is placed on location 9

k. Data '5'
Main Table -> h (x) = 5 mod 10 = 5, location 5 is empty so data '5' is placed at location 5

l. Data '91'
Main Table -> h (x) = 91 mod 10 = 1 -> collision with data '11' the search is continued to Overflow Table
Table Overflow -> g (x) = 91 mod 10 = 1 -> collision with data '21'
G (x) = (1 + 1) mod 10 = 2 -> a collision occurs with data '32'
G (x) = (2 + 1) mod 10 = 3, location 3 is empty so the '91' data is placed at location 3

m. Data '9'
Main Table -> h (x) = 9 mod 10 = 9 -> collision with data '39' the search is continued to Overflow Table
Table Overflow -> g (x) = 9 mod 10 = 9 -> collision with data '59'
G (x) = (9 + 1) mod 10 = 0, location 0 is empty so the data '9' is placed at location 0

n. Data '33'
Main Table -> h (x) = 33 mod 10 = 3 -> collision with data '43' the search is continued to Overflow Table
Table Overflow -> g (x) = 33 mod 10 = 3 -> collision with data '91'
G (x) = (3 + 1) mod 10 = 4, location 4 is emptyso the '33' data is placed at location 4

The following is the order of the numbers in the table hashing:

IOP Conf. Series: Materials Science and Engineering **237** (2017) 012027 doi:10.1088/1757-899X/237/1/012027

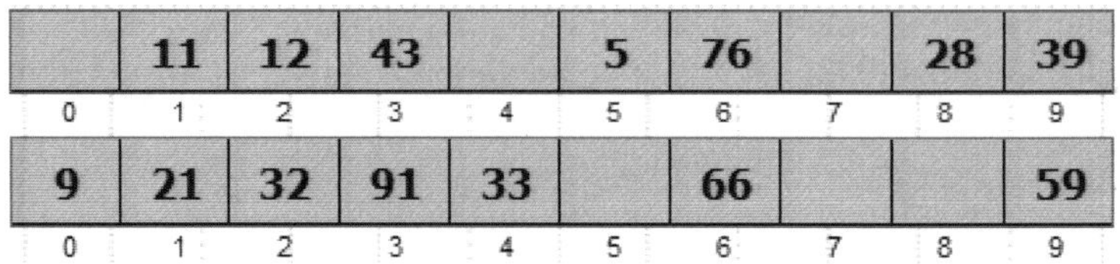

Figure 2. (a) Main Table and (b) Overflow Table

After determining the position of the numbers in the main table and the overflow table, the next test is to perform the following search process:

a. Search Data '91' using Double Hash technique
   Main Table Size = 10, hash Function = h (x) mod 10
   Overflow Table Size = 10, hash Function = g (x) mod 10

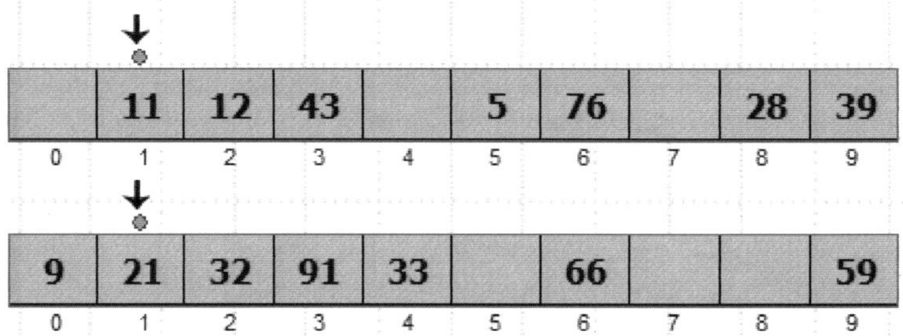

Figure 3. Initial search from position 1 at Main Table and Overflow Table

b. Main Table -> h (x) = 91 mod 10 = 1 -> Data '91' no search found continued to Overflow Table
   Overflow table ->
   G (x) = 91 mod 10 = 1 -> Data '91' not found.
   G (x) = (1 + 1) mod 10 = 2 -> Data '91' not found.

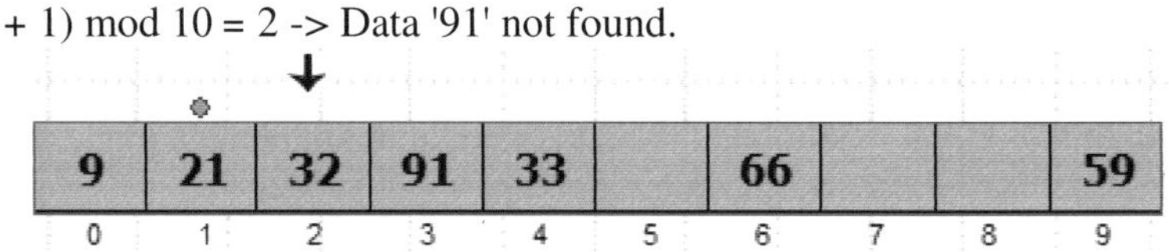

Figure 4. Search Position 2 at Overflow Table

G (x) = (2 + 1) mod 10 = 3 -> Data '91' is found in location 3.

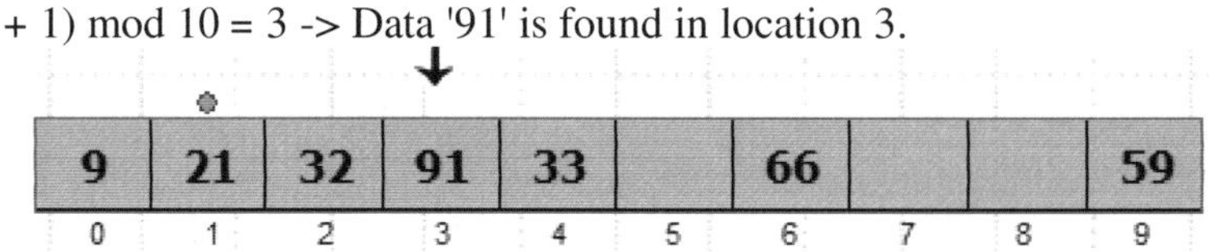

Figure 5. Search Position 3 at Overflow Table

In the search process the number 91 obtained the number is on the overflow table and hashing process stops after the data is found.

IOP Conf. Series: Materials Science and Engineering **237** (2017) 012027 doi:10.1088/1757-899X/237/1/012027

4. Conclusion

In the process of searching by double hashing technique can be concluded that the value will be divided into two tables to place the value in the main table and overflow, and for the search process is done first once in the main table and continued on the table overflow until all the numbers listed in the table are processed.

References

[1] N. Arora, G. Bhasin and N. Sharma, 2014, "Two way Linear Search Algorithm," *International Journal of Computer Applications,* vol. 107, no. 21, pp. 6-8.

[2] D. Kumar and S. Manish , 2014, "BINARY SEARCH IS FASTER THAN THE LINEAR," *INTERNATONAL JOURNAL OF INNOVATIVE RESEARCH IN TECHNOLOGY,* vol. 1, no. 5, pp. 796-799.

[3] D. Jia, W. Xie and Y. Jiang, 2012, "Research of database query technology based on hash function," in *IEEE 8th International Conference on Information Science and Digital Content Technology (ICIDT)*, Jeju, South Korea.

[4] E. Lodi and F. Luccio, 1985, "SPLIT SEQUENCE HASH SEARCH," *Information Processing Letters* , vol. 20, pp. 131-136.

[5] J. N, V. V, K. M and K. A, 2012, "Frequent Itemset Generation using Double Hashing Technique," in *International Co\nference on Modelling Optimization and Computing*.

[6] K. K. Pandey and N. Pradhan, 2014, "A Comparison and Selection on Basic Type of Searching Algorithm in Data Structure," *International Journal of Computer Science and Mobile Computing,* vol. 3, no. 7, pp. 751-758.

[7] Y. M. K. Omar, H. Osama and A. Badr, 2017, "Double Hashing Sort Algorithm," *Computing in Science & Engineering,* vol. 19, no. 2, pp. 63-69.

[8] L. J. Guibas and E. SZEMEREDI, 1978, "The Analysis of Double Hashing," *JOURNAL OF COMPUTER AND SYSTEM SCIENCES* , vol. 16, pp. 226-274.

[9] A. Mehta, A. Saxena, J. Patel and A. Thanna, 2015, "A REVIEW ON COMPARISION OF BINARY SEARCH AND LINEAR SEARCH," *International Journal OF Engineering Sciences & Management Research* , vol. 2, no. 10, pp. 85-89.

IOP Conf. Series: Materials Science and Engineering **237** (2017) 012028 doi:10.1088/1757-899X/237/1/012028

Comparative analysis of e-commerce and social media based trading in Indonesia

Sfenrianto[*]**, Wang Gunawan**[**]**, Halim Abdul, Fajar Ahmad Nurul**
Master in Information Systems Management, Bina Nusantara University, Jakarta, Indonesia.

[*] sfenrianto@binus.edu, [**] gwang@binus.edu

Abstract. The growth of social media and e-commerce recently has caused the unique phenomenon in Indonesia. That phenomenon is online trading using social media like Facebook, Blackberry messenger, Whatsapp, etc. This phenomenon attracts the researcher to know how to use social media to trade, not using e-commerce marketplace like Lazada, Tokopedia, Bukalapak. The aim of the research is to know the profiling of product and buyer in e-commerce and social media and to do the comparison of that 2 channels based on eight purchase decision factors. Research conducted by taking surveys involving 306 respondents. For product and buyer profiling, this study identifies profile base on sex, education, area, jobs and online trading channel. For e-commerce & social media comparison, this study use 8 factors made based on previous related studies. Finding related purchase decision comparisons are: online trust and usability are the factors influencing people to buy the product via e-commerce. While, marketing mix, and interactivity are the factors why people buy the product via social media.

1. Introduction

The growth of the Internet in Indonesia, coupled with the rapid development of the number of users of e-commerce. The internet also affect to the rapid growth of social media in Indonesia. The growth of e-commerce and social media, led to the phenomenon of buying and selling through internet technology.

In recent years, e-commerce sites have sprung up recently as Lazada.co.id, Olx.co.id, Tokopedia.com, Elevenia.com, Blibli.com, Traveloka.com, Bhineka.com, Pegipegi.com, Qoo10.com, Kaskus.com, Mataharimall.com and others. In fact, based on the rating agency website, Alexa.com, 5 of the 20 best-selling sites in Indonesia is an e-commerce site, namely: Lazada.co.id, Bukalapak.com, Onclickads.net, Tokopedia.com, Olx.co.id.

That social media are: Facebook, Twitter, Instagram and also chatting group (Blackberry group, WhatsApp group). The sellers and buyers trade with semi-manual system. There is no "buy button", so communication and transactions are not supported by the system. Payments are made without going through the payment gateway. Buyers could not check stock directly. Surprisingly seems many people are comfortable with this trade models. Even, often encountered sellers take advantage of e-commerce marketplace only as a means of advertising. The stock availability is not updated, buyers and sellers communicate through the chat group or other social media. This is a unique thing that is rarely found in other countries.

Thus the first research objective are knowing the customer and product profiling for buying and selling through social media and e-commerce in Indonesia. The second objective is analysing the factors that distinguish buying decision between e-commerce and social media.

2. Basic Theoretical

Social media is a tool/technology that allows us to communicate with others in a more effective, such as Blog, Twitter, Facebook, Linkedin, YouTube, Instagram, etc [1] [2]. Facebook, Instagram, and Twitter are some of the more popular social media in Indonesia.

Whereas, E-Commerce is use of ICT technology to support the exchange and transactions both in internal and external organizations [3]. External organizations include customers, suppliers, partners, shareholders and other stakeholders. There are many models of e-commerce in foreign countries, but in Indonesia, there are several popular e-commerce: Bukalapak, Tokopedia, Lazada, etc. The access rank of e-commerce in Indonesia is described in table 1.

Table 1 Top E-Commerce in Indonesia (Source: http://www.alexa.com/siteinfo)

E-commerce	Category	Alexa Rank - Indonesia
Bukalapak.com	C2C	10
Tokopedia.com	C2C	12
Lazada.co.id	B2C, C2C	22
Olx.co.id	C2C, Classifieds	28
Elevenia.co.id	C2C	29
Mataharimall .com	B2C,C2C	105
Blibli.com	C2C	114
Qoo10.co.id	C2C	214

3. Research Methodology

This study has 306 respondents from all around Indonesia consist of 167 male respondents and 139 female respondents. Judging from the level of education, respondents with undergraduate education is the highest respondents, which is 232 people. The number is difference significantly compared to the number of respondents with another education level. Then respondents with education level Master upwards are 60 people. High schools are 14 people. Most respondents are private employees that are 180 people. The second largest number is Housewife, 41 people. Then followed by civil servants/military/police/other Government Employees 36, entrepreneur/self-employed 29 people, others are 17 people, and students only 3.

3.1 Research Model

Those eight factors are used to build research model as showed in figure 1.

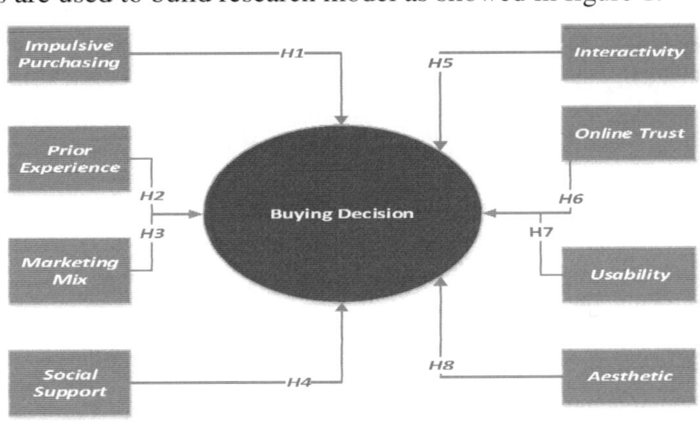

Figure 1. Research Model

Factors that influence of buying online are the impulse purchase, prior experience, and online trust [4] [5] [6]. Impulse purchasing factor has a positive correlation to shop online [4]. Prior experience factor has a positive correlation to the decision to shop online [5]. Another consideration factor is the online trust of e-commerce, and social media [6].

Factors that influence of online purchasing are interactivity, aesthetic, usability, and marketing mix [7] [8] [9]. Interactivity factor is adopted for buying decisions variable in social media because of one benefit in selling through social media is the interactivity between buyers, sellers and between fellow shoppers [7] [8]. Aesthetic factor has a positive correlation to shop commerce and social media [9]. Usability factor has the highest influence in the virtual buying behaviour [7] [8]. Marketing mix factor has the 2nd position in virtual buying behaviour influence [8].

Social support factor has a significant effect on intention to trade via social commerce [10] [11]. The factor is used in this study because people in Indonesia known as a socialize community that emphasizes on social factors, solidarity, mutual help, mutual aid. Scientific research also shows the influence of social support in the development and selling through social media [12] [13].

3.2 Hypothesis

There are 8 hypothesis made in this study. The number of hypothesis is the same with the number of purchase intention factors discussed above. Each hypothesis is comparing each factors in social media and e-commerce. The hypotheses are (1) H1: Impulse purchasing factor in e-commerce is more influence than in social media; (2) H2: Prior experience factor in social media is more influence than in e-commerce; (3) H3: Marketing mix factor in social media is more influence than in e-commerce; (4) H4: Social Support factor in social media is more influence than in e-commerce; (5) H5: Interactivity factor in social media is more influence than in e-commerce; (6) H6: Online trust factor in e-commerce is more influence than in social media; (7) H7: Usability factor in e-commerce is more influence than in social media; (8) H8: Aesthetic factor in e-commerce is more influence than in social media.

4. Results

4.1 Validity Test

Validity test is done with the help of SPSS Software. The correlation method used for validity test is Pearson Correlation. For e-commerce & social media data, with the number of 306 responses (N = 306), and a significant level of 5%, a score r table 0,113. All results obtained count r value over 0,113. Because of that, the data from questionnaires for all indicators are passed the test of validity. The comprehensive validity test result is listed in table 2.

Table 2. Validity test result

E-Commerce		**Social Media**	
Indicator	**r Calculation**	**Indicator**	**r Calculation**
EIP1	,443**	SIP1	,481**
EIP2	,472**	SIP2	,531**
EPE1	,717**	SPE1	,723**
EPE2	,699**	SPE2	,682**
EMM1	,677**	SMM1	,627**
EMM2	,513**	SMM2	,392**
EMM3	,550**	SMM3	,482**
EMM4	,740**	SMM4	,715**

ESS1	,601**	SSS1	,587**
ESS2	,407**	SSS2	,522**
EIT1	,690**	SIT1	,717**
EIT2	,623**	SIT2	,663**
EOT1	,694**	SOT1	,630**
EOT2	,516**	SOT2	,582**
EUS1	,699**	SUS1	,684**
EUS2	,655**	SUS2	,724**
EUS3	,741**	SUS3	,718**
EAS1	,539**	SAS1	,590**
EAS2	,683**	SAS2	,721**

4.2. Reliability Test

There are 2 factors that have Cronbach's Alpha values below 0,7. That unreliable data are from factor Aesthetic & Social Support. For the rest of chapter, that factors are excluded from analysis.

Table 3. Reliability test for each factor.

Factor	Channel	Cronbach's Alpha	N of Items
Impulse Purchasing	E-Commerce	0,786	2
	Social Media	0,762	2
Prior Experience	E-Commerce	0,741	2
	Social Media	0,785	2
Marketing Mix	E-Commerce	0,742	4
	Social Media	0,711	4
Social Support	E-Commerce	0,59	2
	Social Media	0,626	2
Interactivity	E-Commerce	0,839	2
	Social Media	0,885	2
Online Trust	E-Commerce	0,71	2
	Social Media	0,738	2
Usability	E-Commerce	0,859	3
	Social Media	0,826	3
Aesthetic	E-Commerce	0,525	2
	Social Media	0,632	2

4.3 Multi Linier Regression Test

With multi linier regression analysis, measured the relationship between each factor on purchase decisions in social media and e-commerce. The p-value (sig value) margin is set to 0.05 or 5%. The factor which has sig value lower than 0, 05 indicating the influence in online buying decision for related shopping channel (e-commerce or social media).

Table 4. Multi linier regression test.

Factor	Sig Value	
	E-Commerce	Social Media
Impulsive Purchase	0,32	0,908
Prior Experience	0,18	0,188
Marketing Mix	0,369	0,041
Interactivity	0,048	0,001
Online Trust	0,022	0,343
Usability	0,002	0,59

Table 4, described that in e-commerce, the buying decision influencing factors are Interactivity, Online Trust, and Usability. Then for social media, the influencing factors are marketing mix and interactivity.

4.4 Hypothesis Test
Based on result of multi linier regression test, this study compare it with the hypothesis. The comparison result is in table 5.

Table 5. Hypothesis test.

Hypothesis	Buying Decision Result		Hypothesis Test
	E-Commerce	Social Media	
H1	Not Influenced	Not Influenced	Not Supported
H2	Not Influenced	Not Influenced	Not Supported
H3	Not Influenced	Influenced (sig = 0,041)	Supported
H4	*Not Applicable*	*Not Applicable*	*Not Applicable*
H5	Influenced (sig = 0,048)	Influenced (sig = 0,001)	Supported
H6	Influenced (sig = 0,022)	Not Influenced	Supported
H7	Influenced (sig = 0,002)	Not Influenced	Supported
H8	*Not Applicable*	*Not Applicable*	*Not Applicable*

H1: *Impulse purchasing factor in e-commerce is more influence than in social media.* Impulse factor does not influence the decision to buy in e-commerce and social media. Therefore, this study cannot compare the impulse purchasing factors influent in the decision to buy in e-commerce and social media. So the hypothesis is not supported. This is a contrast in [4], which states impulse purchasing influence in purchase intention. One possible reason for this difference is related to credit card usage in [4], which reach 53.72%. Contrary with this study, respondents who use a credit card reached 12%. By paying in cash, will make people think longer to make buy decision.

H2: *Prior experience factor in social media is more influence than in e-commerce.* These factors were not affected in the decision to buy. Both decision to buy in e-commerce and social media. Therefore, this study cannot compare the prior experience factors influent in the decision to buy in e-commerce and social media. So the hypothesis is not supported. This is a contrast with research in [4], which states prior experience affects purchase intention. The difference is probably because of this study did not distinguish between new and existing online shoppers respondent. And also did not aware about how often they do online shopping.

H3: *Marketing mix factor in social media is more influence than in e-commerce.* The marketing mix factors influence the buying decisions in social media, whereas in e-commerce is not very influential. So the hypothesis is accepted. For social media, this is in line with research in [4] [9] that conclude the marketing mix influence on buying decisions online. This factor gets second ranks in buying decision influence, after usability factor. Marketing mix has four components commonly called 4P (product, price, promotion, and place). In this study, three of the four components was taken as indicators. Product components were not included because this study focuses on store/channel, not products. Actually, if only the price indicators are used, e-commerce is more attractive than social media, because, during research time, there were many discounted price from e-commerce marketplace new players. But since this indicator merged with other indicators, the results are different. The directions for future research, it is advised to use a more specific factor.

H5: *Interactivity factor in social media is more influence than in e-commerce.* Interactivity factor is more prominently found in social media. So the hypothesis is accepted the results are consistent with the hypothesis baseline in [5] [6] that interactivity influence the buying decisions online. But it is a contrast to the final results of that research. It could be due to time differences is far enough, t study the study conducted in 2004 at which time social media has not been as popular as now. And also to e-commerce, communications facilities have not been as good as today.

H6: Online trust factor in e-commerce is more influence than in social media. This factor has a sig value 0.022 in E-Commerce and 0.343 in social media. So the greater value of Online Trust found in E-Commerce. So the hypothesis is accepted. The results are consistent with research Kwek (2010) [4] [6], that state online trust factor is influential in the decision to buy in the online shop.

H7: *Usability factor in e-commerce is more influence than in social media.* Usability factor is found in E-Commerce. So the hypothesis is accepted. This study is consistent with research by Constantinides (2004) [5] [6], this state usability is the most influential factor on purchase decisions online.

5. Conclusion

Based on the research result, this study can have several conclusions. First, the factors that influence to purchase decision in e-commerce are usability and online trust. Buyer buys a product in e-commerce due to its easiness, like easy to search, easy to pay, easy to be delivered, etc. E-commerce is trusted channel, since it's can detect the historical record of buyers, using the secure line for transaction etc. Second, the factors that influence to purchase in social media are interactivity and marketing mix. Using social media, the buyer can chat intensively with the seller. This condition makes buyer feel comfortable. In Indonesia, social media based trading is older than e-commerce, for some product are only able to be found in social media. But in the future, this condition might be changing. Third, top 3 social media channel used for trading are facebook, Instagram, Whatsapp group. Fourth, top 3 e-commerce used for trading in Indonesia are lazada, tokopedia, bukalapak. Fifth, usage of the credit card for a transaction is low that is 12%. This number is very low compared to other countries.

Recommendation for future research, in modelling the decision to buy online, for the marketing mix factors should be broken down into factors more specific. Factors such as price, promotion, place. Related to social support factor in a decision to buy online, should be broken down into more specific factors, for example as informational support and emotional support. And also during the date gathering via questionnaire, respondent should also express how log & how often they do online shopping.

Reference

[1] Safko L 2009 *The Social Media Bible: Tactics, Tools, and Strategies for Business Succes* (John Wiley & Sons)

[2] Zarella D 2010 *The Social Media Marketing Book* (United States: O'Reilly Media)

[3] Currie W L 2010 *Value Creation of E-Business Model* (Warwick Business School, UK)

[4] Ling K C, Chai L T and Piew T H (2010) The effects of shopping orientations, online trust and prior online purchase experience toward customers' online purchase intention. *International Business Research*, *3*(3), 63

[5] Karbasivar A and Yarahmadi H (2011) Evaluating effective factors on consumer impulse buying behaviour. *Asian Journal of Business Management Studies*, *2*(4), 174-181

[6] Thaw Y Y, Mahmood A K and Dominic P (2009). A Study on the Factors That Influence the Consumers Trust on Ecommerce Adoption. *arXiv preprint arXiv:0909.1145*
[7] Constantinides E and Geurts P (2005). The impact of web experience on virtual buying behaviour: an empirical study. *Journal of customer behaviour*, *4*(3), 307-335
[8] Constantinides E (2004). Influencing the online consumer's behavior: the Web experience. *Internet research*, *14*(2), 111-126
[9] Liang T P, Ho Y T, Li Y and Turban E (2011). What drives social commerce: The role of social support and relationship quality, *International Journal of Electronic Commerce*, *16*(2), 69-90.
[10] Jarvelainen J (2003). The impact of prior online shopping experience on future purchasing channel choice. ECIS 2003 Proceedings, 58
[11] Thamizhvanan A and Xavier M J (2013). Determinants of customers' online purchase intention: an empirical study in India. Journal of Indian Business Research, 5(1), 17-32
[12] Koufaris M (2002). Applying the technology acceptance model and flow theory to online consumer behavior. Information systems research, 13(2), 205-223
[13] Junadi and Sfenrianto (2015) A model of factors influencing consumer's intention to use e-payment system in Indonesia. Proc Comput Sci 59:214–220.

IOP Conf. Series: Materials Science and Engineering **237** (2017) 012029 doi:10.1088/1757-899X/237/1/012029

Stabilization of nonlinear system using uniform eigenvalue assignment in linear model

Sahat P Siahaan[1], Timbang Pangaribuan[1,2]
(1) Department of Electrical Engineering, Faculty of Engineering, HKBP Nommensen University, Jl. Sutomo No. 4A, Medan , 20234
(2) Department of Physics, Faculty of Mathematics and Natural Science, University of North Sumatera, Jl. Bioteknologi No.1, Medan, 20155

spasi590817@gmail.com, timbang.pbn@gmail.com

Abstract. Some plant in control system has nonlinear dynamic, so it is not easy to do in analysis to see its behavior using eigenstructure assignment. From many observations which have been made, some literature give methods to design nonlinear control system. The modern control theory uses state-space method to explain the behaviour on stability of a plant. To improve the stability of the closed-loop system, designer commonly use the state feedback control law. For the case inverted pendulum plant with the nonlinear dynamics, its need to perform the nonlinear control law with the concepts of modern control theory to satisfy the closed-loop system characteristic, and all the behaviour of the closed-loop system only determined from the given linear pole specifications.

1. Introduction

For the class of linear control systems, the performance of the systems is essentially determined by using method the eigenstructure assignment in the closed loop systems. It is also empirically true that the performance of a non linear system, is determined by the eigenstructure of the linearized system around certain nominal conditions. In this paper, the closed loop system will solve the uniform eigenvalue placement problem. That is, designing a single nonlinear state feedback or output feedback control law such that the linearized closed-loop system around the whole equilibrium manifold has the given set of eigenvalues [1]. Clearly, this objective is motivated by the intuition that performance of the resulting closed-loop system is less sensitive to variations of the equilibrium point.

2. System Details

Consider a general nonlinear system described [1] by equations,

$$\dot{x}(t) = f\{x(t), u(t)\}$$
$$y(t) = h\{x(t), u(t)\} \quad (1)$$

where $x(t)$ is the n-dimensional plant state, $u(t)$ is the m-dimensional plant input, $y(t)$ is the p-dimensional plant output. The equilibrium set of system (1) is defined as,

$$E = \{(x,u) \in R^n \times R^m \mid f(x,u) = 0\}$$

For x is the function of α as below,

$$f\{x(\alpha), u(\alpha)\} = 0 \quad (2)$$

then the equilibrium set given by,

$$E = \{(x(\alpha), u(\alpha)) \in R^n \times R^m \mid f\{x(\alpha), u(\alpha)\} = 0\} \quad (3)$$

Now consider the following parameterized system that is obtained by linierizing the plant around each point of the equilibrium manifold E [2].

$$\dot{x}_\delta = A(\alpha)\, x_\delta + B(\alpha)\, u_\delta$$

$$y_\delta = C(\alpha)\,x_\delta \tag{4}$$

where

$$x_\delta = x - x(\alpha)$$
$$u_\delta = u - u(\alpha)$$
$$y_\delta = y - h\{x(\alpha), u(\alpha)\} \tag{5}$$

and

$$A(\alpha) = \frac{\partial f}{\partial x}\{x(\alpha), u(\alpha)\}$$
$$B(\alpha) = \frac{\partial f}{\partial u}\{x(\alpha), u(\alpha)\}$$
$$C(\alpha) = \frac{\partial f}{\partial x}\{x(\alpha), u(\alpha)\} \tag{6}$$

The fundamental idea can be well characterized in terms of the state feedback control and the design of nonlinear state feedback control law [4] of the form,

$$u(t) = K\{x(t), v(t)\} \tag{7}$$

where $v \in R^s$ represent an exogenous input, and $K\{x(t), v(t)\}$ is a smooth function satisfying $K(0,0) = 0$ such that the closed loop system has the following properties:

(1) There exists a smooth function $v(o)$ satisfying $v(o) = 0$ such that,

$$f[x(\alpha),\ K\{x(\alpha), v(\alpha)\}] = 0 \tag{8}$$

(2) The linearized state feedback control law around the whole equilibrium manifold of the closed loop system [3] is given by,

$$u_\delta = K_1(\alpha)\,x_\delta + K_2(\alpha)\,v_\delta$$
$$v_\delta = v - v(\alpha) \tag{9}$$

where $K_1(\alpha)$, $K_2(\alpha)$ are smooth function synthesized according to the specific task such as output regulation. The above idea can be illustrated using the block diagrams shown in Figure 1.

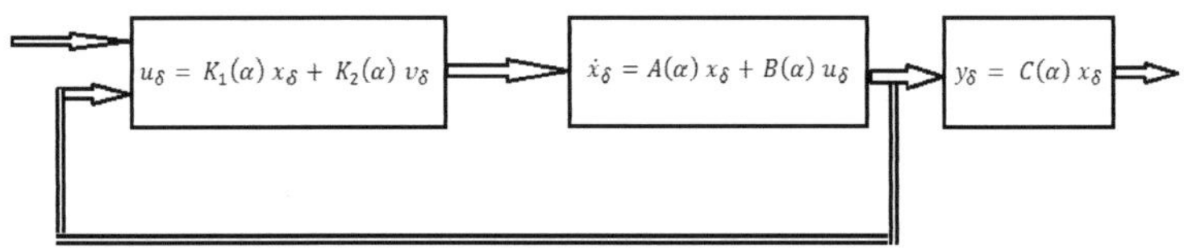

Figure 1. Parameterized Linear System

3. Design Of Nonlinear Controller For An Inverted Pendulum

For the asymtotic tracking, supposefor each α that the pair of equation below:

$$\frac{\partial f}{\partial x}\{x(\alpha), u(\alpha)\}\ ,\ \frac{\partial f}{\partial u}\{x(\alpha), u(\alpha)\} \tag{10}$$

Is controllable and has constant controllability indexes and has constant observability indexes. Under the additional requirement that the equilibrium manifold satisfy the static decoupling condition below for convenience.

$$f\{x(\alpha), u(\alpha)\} = 0$$
$$h\{x(\alpha), u(\alpha)\} = 0 \tag{11}$$

For under the class of either state feedback control laws, the closed loop satisfies, in addition to the local stability around each point in the equilibrium manifold,

$$\alpha = h\,[\{x(\alpha), u(\alpha) + K_1(\alpha)\langle v(\alpha) - x(\alpha)\rangle\}] \tag{12}$$

which complies, in term of state feedback for simplicity, that, given any constant exogenous input $v(t) = \alpha$. The output trajectory $y(t)$, asymtotically approach $v(t) = \alpha$; that is, we have,

$$\lim_{t\to\alpha} y(t) = v(t) \tag{13}$$

Consider a problem of balancing an inverted pendulum on a cart as a nonlinear plant. The dynamics equation of motion for the inverted pendulum shown in equation (14).

$$\begin{bmatrix}\dot{x}_1\\ \dot{x}_2\end{bmatrix} = \begin{bmatrix} x_2 \\ \dfrac{g\ \sin(x_1) - a\,m\,L\,x_2^{\,2}\ \sin\frac{2\,x_1}{2} - a\ \cos(x_1)\,u}{\frac{4}{3}L - a\,m\,L\ (\cos\ (x_1))^2}\end{bmatrix} \tag{14}$$

where x_1 is the angle (in radians) of the pendulum from vertical axis, and u is the force applied to the cart (in Newtons). The various parameters and values are,

m = pendulum mass = 2.0 kg.
M = cart mass = 8.0 kg.
2L = pendulum length = 1.0 m
G = 9.8 *m/sec*^{2}
$a = \frac{1}{m+M}$

It is straighforward to verify that equilibrium manifold,

$$x(\alpha) = \begin{bmatrix}\alpha\\0\end{bmatrix}\ ,\ \ u(\alpha) = \frac{g}{a}\tan\alpha \tag{15}$$

Is such that equation ( ) we take calculation and gives two parameters below.

$$A(\alpha) = \begin{bmatrix} 0 & 1 \\ \dfrac{g}{\left[\frac{4}{3}L - a\,m\,L\ \cos\alpha^2\right]\ \cos\alpha} & 0\end{bmatrix}\ ,\ \ B(\alpha) = \begin{bmatrix} 0 \\ \dfrac{-a\ \cos\alpha}{\frac{4}{3}L - a\,m\,L\ \cos\alpha^2}\end{bmatrix} \tag{16}$$

To satisfy the closed-loop system specification according to Figure 1, the effect of feedback controller from (9) and (4) will give equation below.

$$\dot{x}_\delta = A(\alpha)\,x_\delta + B(\alpha)\,[K_1(\alpha)\,x_\delta + K_2(\alpha)\,v_\delta]$$

This equation give result,

$$\dot{x}_\delta = [A(\alpha) + B(\alpha)\,K_1(\alpha)\,]\,x_\delta + B(\alpha)\,K_2(\alpha)\,v_\delta \tag{17}$$

The solution of equation (17) will give the linear model according to equation below [5].

$$\frac{Y(s)}{V(s)} = \frac{w_n^{\,2}}{s^2 + 2\,z\,w_n\,s + w_n^{\,2}} \tag{18}$$

dimana z adalah faktor redaman dan w_n adalah frekuensi alamiah, serta s adalah variabel Laplace. Given any second order polinomials, the requirement is,

$$F(s) = s^2 + 2\,z\,w_n\,s + w_n^{\,2} = 0 \tag{19}$$

The closed-loop system (17) has linear system defined below.

$$\begin{aligned}\dot{x}_1 &= x_2\\ \dot{x}_2 &= -w_n^{\,2}x_1 - 2\,z\,w_n\,x_2 + b\,v\end{aligned} \tag{20}$$

With combination of equation (17) and (20) we found the parameter of controller $K_1(x)$ and $K_2(x)$ below [6],[7] :

$$\frac{g\ \sin(x_1) - a\,m\,L\,x_2^{\,2}\ \sin\frac{2\,x_1}{2} - a\ \cos(x_1)\ [K_1(\alpha)\,x_\delta + K_2(\alpha)\,v_\delta]}{\frac{4}{3}L - a\,m\,L\ (\cos\ (x_1))^2} = -w_n^{\,2}x_1 - 2\,z\,w_n\,x_2 + w_n^{\,2}\,v \tag{21}$$

This equation (21) gives two parts equation for $K_1(x)$ and $K_2(x)$ as follows:

For $K_1(x_1, x_2)$ the equation is,

$$\frac{g\ \sin(x_1) - a\,m\,L\,x_2^{\,2}\ \sin\frac{2\,x_1}{2} - a\ \cos(x_1)\,K_1(\alpha)\,x_\delta}{\frac{4}{3}L - a\,m\,L\ (\cos\ (x_1))^2} = -w_n^{\,2}x_1 - 2\,z\,w_n\,x_2$$

and give the result,

$$K_1(x_1, x_2) = \frac{\frac{4}{3}L - a\,m\,L\ (\cos\ (x_1))^2}{a\ \cos(x_1)}\left[\frac{g\ \sin(x_1) - a\,m\,L\,x_2^{\,2}\ \sin\frac{2\,x_1}{2}}{\frac{4}{3}L - a\,m\,L\ (\cos\ (x_1))^2} + w_n^{\,2}x_1 + 2\,z\,w_n\,x_2\right] \tag{22}$$

Next, for $K_2(x_1, x_2)$ we make,

$$\frac{a\ \cos(x_1)\, K_2(\alpha)}{\frac{4}{3}L - a\, m\, L\ (\cos\ (x_1))^2} = {w_n}^2$$

and give the result,

$$K_2(x_1, x_2) = \frac{\frac{4}{3}L - a\, m\, L\ (\cos\ (x_1))^2}{a\ \cos(x_1)}\ {w_n}^2 \tag{23}$$

Both equations (17) and (18) show that performance of the closed-loop system only determined by damping factor of z and natural frequency of w_n.

4. Simulation Result

From the dynamics of inverted pendulum equation (14), it needs to generate the linear responses of nonlinear plant with nonlinear control system. For thoose equation from (14) to (23), the dynamics of inverted pendulum have changed into the equation (21). The response of the system before controlled, can be seen in Figure 2.

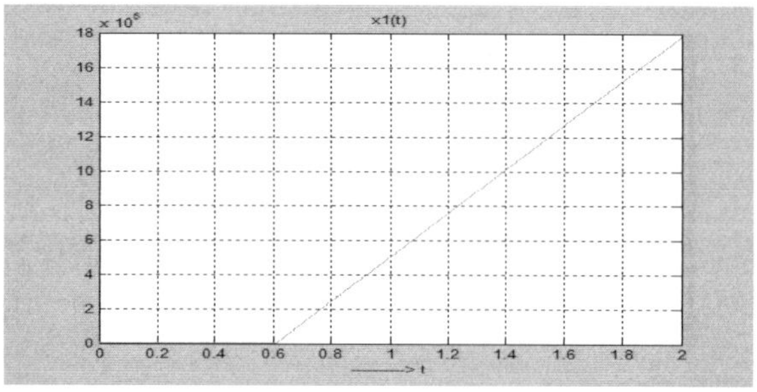

Figure 2. Initial Responses of Inverted Pendulum's Positions

From the dynamics result as shown in Figure 2, the initial position is given with value $\theta = \pi/4$ rad or 22.5°. The position response of that inverted pendulum is not stable, so its need to stabilize the system.

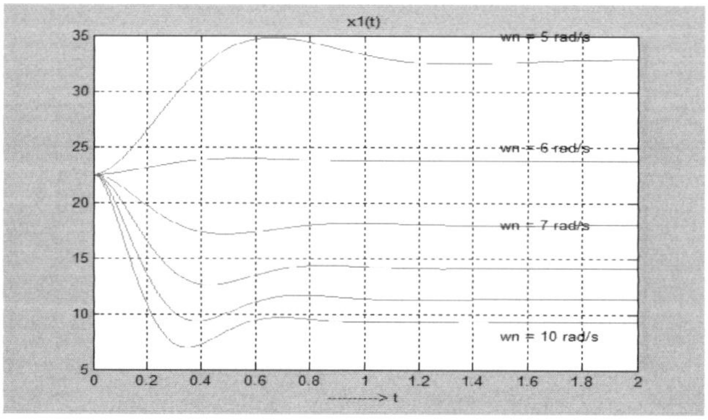

Figure 3. Positions Responses of Inverted Pendulum with Different w_n

With apply the feedback control law equation (22) and (23), first we have to test which value of w_n can satisfy the response needed. The natural frequencies w_n is varied from 5 to 6, 7, 8, 9 and 10 rad/s, and all the responses is shown in Figure 3. The good response happen if the response decrease from the given set point toward steady state position. From the result given in Figure 3, the good response start when w_n = 7 rad/s, 8 rad/s, etc. Furthermore, in this case we choose that w_n = 12 rad/s, and the response is shown in Figure 4. For all response in Figure 3 and Figure 4 we choose damping factor z = 0.5, its mean that maximum overshoot is below 20%.

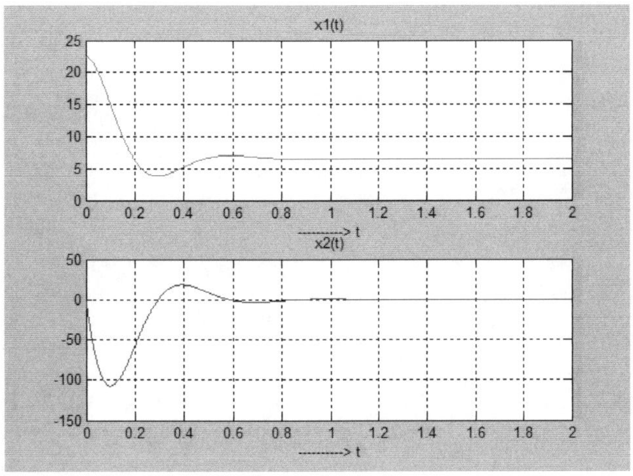

Figure 4. Positions Responses of Inverted Pendulum with v = 0.000 rad

The next problem appear from the result in Figure 4. The movement of the position starts from π/4 rad (22.5°) to 6.56° (not zero) , so the response of inverted pendulum still not adequate.

In the next step of the simulation, we need to satisfy the goal to find the zero position on state state, we try to give the set position from $v(t)$. We choose v(t) = -0.2 rad, and the response its shown in Figure 5. The movement of the position starts from π/4 rad (22.5°) to -4.90° (not zero) , so the response of inverted pendulum still not adequate.

If we do trial and error to find the match set point, we choose v(t) = -0.115 rad. The movement of the position starts from π/4 rad (22.5°) to -0.049° (near zero or ~0) , so the response of inverted pendulum is adequate, as shown in Figure 6.

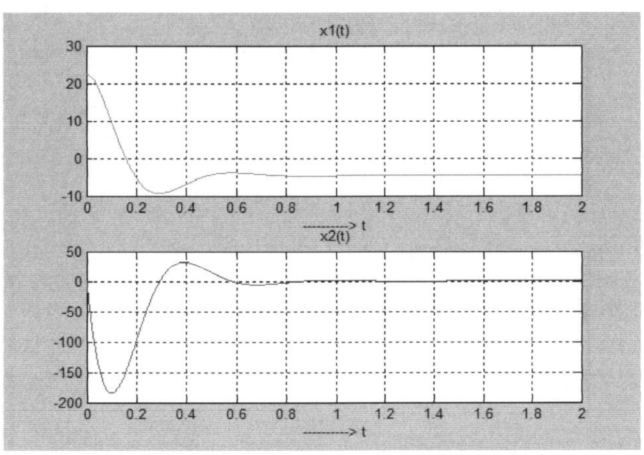

Figure 5. Positions Responses of Inverted Pendulum with v = -0.2 rad

IOP Conf. Series: Materials Science and Engineering **237** (2017) 012029 doi:10.1088/1757-899X/237/1/012029

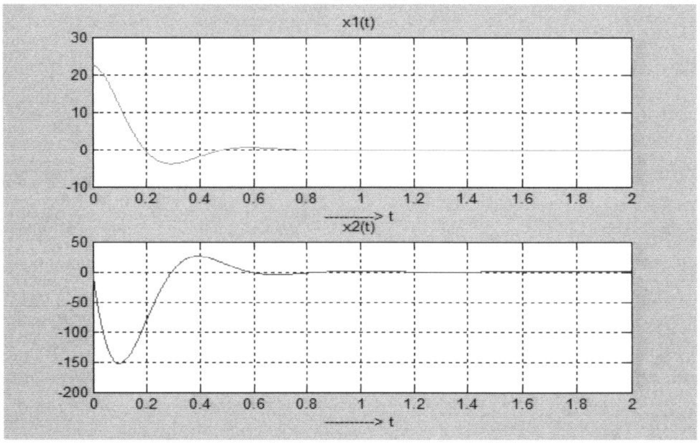

Figure 6. Positions Responses of Inverted Pendulum with v = -0.115 rad

From the last results in Figure 6, we saw the response of the pendulum will be continued toward zero. These all responses mean, the system has stable, the both responses toward equilibrium pont.

5. Conclusion

As we have described in the introduction, we need to observe how to choose feedback control law $\boldsymbol{K}_1(\alpha)$ and $\boldsymbol{K}_2(\alpha)$, and how to choose the eigenvalue needed by choosing z and $\boldsymbol{w_n}$.

Lastly, the controller has been designed to the linear system by using the state feedback nonlinear techniques using eigenstructure assignment for the nonlinear plant.

References

[1] Chiang and Alberto 2015 Stability regions of Nonlinear Dynamical Systems (Cambridge University Press: United Kingdom)
[2] Ching-Fang Lin 1994 Advanced Control Systems Design (New Jersey : Prentice Hall)
[3] Henders M G, Soudack A C 1996 Dynamics and Stability State Spaceof a Controlled Inverted Pendulum (Int. Journal of Non-Linear Mechanics Vol 31) (Vancouper:Elsevier) p 215-227
[4] Katsuhiko Ogata 2010 Modern Control Engineering Fifth Edition (Boston: Prentice-Hall)
[5] Andrey D Polyanin 2015 The Generating Equation Method Construction Exact Solution Other Non-Linear Coupled Delay PDE's Institute of Problem in Mechanics (ELSEVIER:Rusia)
[6] Ridong Zhang 2014 Design of State Space Linear Quadratic Tracking Control Hongkong Univerwsity (ELSEVIER)
[7] Yao Yan 2014 Non-Linear Analysis and Quench Control International Journal of Nonlinear Mechanics (ELSEVIER : Scotland)

The analysis of waste treatment methods and managerial skills towards the effectiveness of CO_2 emmissions (an *ex post facto* study at TPA Bantar Gebang Bekasi)

Jenni Ria Rajagukguk[1,*], Lestina Siagian[2]
[1]Mechanical Engineering Department, Krisnadwipayana University, Jakarta, Indonesia
[2]Electrical Engineering Department, Nommensen HKBP University, Jl. Sutomo No. 4a Medan 20234, Indonesia

* jenniria_rajaguguk@yahoo.com

Abstract. In the last three years, Java Island produces 29.413.336 m^3/year of waste, coming from settlement (house hold) and non-settlement waste. Recently, this waste is managed with conventional technology, composting and recycling. Based on law No. 18 of 2008 on Waste Management, Chapter III Article 5, it is firmly stated that the government and regional governments are responsible for ensuring proper and environmentally sound waste management in accordance with the objectives. The observation of managerial skills is highly needed to investigate the operation of waste management at TPA Bantar Gebang towards the effectiveness of CO_2 emissions.The problems are (1)Whether there is any influence between the method of waste management through Biogas Technology to the effectiveness of CO_2 emissions. (2) Whether there is any influence between managerial skills to effectiveness of CO_2 emission. (3) Whether there is any simultaneous influence between waste management method and managerial skill to CO2 emission effectiveness and (4) how is the method of waste management. Quantitative and egineering method were used to process the data.Biogas Technology variables and Managerial Skill are simultaneously and significantly influenced to CO_2 Emission Effectiveness, this is based on Fh > Ft value of 168,453 > 3.072467) and its significance is 0.000 < 0,05. Then Ho was rejected and Ha was accepted which means that variable of Managerial Skill have influence or very big influence to Effeciveness of CO_2 Emission, Correlation coefficient value 94,1% which means there is very strong relation between variable of Biogas Technology, Managerial Skill to Effectiveness of CO_2 emission. Then Technology management through Biogas Technology is anaerobic biology.

1. Introduction

In the last three years, landfill/waste disposal coming from settlement waste (house hold) and non-settlement of waste volume for Java Island reaches 29,413,336 m^3 / year or 44%. In 2008 the volume of kitchen waste was 22,2 million tons/year (58%) and was increasing in 2009 to 6.594,72 tons/day. It is estimated that it will become 7200 tons/day in 2020 (BSN,2008). Waste management methods through conventional technology cannot be avoided, so it was caused adverse effects for human health. Conventional technological methods produce very high levels of pollutant gas emissions that are harmful to human health, especially respiratory infections (Respiratory) and skin irritation. One of the problems that arises with the existence of TPA Bantar Gebang Bekasi is the presence of solid waste associated with environmental pollution, such as foul odors that can interfere with human health(Daryanto, 2004).

The government's role in waste management at TPA Bantar Gebang cannot be separated from the participation of the community, operational employees and businessmen/ stakeholders. Especially, the management of companies involved in waste management in Bantar Gebang was handled by DKI government. The role of government from the aspect of the regulation can be seen from various policies that have been established, based on the policy and mandate of the law mentioned that in carrying out the activities, the entrepreneur was obliged to prevent and overcome the occurrence of interference or pollution to the environment.(BSN, 2008). Up to now, it is uncertainly known how much managerial skills of employee in operational waste management at TPA Bantar Gebang relate to the effectiveness of CO_2 emissions.

From the description above, an analysis is needed to investigate the solution of the effectiveness of CO_2 emission with Biogas Technology and managerial skill of operational employees in the company. In the regulatory context, TPA Bantar Gebang Bekasi location was expected to meet health requirements, especially in terms of effectiveness of CO_2 emissions. The formulation of the problems: (1) whether there is any influence between the method of waste management with Biogas Technology to the effectiveness of CO_2 emission in Bantar Gebang Bekasi, (2) whether there is any influence between managerial skill to effectiveness of CO_2 emission in Bantar Gebang Bekasi, (3) whether there is any influence between waste management method with Biogas Technology and managerial skill to the effectiveness of CO_2 emission in Bantar Gebang Bekasi and (4) how is the method of waste management.

The effectiveness of CO_2 emissions by the method of waste management through the use of Biogas Technology is the achievement in the use of waste processing system through the use of Biogas Technology shows the achievement of CH_4 (Methana) and CO_2 value and can influence its environment naturally (Scott JC and Janet M. Thomas 2000: 46).

CO_2 is a heavy, colorless gas produced by the combustion and decomposition of organic substances and as chemically processed goods. This gas comes from inhaled and exhaled smoke which will not burn and is relatively non-toxic and is not reactive.CO_2 can cause inefficient environments in large concentrations (Christian L. Hackman, 2000).Air pollution or CO_2 emissions can be caused by natural sources or as a result of human activities.

Pollutants are air or components imposed from a combustion activity (a particular process) issued directly from the source. In general, pollution which was caused by natural sources is difficult to know the magnitude (Robert Noyes, 2001).

The effectiveness of CO_2 emissions is synonymous with the term eco-efficiency based on the concept of creating goods and services with the least possible resources and creating little waste (Surna T. Djayadiningrat and Melia Famiola, 2004). In addition, improvements as efficient as possible in energy usages are increased productivity, reducing pollution, and improving company performance. If the allocation of resources does not take into account the environmental impact then the allocation is considered inefficient (Adinul Yakin, 2004).

Waste management is the collection, transportation, processing, recycling or disposal of waste materials (KNLH, 2008). Based on Law No. 18 of 2008 on waste management, in Chapter I in the general provisions of article 1 paragraph 5 the definition of waste management is a systematic, comprehensive, and continuous activity that includes the planning, reduction, and handling of waste. Municipal solid waste management is part of urban sanitation management.The net notion actually means not only the absence of waste, but also the sense that leads to an aesthetic review (Mr. Pankaj Patel, 2000).

Biogas technology is a modern technology, the decay technology of organic material formed by putrefactive bacteria with stable temperature. Combustion of CH4 and CO_2 generally form Biogas and produce slurry in Digester (Dieter Deublein and Angela S., 2008). Martin Von Lampe and Organization for Economic Co-Operation and Development or OECD Secretariat, defines Biogas as a methane or gas commonly generated from the formation of anaerobic digester biomass. Biogas can be used to fuel transportation on gas fuel in the engine, generate heat, controll industrial machinery and warm water heating (Martin Von Lampe and OECD, 2010)

Biogases organic material in this case livestock waste in the form of livestock gonad and urine processed by microorganisms in rare condition oxygen, anaerob and sludge, biogas Consisting of 60% CH_4 (methane), 38% CO_2, 2% N_2 and Hydrogen and other gases. Biogas is having the main elements of methane (35 - 70)% CH_4, (30 - 45)% CO_2 and added other gases (L.Widarto & FX S.C., 1997).

Biogas technology is collecting methane gas. Methane is a flammable gas which can be used for cooking, lighting gas inside the tool, run the combustion engines and create electricity. This gas is created naturally from the dung, human waste, rice husk, foliage, water plants and rot grass. This gas has many benefits including creating high-quality fertilizers from materials that have been used. The advantage of appropriate technology or technologies of biogas is flammable and can be a substitute fuel alternative energy gas and the rest of the process can be used as organic fertilizer (Widarto and Sudarto, 1997). It is synthesized that Biogas Technology is the technology of sewage treatment or waste from organic materials that generate gases, with the goal of eliminating or reducing air pollution. Some understandings of managerial skills, are shown in table 1.

Table 1. Definition of Managerial Skills

No	Source	Definition of Managerial Skills
1	M. Ivanc and James H (1989).	An ability or proficiency in carrying out specific tasks
2	A. V. and Eugene (1981)	Skill has three expertises i.e. technical, human and Conceptual
3	Chuck Wiliam (2001)	*Skill has three expertises i.e. technical, human & conceptual *The Manager of the top, middle and first l line and group leader.
4	Freeman S. and Gilbert (2003)	Skill has three expertises i.e. technical, human & conceptual
5	Richard L. Daft. (1990)	Technical expertise also includes specialized knowledge, the ability of analysis and the use of appropriate technical d to solve problems in a specific field of discipline

From theories above, it can be synthesized that managerial skill is an ability in carrying out operational duties with regard to the ability to use the tools, procedures and techniques work together, motivating others, and understand the importance of the activities of all the organizations concerned.

2. Research methods

This study employs expost facto method to examine whether there is influence of the variable bound against free. The measurement at the occurrences of events already in progress but seen in the context of the current time without doing the manipulation of the variables examined (Putrawan, 1990). Occurences that have been unfolding as measured data in this study are the effectiveness of CO_2 emissions from landfill garbage in Bantar Gebang, Bekasi as a bound variable (Y) is a free variable and X1 (composition of NH_4) and X2 (Managerial Skills). The value of the effectiveness of the CO2 emission in size with use of Gas Analyzer, then the research results as shown in table 2.

Table 2. Measurement of Data & Questionnaire Results

No	NH_4	Resp.	CO_2
1	47.7	5	31.5
2	46.1	5	30.5
3	47.4	5	29.1
4	48.3	5	29.8
5	47.9	5	30.1
6	48.5	5	30.2
7	49.3	4	29.7
8	49.4	4	29.1
9	48.7	5	31.2
10	49.2	5	31
11	49.8	5	30.3
12	47.9	4	30.5
13	47.4	5	29.5
14	47.5	5	28.9
15	48.6	3	30.6
16	46.4	5	38.4
17	46.9	5	37.8
18	47.3	3	38.1
19	47.3	5	38.3
20	46.3	3	38.7
21	49.9	5	45.5
22	45.2	5	44.8
23	45	5	44.3
24	45.3	5	44

Data collection techniques in the study include: (1) collection of data the results of measurements of the methane CH4 (2) measurement of the effectiveness of CO_2 emissions (3) data questionnaire consists of 24 questions with answers used Likert scale. Analysis of the data, from the equation: $Y = b_o + b_1X_1 + b_2 X_2$, Where: b_o = coefficient, b_1 = coefficient for NH_4 composition, b_2 = coefficient for managerial skills, X_1 = composition for NH_4, X_2 = for managerial skills and Y (CO_2emissions effectiveness). Next, to find out the significant regression mutiple compare between 0.05 probability values with significant probability values as follows: If the value of the probability of significant probability Sig $\leq$ 0.05, Ho and Ha accepted rejected it means not significant.

3. Results and Discussion

The data was calcullated by Program SPSS Statistics with SPSS programs 21.0, provides a general overview of the distribution of the normal, regression, linearity and others. The classical linear regression analysis test compounds were obtained, good regression model is the distribution of the data is normal or close to normal, residual values and it indicates that the model has a zero distribution. In the Normal diagram of P-P Standardized regression plot indicates the existence of points around the line, it is concluded that the model is normally distributed.

Autocorrelation was used to test the presence of internal correlations among the variables arranged in the time series. To detect the presence or absence of autocorrelation in a regression model, Durbin-Watson method was used.Regression model can be expressed no symptoms, if probability Durbin-Watson> 0, 05 and from the above table Durbin-Watson value is 2,118. So that shows 2,118> 0,05 concluded regression model did not experience autocorrelation symptoms.

On the equations of the regression may not be occurring multicolinearity meaning there should be no correlation or relationship that is perfect or near-perfect free variables that make up between equations. If on a model equation occurs symptoms of multicolinearity that means fellow free variable occurs or correlation similarity. Multicolinearity test results can be detected from values, VIFs greater than 10, when <from 10 it can be said that there is no multicolinearity. Biogas Technology variable and Management Skill has a value of VIF 3.384 <10. It has a value of VIF 3.384 <10.

So it can be concluded that in the regression model, multi colinearity did not occur. Regression model is predicted to have a Heteroscedasticityproblem. It means there are some different variants of variable in the model. This symptom can also be interpreted that in the model there is no similarity of variance of residual on observation of regression model. TheHeteroscedasticitywere tested using Park Gleyser theory.

The Regression coefficient of the test results indicates that the characteristics of probability (significance), if the value is 0.05 < 0.000 it means there is no symptoms of heteroscedasticityIt turns out that the variable of Biogas Technology is 0,000 while the managerial skills is 0,000. Thus, we can coclude that there is no heteroscedasticity symptomps in the regression model.Hypothesis testing in this study was conducted by using statistical analysis tools in the form of the correlation coefficient, coefficient of determinants, F-test and t-test significance level. Where the level of significance orprobability (α) is 0.05 or 5%. The correlation coefficient measures the strong relations. Thestrong relations is stated by a quantity value of the correlation (r) whose value is in the range-1 to 1. If the value obtained is increasingly close to the numbers-1, this means the relationship is getting stronger and the direction of the relationship is reverse. It figures out that there is a weak relationship between managerial skills and Biogas Technology towards the effectiveness of CO_2 emissions, from table 6. The value of the correlation coefficient of 0.97

The determinant coefficient was used to determine the number or presentation of the contribution of independent variables in the regression model simultaneously or synchronously giving effect to the dependent variable. Table 6 shows that the value indicates that the value of the Adjusted R Square r^2 was 0941, it means E simultaneously influenced Biogas Technology and Managerial Skills 94.1%, while the rest of 5.9% influenced by other factors were not examined in this study.

The coefficient results of a determinant shows, for testing the equations simultaneously using between countdown with a comparison between the calculated F > F table, or by using the value of significance. The basis of decision-making was Ho will reject or accept Ha. Consideringthe value of the significance is less than 0.05 and the value of F is F count (168.453 > 3.072467) value of significance was 0.05 < 0.000. It inferred that Ho is denied and Ha is accepted. Thus, the Managerial skills and Biogas Technology simultaneously affect the effectiveness of the company's significant CO_2 emissions.

T-test was conducted to find out whether in partially or individually affects the dependent variable. The t-test can be done by comparing the t with a t > count table (15,586 > 2.079614), or by using the value of significance. The basic decision-making is Ho will be rejected or Ha will be accepted if the significance value t 0,000 < 0,05. Table 8shows that: (1) Variable X_1 towards Y: the significance value of TB variable towards the effectiveness of CO2 emissions was 0,000 < 0,05. It concludes that Ho was rejected and Ha is accepted. Thus, Biogas Technology variable affects the effectiveness of CO2 emissions(15.586>2.079614).(2) X_2 variable towards Y (-7.811 <2.079614), the significance value of managerial skills variable towards the effectiveness of CO2 emissions is 0,000 < 0,05 significant. It concludes that managerial skills variable have no (or small) effect to the effectiveness of CO2 emissions.

In multiple linear regression analysis (see table 8),the correlation coefficients showed that the values of the constants b_1 = -232,961; b_1 = 6.202 b_2 = -6.157. Linear regression of equations, so is Y = -232,961 + 6.202 X_1- 6.157X_2. For multiple regression equations above can be deduce through statements as follows: If the Biogas Technology variable rise one unit the scale ratio, then the effectiveness of CO_2 decreases and ifthe Managerial Skills variable rise one unit the scale ratio, then the effectiveness of CO_2 decreases.

Management technologies through Biogas Technologies in Anaerobic Biological, according to (Ketut Sumada, 2012): In anaerobic biological treatment process occurred four (4) stages of the processes involved include: 1. hydrolysis Process, 2 Acidogenysis Process, 3. Acetigenysis Process4. Methanogenysis Process. Based on the model of the growth of microorganisms, waste water treatment in anaerobic biological is divided into 2 (two) model: (1) Suspended Microorganisms growth Model and (2) of the microorganism growth Model is attached. On waste water treatment in anaerobic biological material organic (COD) converted to gas produced 90% of CH_4 and CO_2, and 10% of his mud. The resulting gases can be purified by a process of absorption of CO2 gas, so the resulting pure CH_4 gas that can be used as fuel.

4. Conclusions

- Biogas Technology variable significantly influence the effectiveness of CO_2 emmissions in TPA Bantar Gebang Bekasi.
- Managerial skills variable significantly influenced the effectiveness of CO_2 emmissions in TPA Bantar Gebang Bekasi.
- Biogas Technology and managerial skills variables simultaneously and significantly influence CO_2 Emissions in TPA Bantar Gebang Bekasi.
- Management technology through biogas technology is done by anaerob biology (hydrolysis, acidogenysis, acetogenysis and methanogenysis process).

References

[1]. C. Hackman, L. (2000). Hazardous Waste Operations & Emergency.New York: Mc Graw-Hill.
[2]. Chuck Wiliam (2001). Management. New Jersey: PrenticeHall
[3]. CCME. (2006). National Guidelines for Hazardous Waste Landfills

[4]. Daryanto. (2004). Problem of Pollution. Bandung: Transito
[5]. Daft, Richard L. (19900. Management, 6th Edition Philadelpia: Dryden Press
[6]. Deublein, Dieter and Angela S. (2000). Biogas from Waste and Renewable Resources. Singapore: Wiley VCH Verlagin GmbH & Co.
[7]. Djayadiningrat, Surna T. and Melia Famiola. (2004) Environmentally Friendly Industrial Area. Bandung: Engineering Science.
[8]. Ivancevich, John M. and James H. Donnelly. (1989). Management Principle and Function. New York: McGraw-Hill
[9]. Journal homepage. (2008), www.elsevier.com/locate/wasman, Waste Management. China.
[10]. KNLH. RI Law, No.18 of 2008, on Waste Management. Jakarta.
[11]. Kostova, Irina. (2006). Leachate from Sanitary Landfills – Origin Characteristics, Treatment, Borovetz.
[12]. Ministry of Environment (1995). Deputy Activity Report of Control Field Pollution. Jakarta: Deputy II KNLH.
[13]. MNLH Decision No. 295 Year 2007. The team's decision on waste management, Jakarta
[14]. Lampe, Martin Von. (2010). Organization for Economic Co-Operation, and Development or OECD Bioheat, Biopower and Biogas, Development and Implications for Agriculture. Netherland G.
[15]. Madibrata, Roni and Ade Moetangad Kramadibrata. (2007). Waste Reactor Management System Integrated. Jakarta: Silar satu.
[16]. Mc Dougall, Forbes R., Peter R. White., (2003). Marina Franke, and Peter Hindle. Integrated Solid Waste Management a Life Cycle Inventory, Published by Blackwell Science.
[17]. Michael, H. (2007). Waste Processing Technology. Jakarta: PT. Gramedia.
[18]. Noyes, Robert. (2001). Hand Book of Pollution Control Processe ,. New Delhi: Jaico Publishing House.
[19]. Patel, Mr. Pankaj, et al. (2000). Municipal Solid Waste Management. New York: WAVE.
[20]. Minister of Home Affairs Regulation No.33 of 2010. (2010). About the Management Manual Trash Consisting of 45 Articles. Jakarta
[21]. Robbin, Stephen P. and Mary Coulter. (2002). Management, 7th Edition. New Jersey: Prentice Hall.
[22]. Scott, J. C. and Janet M. Thomas. (2000). Environmental Economics and Management, Theory, Policy and Apllication.Harcourt: College Publisher, Printed in the United States of America, 46.
[23]. Sink, RK and Misra. (2005). Biofels fromt Biomass, Department of Chemical Engineering National Institute of Technology. Roukela.
[24]. Soemarwoto, Otto. (2004). New Paradigm of Environmental Management. Yogyakarta: Gadjah Mada University Press.
[25]. Stoner, James A. F. (2003). Jersey Management.New: Prentice Hall.
[26]. Supardi, U.S., et al. (2012). Statistics In Research Applications, Most Statistics Books Comprehensive, First Print. Jakarta: Ufuk Press.
[27]. Stohr, Uli Werner Ulirich and Nicolai Hees., (1989). Biogas Plant in Animal Husbandry, Humanity Develomenty Library Document Text. Vierg.

[28]. Stoner, James A. F. and Charles Wankel. (2003). Management, 2nd Edition. Jakarta: Rineka Cipta.
[29]. Tammega, Hans. (1999). The Waste Crisis, Landfill, Incinerators, and the Search for a Sustainable Future. New York: Oxford University Press.
[30]. TPST. (2008). Company Organization. Jakarta: PT GTJ and PT NOEI
[31]. Widarto, L. and FX. Sudarto C. Ph (1997 dan 2010) Appropriate Technology to Make Biogas, Jakarta.
[32]. Wignyosubroto, Sritomo. (2003). Introduction to Industrial Management Techniques. Surabaya
[33]. Wiliam, Chuck. (2001). Management. New Jersey: PrenticeHall.
[34]. Winardi, J. (2006). Organizational Theory and Organizer. Jakarta: Raja Grafindo Persada

Stabilization / solidification of polluted marine dredged sediment of port en Bessin France, using hydraulic binders and silica fume

Ernesto Silitonga
Universitas Negeri Medan, Jl Willem Iskandar Pasar V Medan Estate 20221
North Sumatera, Indonesia

ernestosilitonga@unimed.ac.id

Abstract. A large amount of sediment is dredged in France every year. Due to the increase of the amount of marine dredged sediments, environmentally reuse of dredged sediment is urgently needed in France. The first objective of this study is to find an application for reuse of marine dredged sediments materials, as new material for road construction. Hence, serial tests need to be realized to identify if marine dredged sediment could be utilized for road construction. The second goal is to enhance the physical, mechanical and chemical characteristics of the mix, by incorporating binders and sediments, and revealed the identification of the mechanical characteristics measured on the mixes is compatible with their use as a base course material. The results show that the treatment by hydraulics binders could satisfy the needed mechanical characteristics. The present of Silica Fume is aimed to reduce the pollution level, especially the heavy metal content. However, the proportion of hydraulics binders and silica fume needed to meet prescribed specification is important, so the reuse of the marine dredged sediments of Port-en-Bessin, France in road construction, as an alternative material could be achieved. After the geotechnical study in laboratory results shown as expected than the study to identify the chemical characteristic realized. To evaluate the environmental impacts of the used material, leaching test is performed. The leaching test was performed to verify the predicted release of pollutants based on total dissolution. And for the final part, the test results show that the polluted marine dredged sediments could be safely used (in term of environmental impact) as a new material in road construction.

1. Introduction

In France, the Port Authority is encouraged to find the application domain for the reuse of dredged sediments. Furthermore, with European regulation [1], the classical method (rejecting the contaminated dredged sediment in to the sea) is not an option anymore, regardful to the environmental issues. Available or attainable, there are many constraints on policy implementation need to be designed. However, before deciding if the dredged material is acceptable for a beneficial use, it is necessary to be realized serial tests to evaluate the capability of marine dredged sediment in its utilization domain. Previous studies with different alternative treatment method to find the domain application for the reuse of marine dredged sediments have been investigated in Europe. The previous studies utilized Marine dredged sediment, Pempe Ozer-Erdogan et al [2]worked on marine dredged sediments in Turkey, Benyerou et al [3]utilized marine dredged sediments as a base material for produce the brick, Couvidat et al [4]worked with marine dredged sediment in Marocco, Hurel et al [5] and Levacher [6] stabilized marine dredged sediment in France.This paper concentrate to determine the type and composition of binder to stabilize marine dredged sediments. The ideal percentage of hydraulic binders needed to meet prescribed specifications is very important. Silitonga in his research using dredged sediments from Cherbourg (France) [2] claimed that the ideal amount of hydraulic binder needed to stabilized dredged sediment was between 7%-9%. Behmanesh[7] stabilized dredged sediment of Port du Havre. The amount of hydraulic binder utilized to stabilized dredged sediment was 15%. Silitonga E., [8] needed 11% of pozzolanic binders (fly ash) to stabilized dredged sediment from Port de Cherbourg (France). Zhibo[9] in his research stabilized marine dredged with cement and lime, the results show that the ideal percentage of hydraulics binder is 15%. The primary goal of this research is to investigate the potential reuse of marine dredged Sediment of Port-en-Bessin as material in road construction. Road construction needs to meet the requirements. In order to fulfill the requirements, the material has to have mechanical characteristics, furthermore if the material polluted. Consequently, the addition of binders that capable of fulfilling the mechanical requirements and environmental regulation is needed. In this study, the binder utilized to enhance the mechanical characteristics and to prevent the pollutants contaminate environment is Silica fume. Silica fume is a by-product of producing certain metals in electric furnaces. Silica-fume, also known by other names such as volatilized silica, micro silica, or condensed silica fume, is a by-product of the induction arc furnaces in the silicon metal industries.

2. Material and Methods

2.1. Dredged Sediment and Silica Fume

This research is realized using marine sediments, which were dredged from Port-en-Bessin, Basse Normandie France. Theharbour Port-en-Bessin (PEB) is located in the Northwestern France. The sediments were dredged from two different, Bassin no.1 (PeB-1) and Bassin no.2. (PeB-2). The binder utilized in this study is Cement and Lime and Silica Fume. Hydraulic binders such as cement and lime utilized are typical hydraulics binders used in soil stabilization and road construction work. In this study, two different types of Silica Fume (FS) were utilized in this research (FS1and FS2). The Silica Fumes utilized are not commercialized yet. In fact, the Silica Fumes used in this study are a prototype of low-cost Silica Fume. The particle size distribution of this Silica Fume is not as fine as Silica Fume that commercialized in the market.

2.2. Index properties.

2.2.1. Particle size distribution

Due to its particle size distribution, a laser diffractometer Coulter LS200 was used to identify the particle size of marine sediments utilized,

Table 1. Particle size distribution of utilized materials

	PeB 1	PeB 2	PeB 3	PeB4	SF1	SF2
<2 μm (%)	10.7	10.6	13.8	9	2.3	13.8
2 - 63 μm (%)	77.2	78.8	74.8	78.8	18.9	68
> 63 μm (%)	12.6	10.6	11.5	12.2	78.8	18.2

The results as shown in Table 1, confirm that the particle distribution of marine dredged sediments that were taken from four different locations (PeB-1, PeB-2, PeB-3 and PeB-4) is almost similar, this means the homogeneity of the marine sediment of PeB.The result shows that dredged marine sediments constituted of a large amount particle with the size between 66,7- 78,4 μm. The Silica Fume type 2 (SF2) has a finer diameter of particle compared to

Silica Fume type 2 (SF2) and Silica Fume type 1 (SF1). According to Silitonga [10], in the previous study on stabilization dredged sediment with fly ash, showed the Unconfined Compressive Strength (UCS) value tends to reduce as the mean particle size increase. From a study of the strength contribution potential of Silica fume Gleize*et al*[11] reported that the particles larger than 45 μm show little or no reactivity under normal hydration conditions and the pozzolanic activity was directly proportional to the amount ofparticles under 10 μm. From this theory, we can observe that SF1 possess coarse particles (>63 μm) more important than SF2 and SF2 has a more important quantity of fine particles (<10 μm) than FS1. From this theory, we could expect that SF2 will be more reactive than SF1

2.2.2. Chemical Characteristics (dredged sediments and Silica Fume)

In order to identify the level of pollution of Marinedredged sediment it is important to assess the environmental risk of the dredged sediment before its land application. The level of pollution is identified using Leaching Test. The results of chemical characteristics are shown in Table 2. The results presented were taken from several different locations in Port en Bessin area. The amount of Micropollutants measured from 2different locations of time was presented in Table 2.In order to identify the level of pollution of the marine sediment, the researcher used the reference established by European Council 2003/33/EC. This reference divided in 3 classes; inert waste, non-hazardous and hazardous waste.

Table 2. Leaching test result of dredged sediment from Port en Bessin, France

Pollutants	PeB 1	PeB 2	European Council Reference		
			Inert Waste	Non Hazardous Waste	Hazardous Waste
As (mg/kg)	14	8.9	0,5	2	25
Cd (mg/kg)	1.33	1.59	0,04	1	5
Cr (mg/kg)	77	58	2	50	100
Cu (mg/kg)	1.41	0.63	0,01	0,2	2
Hg (mg/kg)	0.63	0.73	0,01	0,2	2
Pb (mg/kg)	20.2	17.9	0,5	10	50
Ni (mg/kg)	29.4	30.2	0,4	10	40
Zn (mg/kg)	137	148	4	50	200

These reference values relate to the elements contained in the leachate and not in the raw material. According to European reference, all amount of micropollutants from 2 different locations are categorized as non-hazardous waste.From the result shown in table 3, we can confirm that Silica Fume 1 and 2 (SF1 and SF2) constituted a slightly different percentage of SiO_2, Al_2O_3, MgO, K_2O and free Si. Silitonga [8] in his research confirmed that the chemical properties have an important effect on the mechanical properties of the marine sediment stabilized. As shown in table 3, SF2 is characterized by a higher content of SiO2, followed by SF1.According to this result, we can assume that SF2 will be more reactive than SF1. Silitonga [10] on his research noticed that type of Silica Fume that possesses higher content of SiO_2, Fe_2O_3 or Al_2O_3 produces higher resistance on Unconfined Compressive Strength Test. Silica Fume type 2 (SF2) has a higher total amount of free CaO.

Table 3. Chemical characteristic of Silica Fume utilized

Parameters	SiO_2	Fe_2O_3	Al_2O_3	CaO	MgO	Na_2O	K_2O	C	free Si	Free CaO	SO_3
Silica Fume 1(SF1)	90-92	1,5-2	1	0,5-1	1-1,5	0,5-1	1-1,3	0,5-1	< 0,2	< 1	< 1
Silica Fume 2(SF2)	90-95	1.5-2	1-1.5	0,5-1	1	0.8-1	1.3-1.5	1-1.5	<0.4	<1	<1

2.3.Methods

2.3.1. Preparation of samples and mix design

After being dredged, physical characteristics of marine sediment were measured. The measured initial water content is about 169%. The sample was realized by using cylindrical specimens (ø = 40mm, h = 80mm). Samples with Silica Fume type 1 (SF1) and Silica Fume type 2 (SF2)addition were set in order to investigate the influence of the Silica Fume on the strength gained, the mixtures wererealized with various percentage of Silica Fume content. The mixture without any Silica Fume content (LIM-CEM) was realized as a reference to identify the effect of Silica Fume addition.

3. Results and Analysis

3.1 Tensile Strength

The Tensile Strength is the most important test to identify the capability of a material to be used in road construction. In this study the Tensile Strength was realized beside to identify the tensile strength gained is to calculate the elastic modulus obtained. The elastic modulus is obtained using a stress–strain curve of an unconfined compression test. Firstly, the maximum compression strength of the mixture is measured. Then the elastic modulus, E, is measured on another sample. In order to not damage the cylinders, the maximum effort applied corresponds to 30% of the maximum compression strength. The results of tensile test are presented in figure 1. As shown in the figure, only the samples with 7% Silica Fume located in class S3 (G and F). According to Scheme of road section and recommended engineering characteristics, the samples with 7% Silica fume are secure to be utilizedas material on sub base and Foundation.

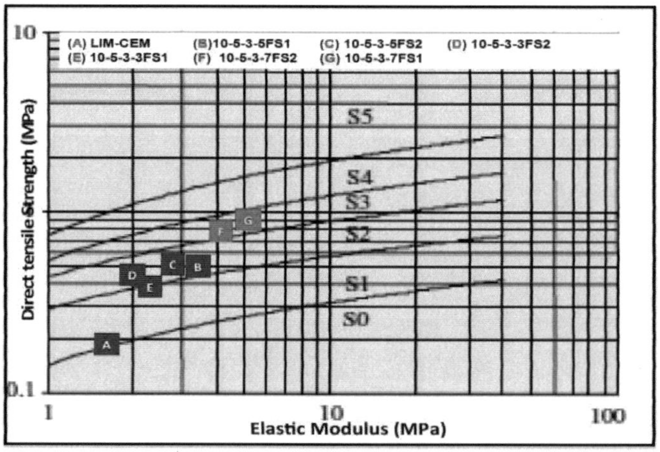

Figure 1.Tensile test result for sample realized.

The result showed that, the sample with Silica Fume addition, produces ahigher Elastic Modulus and higher Tensile strength value than sample without Silica Fume (LIM CEM). The

IOP Conf. Series: Materials Science and Engineering **237** (2017) 012031 doi:10.1088/1757-899X/237/1/012031

sample without Silica fume LIM-CEM (A) is situated within the limit of S0 and S1 classes, it means the reuse for sub base need to be considered. On the other hand the samples with 3%and 5% Silica fume (B, C, D and E) are situated in the same class (S2) yet the Elastic Modulus values of sample with 5% of Silica Fume is higher than samples with 3% Silica Fume. The presents of Silica Fume with its filler effect, due to its fineness can fit into space and fill the empty space between sediment grains and produces more solid bonds between sediment grain, this is the reason the enhancement of Elastic Modulus because a stiffer material will have higher Elastic Modulus. The results show that the samples with 3% and 5% of Silica fume (B, C, D and E) could be used as material only in sub base. In the other hand the samples with 7% of Silica fume (H and G) could be reused in sub base and foundation.

3.2 Leaching Test

The main objective of this study is to reuse the marine dredged sediment in road construction as a new material which is fulfil the road regulation criteria and harmless to environment. However, on its application, due to drying and oxidation process of polluted marine dredged sediment may affect mobility and bioavailability of the land utilized. Therefore Lixiviation Test should be realized to identify the strength or reaction of sample to aggressive environment, such as the extreme change of pH in environment.The Leaching Test refers to the solution containing the solubilized elements during the test, which are performed on the analytical characterization. In order to identify the pollution level of dredged sediment of Port en Bessin, the researcher decided to used reference values established by European Council No. 2003/33 / EC. According to the result (Figure 2), the amount of Cadmium (cd) is 1.33 mg/kg, according to European reference; the quantity of Cadmium categorized as non-hazardous waste.The result of leaching test (figure 2) shows that, in the case of Cadmium (Cd), sample with or without Silica fume show a reduction of Cadmium content. The sample without Silica fume; LIM_CEM (Bin Figure 2), show a slight reductionof amount of Cadmium (from 1.33 to 1.25) but it is still in the same level of pollution as the original amount of pollutants (non-hazardous waste level).

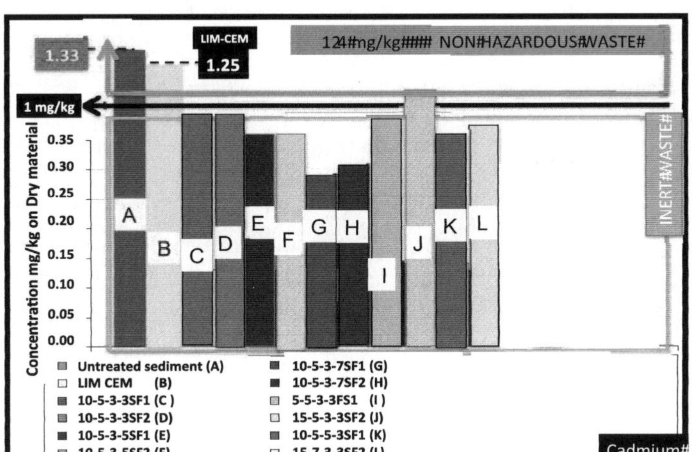

Figure 2. Leaching Test result for marine dredged sediment for Cadmium.

All the samples mixed with Silica fume established a major reduction on amount of Cadmium; all the samples are situated in Inert Waste. The effect of different quantity of Silica Fume can be noticed in Figure 2.The samples with 3% percentage of Silica fume (C and D) showminimal reduction of Cadmium content compared to samples with 5% (E and F) and 7% Silica Fume (G and H). The samples with 7% of Silica fume show a very remarkable reduction of Cadmium content and established a group with lowest amount of Cadmium. One of the principle physical effect of silica fume is the filler effect, because of its fineness, Silica Fume can fit into space between sediment grains in the same way that sand fills the space between particles of coarse aggregates and cement grains, fill the space between sand grains. After the pozzolan reaction, Silica Fume iscapable to trap the pollutants, so that the pollutants could not escape from the matrix and pollute the environment. This is the main reason why the Silica Fume is being used in this study. The different quantity of sand also realized and could be noticed in the result. As shown in Figure 2, the raise of sand percentage to 15% (J) the result show an increase in Cadmium content if compared to 10% sand (D). The addition of 15 % of sand, make its position located from inert waste to non hazardous waste (European Council reference). The result shows that the raise of cementpercentage from 5% to 7% (L), and limefrom 3% to 5% (K),can enhance the reduction of Cadmium content.This results is due to the hydration of cement which also produces $Ca(OH)_2$, required for the pozzolanic reaction in Silica Fume and for ion exchange in clays, which accelerates the cement and pozzolanic reaction of Silica fume. The enhancement of cement/ lime reaction can promote the pozzolanic reaction of Silica thus produce the reduction of Cadmium content. According to European reference value for dredged sediment, from the environment point of view, after stabilization process with 3% to7% of Silica fume, the marine dredged sediment categorized as inert waste, hence harmless to environment and could be reused as a material in road construction

4. Conclusion

The main objective of this research is to fulfil the criteria (geotechnical and environment impact point of view) for the reutilization of polluted dredged sediment as material in Road construction. The granulometric test shows that dredged sediment from Port en Bessin constituted a large amount particle with silt fraction (2 - 63 μm). The leaching test result state that dredged sediment of Port en Bessin content amount of pollutants is categorized as non-hazardous waste (according to European Reference).Test on tensile strength confirmed that the sample with addition 7% Silica fume could to be used as material in sub base and foundation. In the other hand sample with 3% and 5% Silica fume only can be used as material in sub base. Leaching test determine that the present of Silica Fume in the specimens (from3% to 7%) remarkably capable to reduce the amount of pollutant (cadmium) and passed from level non-hazardous waste to categorize as inert waste. Hence the marine sediment with 3%-7% Silica fume can be used in road construction and harmless to the environment. On the other hand, the sample without Silica Fume (LIM_CIM)could not reduce the Cadmium content and still categorized in Non-Hazardous waste. This result confirms the main reason why the researcher used Silica Fume as binder. Although the use of Silica Fume cost additional price the result showed that Silica Fume capable to reduce the pollutant content in the sediment.

References

1. OSPAR Convention, Convention for the Protection of the Marine Environment of the North-East Atlantic, Paris (1992)
2. PembeOzer-Erdogan, Ibrahim Erden, LeylaTolun,Beneficial use of marine dredged materials as a fine aggregate in ready-mixed concrete: Turkey example. Construction and Building Material, Volume 124, 15 October 2016, p 690-704
3. D. Benyerou, N. Boudjenane, M. Belhadri, Caractérisation et valorisation des sediments marins dans la fabrication des briques (Characterization and Valorization of marine sediments in the manufacture of bricks) Journal of material and environmental science(2016) p 1317-1327
4. Julien Couvidat, Mostaba Benzaazoua, Vincent Chatalin, Hassan Bouzahzah, Environmental evaluation of dredged sediment submitted to a solidification stabilization process using hydraulic binders, Environmental Science and Pollution Research, May 2016.
5. C Hurel, M Taneez, A VolpiGhirardini, G Libralato, Effects of mineral amendments on trace elements leaching from pre-treated marine sediment after simulated rainfall events, Environmental Pollution 2017, p363-374
6. Daniel LEVACHER, Dahbia BERKAINE, Consolidation parameter of dredged marine sediments: case study of Dunkirk port sediments, Proceeding of Geo-Environmental Engineering 2016, Nantes, France, 2016, pp 41-48
7. J. Behmanesh.; Etude de la durabilité d'un sédiment, traité au ciment et additifs, Thèse de doctorat de l'Université de Caen. 214p (2008)
8. E. Silitonga, D. Levacher, S. Mezazigh, Utilization of fly ash for stabilization of marine dredged sediment (2010), Revue Européenne de Génie Civil EnvironnementVol 14, p 253-265, (2010)
9. Zhibo D. Caractérisation, stabilisation et solidification de sédimentmarin. Thèse dedoctorat, Université de Caen, 145p (2008)
10. Silitonga E., Levacher D., Mezazigh S., Effects of theuse of fly ash as a binder on the mechanical behavior of treated dredged sediments, Environmental Technology vol 30. (2009)
11. Gleize P.J.P., Müller A., et Roman H.R., (2003). Microstructural investigation of a silica fume – cement –lime mortar, Cement and Concrete composites, Vol.25, pp.171-175.

IOP Conf. Series: Materials Science and Engineering **237** (2017) 012032 doi:10.1088/1757-899X/237/1/012032

Extraction of basil leaves (*ocimum canum*) oleoresin with ethyl acetate solvent by using soxhletation method

R. Tambun*, R. R. H. Purba, H. K. Ginting
Department of Chemical Engineering, Universitas Sumatera Utara, Kampus USU Medan 20155, Indonesia

* rondang@usu.ac.id

Abstract. The goal of this research is to produce oleoresin from basil leaves (*Ocimum canum*) by using soxhletation method and ethyl acetate as solvent. Basil commonly used in culinary as fresh vegetables. Basil contains essential oils and oleoresin that are used as flavouring agent in food, in cosmetic and ingredient in traditional medicine. The extraction method commonly used to obtain oleoresin is maceration. The problem of this method is many solvents necessary and need time to extract the raw material. To resolve the problem and to produce more oleoresin, we use soxhletation method with a combination of extraction time and ratio from the material with a solvent. The analysis consists of yield, density, refractive index, and essential oil content. The best treatment of basil leaves oleoresin extraction is at ratio of material and solvent 1:6 (w / v) for 6 hours extraction time. In this condition, the yield of basil oleoresin is 20.152%, 0.9688 g/cm^3 of density, 1.502 of refractive index, 15.77% of essential oil content, and the colour of oleoresin product is dark-green.

1. Introduction

Indonesia is a rich country in resources and natural products. Basil (*Ocimum canum*) is annual plant and herbs, with a height of 20 to 60 cm, have white-purple flowers. This plant comes from Asia, Africa, and South America and have been almost all over the world [1–4]. Basil is one of the resources that has not been used optimally. Basil is usually used as a flavor enhancer of food and as complementary foods [5-6]. One of the valuable contents in basil is oleoresin. Basil oleoresin contains the main component of volatile substances (essential oils) and non-volatile (resin and gum), each of that aims to determine the aroma and flavor [7-8]. Usually oleoresin is used as a flavouring agent in food, in cosmetic and ingredient in traditional medicine [9-10]. Oleoresin is derived from oleo, which means oil and resin, which means amber. Oleoresin is oil and resin that is a mixture of essential oils as an aroma carrier and a kind of resin as a carrier of flavor [7-8]. Basil oleoresin is dark-green oil and contains essential oil about 4 to 12 percents. In the trading, the quality of oleoresin is indispensable. The quality standard of basil oleoresin is dark-green color, typical aroma of basil, oil content of 4 % to 12 %, 1 g/cm^3 of density, and 1,501-1,521 of refractive index [11]. To get the oleoresin contained in the basil leaves, one of methods is soxhletation, which is a standard technique where a fresh solvent is contact with sample periodically [12]. The efficiency of soxhletation method can be determine by several factors, such as the average size of the particles, the ratio of material and solvent, and the usage of polar or non-polar solvents [13-14], such as n-hexane, ethyl acetate, acetone, and ethanol [15]. The high of yield can be influence by extraction time and ratio of material and solvent [16]. Based on the above description, this research aims to produce oleoresin from basil leaves by using soxhletation method and ethyl acetate as solvent by varying extraction time and the ratio of material and solvent.

2. Methods

2.1 Raw Materials and Equipment

The sample in this research is basil leaves (*Ocimum canum*). Ethyl acetate is used as solvent. The equipment in this research is a hot plate, 40 mesh of sieve tray, funnel, condenser, round-bottom flask, blender, electric scales, thermometer, filter paper, rotary evaporator, measuring cups, and glass beaker.

2.2 Basil Leaves Powder Preparation Procedure

Basil leaves are washes using water, and then the samples are drying using the solar heat. After it is dried, the sample are crushed using a blender, and it is sieved using sieve (40 mesh) to obtain a basil leaves powder.

2.3 Basil Oleoresin Extraction Procedure

Extraction of basil leaves oleoresin has done using soxhletation method. In this method, the ethyl acetate is boiled at the boiling point, that is 77 ^{0}C as long as 3 hours, 4 hours, 5 hours, and 6 hours with the ratio of material and solvent (w/v) much as 1:5; 1:6; 1:7 and 1:8. Samples were finely put 50 g into a thimble located in the middle part of the soxhlet equipment. Ethyl acetate solvent used for the extraction process is put in a round bottom flask with a comparison between the material and solvent (w/v) 1:5; 1:6; 1:7 and 1:8. Oleoresin is extracted with soxhlet equipment for 3 hours, 4 hours, 5 hours, and 6 hours at temperature of 77 ^{0}C. Once extraction is complete, oleoresin has separated from the solvent using a rotary evaporator. The oleoresin is stored in the bottles for the analysis.

Analysis of the results in this research includes analysis of oleoresin yield, oleoresin density, oleoresin refractive index, and essential oil content.

3. Result and Discussion

3.1 Analysis of Oleoresin Yield

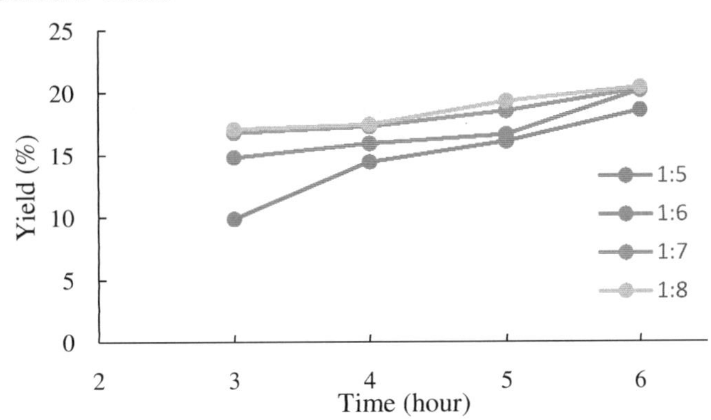

Figure 1. The influence of ratio between material and solvent and extraction time to oleoresin yield

Figure 1 is an overview of extraction, which shows the influence of the ration of material and solvent and extraction time to oleoresin yield. From Figure 1, we can see that in overall, the ratio of material and solvent and extraction time give an increasing effect to oleoresin yield, which oleoresin concentration increases, with the increase ratio of material and solvent and extraction time. It can be explained that extraction time is required to obtain the higher oleoresin yield during prolong contact between basil leaves and solvent and giving basil leaves a chance to contact with solvent so that the oleoresin can be extracted optimally. At the time of extraction time unchanged, but the increase in the ratio of the material to the solvent, this has a major influence in producing oleoresin. To increase the oleoresin extract, much solvent should be available to maximize the diffusion of oleoresin yield. The increasing in the ratio of material and solvent with basil leaves affect the diffusion of oleoresin from basil leaves to the solvent, more solvent making diffusion of oleoresin will be greater, so solvent distribution to basil leaves will be greater. Equitably solvent distribution to basil leaves will increase the yield of oleoresin produced. The more solvents used will reduce the saturation level of the solvent, so that the extracted extracted component can be maximized [16]. The highest yield of oleoresin in this research is 20,434 % with the ratio of material and solvent 1:8 at 6 hours. At the same time, but with the ratio of material and solvent 1:6 obtained yield of 20,152 %, and lowest yield is 9,8676 % with the ratio of material and solvent 1:5 and extraction time at 3 hours. Effect of variations in the ratio of material and solvent and extraction time is different to each oleoresin yield product. The increasing in ratio of material and solvent and extraction time will increasing oleoresin yield.

3.2 Analysis of Oleoresin Density

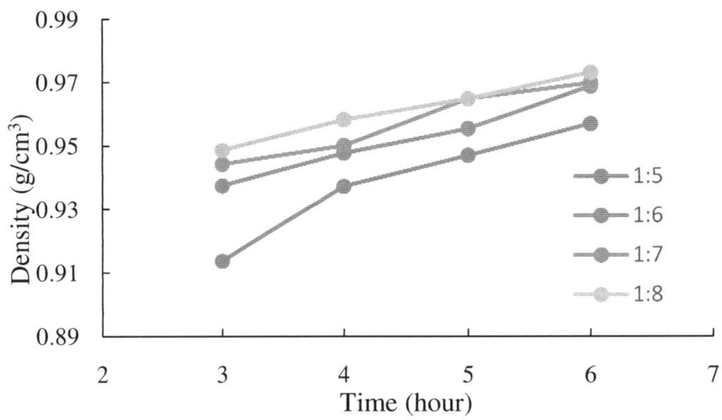

Figure 2. The influence of ratio between material and solvent and extraction time to oleoresin density

Figure 2 is an overview of extraction which shown the influence of the ratio of material and solvent and extraction time to oleoresin density. We can see that in overall, the ratio of material and solvent and extraction time give effect to the density of oleoresin, which density of oleoresin increases with the increase in ratio of material and solvent and extraction time. For the ratio of material and solvent 1:5 at 3 hours has obtained oleoresin density of 0.9137 g/cm^3. At 4 hours has obtained oleoresin density of 0.9373 g/cm^3 and at 5 hours has obtained oleoresin density of 0.9470 g/cm^3. At 6 hours, the oleoresin density has obtained about 0.9570 g/cm^3. For the ratio material and solvent 1:6 at 3 hours has obtained oleoresin density of 0.9375 g/cm^3. At 4 fours is obtained oleoresin density of 0.9477 g/cm^3 and at 5 hours is obtained oleoresin density of 0.9554 g/cm^3. At 6 hours has obtained oleoresin density of 0.9688 g/cm^3. For the ratio of material and solvent 1:7 at 3 hours has obtained oleoresin density of 0.9442 g/cm^3. At 4 hours has obtained oleoresin density of 0.9501 g/cm^3 and at 5 hours has obtained oleoresin density of 0.9649 g/cm^3. At 6 hours, the oleoresin density has obtained about 0.9698 g/cm^3. For the ratio of material and solvent 1:8 at 3 hours has obtained oleoresin density of 0.9487 g/cm^3. At 4 hours has obtained oleoresin density of 0.9582 g/cm^3 and at 5 hours has obtained oleoresin density of 0.9649 g/cm^3. At 6 hours, the oleoresin density has obtained about 0.9731 g/cm^3. At this research, the density obtained are between 0,9137-0,9731 g/cm^3. Oleoresin density obtained in this research is lower when it was compared with the density according to Lluch Essence, that is 1,000 [11]. The density differences caused by oleoresin obtained in this study contains relatively small essential oils. The lower the essential oil content, the content of resins, fatty acids, and irregular compounds will be higher [7].

3.3 Analysis of Oleoresin Refractive Index

Figure 3 is an overview of extraction which shown the influence of the ration of material and solvent and extraction time to oleoresin refractive index. We can see at Figure 3 that with the ratio of material and solvent and extraction time given different of refractive index results. From Figure 3 we can see that more the amount of solvent and longer the time of extraction showed an increasing in the value of refractive index. For the ratio of material and solvent 1:5 at 3 hours is obtained oleoresin refractive index of 1,4938. At 4 hours is obtained oleoresin refractive index of 1,4950. At 5 hours is obtained oleoresin refractive index of 1,4963. At 6 hours is obtained oleoresin refractive index of 1,4977. For the ratio of material and solvent 1:6 at 3 hours is obtained oleoresin refractive index of 1,4992.

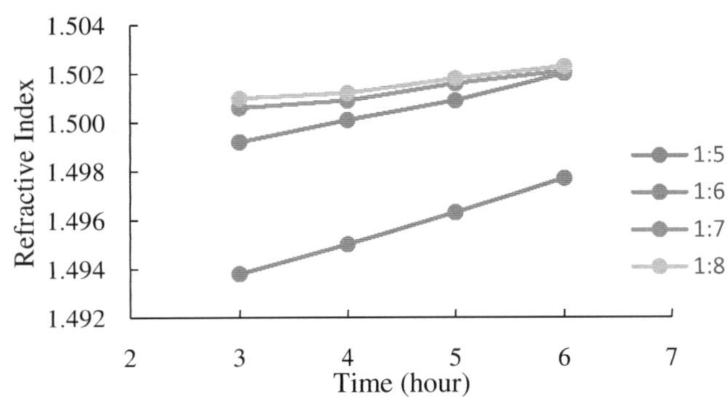

Figure 3. The influence of the ratio between material and solvent and extraction time to oleoresin refractive index

At 4 hours is obtained oleoresin refractive index of 1,5001. At 5 hours is obtained oleoresin refractive index of 1,5009. At 6 hours is obtained oleoresin refractive index of 1,5020. For the ratio of material and solvent 1:7 at 3 hours is obtained oleoresin refractive index of 1,5006. At 4 hours is obtained oleoresin refractive index of 1,5009. At 5 hours is obtained oleoresin refractive index of 1,5016. At 6 hours is obtained oleoresin refractive index of 1,5021. For the ratio of material and solvent 1:8 at 3 hours is obtained oleoresin refractive index of 1,5010. At 4 hours is obtained oleoresin refractive index of 1,5012. At 5 hours is obtained oleoresin refractive index of 1,5018. At 6 hours is obtained oleoresin refractive index of 1,5023. At this research, refractive index are between 1,4938 – 1,5023. According to Lluch Essence, the value of refractive index allowed is 1,5010 – 1,5210 [11]. From the research has been done, it has been fulfil the standards of basil oleoresin refractive index.

3.4 Composition of Essential Oil in Basil Leaves Oleoresin

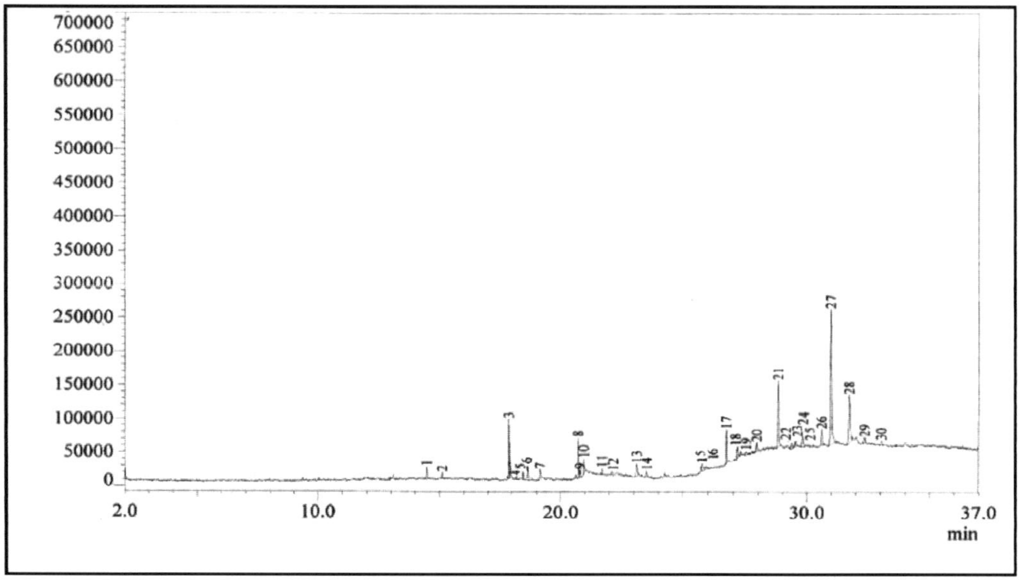

Figure 4. Basil leaves oleoresin chromatogram using GC/MS when extraction time at 6 hours and 1: 6 of raw material and solvent ratio

Identification of essential oil composition in basil leaves oleoresin when extraction time at 6 hours and 1: 6 of raw material and solvent ratio has shown at Figure 4. This result obtained by soxhletation method, and chromatogram obtained by using Gas Chromatography Mass Spectrometry (GC/MS). The components of basil leaves oleoresin are shown in Table 1, and the characteristics of basil leaves oleoresin are summarized in Table 2.

Table 1. Composition of Essential Oil in Basil Leaves Oleoresin at the Ratio of Material and Solvent 1:6 with Extraction Time 6 Hours

Peak	R. Time	Area (%)	Component
1	14,485	1,75	Trans-alpha-bisabolene
2	15,111	0,77	Undetected
3	17,908	6,69	Neophytadiene
4	17,981	1,17	Phytol
5	18,173	0,97	Phytol
6	18,374	1,35	Phytol
7	19,161	1,83	Octadecenoic acid
8	20,725	4,83	Phytol
9	20,794	1,05	Undetected
10	20,958	2,23	Methyl linolenate
11	21,710	0,82	Phytol
12	22,099	0,79	Flavone
13	23,133	2,15	Beta-pinene
14	23,512	1,18	Beta-pinene
15	25,750	1,80	Methyl linolenate
16	25,926	0,78	Undetected
17	26,736	4,12	Farnesol

18	27,181	1,90	Pentatriacontane
19	27,525	0,81	Farnesol
20	27,957	1,40	Tetracosane
21	28,825	9,88	Heptacosane
22	29,379	1,20	Vitamin E
23	29,508	0,97	Undetected
24	29,557	1,09	Octadecane
25	29,821	2,63	Tetratetracontane
26	30,607	3,53	Cholest-5-ene
27	30,986	27,72	Dotriacontane
28	31,742	12,23	Cholest-5-en-3-ol
29	32,367	1,14	Octadecane
30	33,057	1,23	Undetected
Total		**100**	

According to table 2, the characteristics of basil leaves oleoresin in this research meet the standard of basil oleoresin, but few of the value of the density and refractive index are still below the standard value [11], and basil leaves oleoresin contains essential oil about 15,77 %.

Table 2. Basil Leaves Oleoresin Characteristics [11]

Parameter	Research Result	Standard for Oleoresin
Color	Dark-Green	Dark-Green
Appearance	Viscous Liquid	Viscous Liquid
Taste	Characteristic of Basil	Characteristic of Basil
Essential Oil Content (%)	15,77	4 - 12
Density (g/ cm^3)	0,9137-0,9731	1
Refractive Index	1,4938-1,5024	1,5010-1,5210

4. Conclusion

The oleoresin of basil leaves can be obtained optimally at ratio of material and solvent 1:6 and extraction time at 6 hours which contains 20,152% of oleoresin yield, contain 15,77 % of essential oil, 0,9688 g/cm^3 of density, and 1,502 of refractive index and physical characteristics are dark-green color, viscous liquid appearance, and has aroma of basil characteristic.

5. References

[1] Eva K, Katerina H, Jana H, Tomas C, Jan P, and Martin K., 2008, *Aroma Profiles of Five Basil (Ocimum basilicum L.) Cultivars Grown Under Conventional and Organic Conditions)*, Elsevier. Food Chemistry, Vol. 107, pp. 464-472.

[2] Jean C C and Mehmet M O., 2008, *Comparative Essential Oil of Flowers, Leaves, and Stems of Basil (Ocimum basilicum L.) Used as Herb.* Food Chemistry, Vol. 110, No.2, 501-503.

[3] Massimo L, Mariangela M, Bernardetta L, Fabrizio G, Mauro M, and Francesco S.. 2004, *Morphological Characterization, Essential Oil Composition and DNA Genotyping of Ocimum basilicum L. Cultivars.* Plant Science, Vol. 167, pp. 725-731.

[4] Mohammed C, Douniazad E A, Njara R, Xavier F, and Farid C., 2016, *Comparative Study of Essential Oils Extracted from Egyptian Basil Leaves (Ocimum basilicum L.)*

Using Hydro-Distillation and Solvent-Free Microwave Extraction. Molecules. MDPI. Vol.113, No.21.
[5] Philippe B, Patrick A N, Eric M B F, Gisèle A F D, and Joseph L T.. 2013, *Chemical composition and residue activities of Ocimum canum Sims and Ocimum basilicum L essential oils on adult female Anopheles funestus ss*. Journal of Animal & Plant Science. Vol. 19. No. 1.
[6] Mehmer Ö, Jean C C.. 2002, *Essential Oil Composition of Ocimum basilicum L. and Ocimum minimum L. in Turkey*. Czech J. Food Sci, Vol 20, No. 6.
[7] Seema R, Hindole G, and Muddasir B.. 2016, *Phytochemical Characterization and Antioxidative Property of Ocimum canum: Effect of Ethanolic Extract of Leaves and Seeds on Basic Immunologic and Metabolic Status of Male Rats*. Immuno Biol. Vol. 1. No. 2.
[8] Susheela R U.. 2000, *Handbook of Spices, Seasonings and Flavoring*. Lancaster-USA: Techonomic Publishing Co. Inc.
[9] Normalina A, Satriana, and Kiki R.. 2013, *Extraction of Oleoresin from Waste of Nutmeg Oil Refining by Using Ultrasonic*. Vol. 9, No. 4, pp.1412-5064.
[10] Yuva B, Fadila B, Abdelhanine, Zina D, Laid B, Abderrahim N, and Mokrane I O.. 2014, *Antioxidant Activity of the Essential Oil and Oleoresin of Zingiber Officinale Roscoe as Affected by Chemical Environment*. Vol. 16. No.6.
[11] Lluch E.. 2016, *Essential Oils-Aromatic Chemicals-Flavors and Fragrances*. Spain : El Prat de Llobregat-Barcelona.
[12] Sabel, W. & J.D.F. Wamen..1973, *Theory and Practice of Oleoresin Extraction*. In Proceedings at the Conference On Spices. London: Tropical Products Institute.
[13] Jibrin M D, Agus A, and Muhammad A A Z.. 2015, *Characterization and Process Optimization of Castor Oil (Ricinus communis L.) Extracted by the Soxhlet Method Using Polar and Non-Polar Solvents*. Journal of The Taiwan Institute of Chemical Engineers. Vol.47, pp.99-104.
[14] Sara B.. 2004, *Essential Oils: Their Antibacterial Properties and Potential Applications in Foods-A Review*. International Journal of Food Microbiology. Vol.94, 223-53.
[15] Kaibing Z, Hui W, Wenli M, Xiaona L, Ying L, and Haofu D., 2011, *Antioxidant Activity of Papaya Seed Extracts*. Molecules Vol: 16, pp. 6179-6192.
[16] Muhammad D S, Anwar F, Pocut N A, and Normalina A.. 2011, *Solvent Extraction of Ginger Oleoresin Using Ultrasound*. Makara Seri Sains. Vol. 15. No. 2, pp. 163-167.

IOP Conf. Series: Materials Science and Engineering **237** (2017) 012033 doi:10.1088/1757-899X/237/1/012033

Bridging the engineering gap: integrated systems thinking

J R Weintré[1*], **M Delfi**[2]
[1]International Indonesian Forum for Asian Studies, www.iifas.info, Australia
[2]Andalas University, Faculty of Social and Political Sciences, Padang, Indonesia

* johan.weintre@mail.ugm.ac.id

Abstract. On visits to rural Indonesia it is apparent that the advances made possible by technical engineered solutions, are rarely at the same pace as the human captivation of technical development. This uneven pace has limited the application of labour-saving equipment and efficiency. It is suggested to be of primary importance to advance technical application skills among communities as part of the continuous advancement cycle in our human environment. A creative approach to inclusive technology and internal transfer of equipment knowledge in society, reduces barriers and could diminish structural or societal undesired situations. Earlier theoretical concepts provide us a lens for describing the practices of habitus, conceptualization of social capital and integrated systems thinking. The interrelationship and complexities in technical and social systems requires to be investigated. This paper aims to describe those, combined with technological applications in an empirical ethnographic approach. The study analyses the negotiations of community members with the available technology. It intends to foster a better understanding of the various cultural-economic values by exploring the systems thinking theory, with a focus on rice cultivation in Indonesia, Japan and Australia. This research suggests that cultural, economic and technical advances vary considerably and human expectations are strongly influenced by local culture.

1. Introduction

Education is focused on the development of technical and engineering skills, whereas the required application skills have been lagging in communities. Without, a direct link to communities, the engineered accomplishments remain below the optimum utilisation. The general prominence is upon change and adaptability, of which the rate of new inventions appears to be at an increasing speed. We have in fact passed beyond the stage of human organisations in which effective communication and collaboration are secured by the well-established traditional routines of wisdom transfer through relationship in society alone.

For this change, it is suggested that a focus on physio-technical engineering development is useful [1]. It is no longer conceivable for an industrial society to assume that only the technical processes of agriculture will remain unchanged. The challenge is how local cultural elements are incorporated.

The technical skill transfer could both incorporate the technical aspects and elements to enhancethe social connectivity. The skills required in contemporary society have developed in two directions. On the one hand, a higher type of skills is required, based upon adequate scientific and engineering knowledge, and secondly the visions of asocio-culturallydynamic society. This paper is to foster a qualitative analysis of the social and technical advances which show; use of information technology (IT), technical advancement and the social environmental dimensions, including community's affinities. Those elements require multiple

attention at the same time. The field work was limited to the artisan rice agriculture which incorporates a strong social and emotional or spiritual motivation, traditional commercial and the technologically advanced agiculture with a strongerhigher yield motivation.
These variables are discussed in the light of the tension between technical advancement and procedural human action. Agriculture as well as manufacturing are complicated by the interaction of environmental, economic and social factors. This fosters an exploratory of the Integrated Systems Thinking theory and visit the current human adoption in a reality setting of open engineered social and closed sciences systems [2].

2.Integrated Systems Thinking

It flows from a growing awareness that multi disiplenary collaborative approaches are obligatory as a result from the increased complexity in technical and human resource diversity. As well because ofthe positive evidence of enhancement when the participation of engineers, staff, corporate and individual farmer or stakeholders are incorporated. In those cases, the quality of participatory processes or synergy was suggested to become a key feature for success to reach adesired wholenessor optimumyield, subject to technical and cultural elements of limitation [3]. The cognitive basis demonstrates how knowledge supports the monitoring process of equipment employed. This movement relates to what Cook and Brown [4] have named an "epistemology of possession" of knowledge, when knowledge in its stabilised form is measured to be able to circulate from one place to another and stimulatesinnovation.

In the field of engineering and society are facing often issues in which realities are uncertain; values might be disputed, the stakes might be high, and making decisions urgent [5]. In that context, social scientists and engineers like others may have some fear of taking sides on value-laden issues. It is claimed by many; this stalemate position is not defensible. Once scientists and technical experts are asked to define what reality is, this distinction becomes problematic because knowledge carries implicit values and local worldviews: What they say and know, cannot be distinguished from what they perceive reality ought to be with the input of equipment or technical process implied. It is considered, social learning is needed to engage various stakeholders in a problem formulation allowing stabilised agreements and purposeful action. These processes occur through interactions with others, involve social, cognitive, as well as emotional dimensions and could take account of Habermas's proposition: "...the paradigm of knowledge of objects has to be replaced by the paradigm of mutual understanding between subjects capable of speech and action" [6]. In the field of management in which engineers, staff and agriculturalists require joint communication in understanding the process and spheres of thought that can be brought together. A joint vision of what problems were perceived and to be solved, provide an obvious picture of what could be done. Checkland [7] stated that hard systems are goal seeking, with given or assumed goals, whereas in soft systems the goals themselves are the bone of contention, this perspective raises the question of the role of traditional knowledge before engineered processing options are implemented, in setting certain goals in the output [8]. Is it used to force interacting stakeholders to conform in their thinking and doing, to problems defined by others in others place. Perhaps we could say it is used to help engineered farm equipment in the bio-socio-physical environmentto be complementary in a process, and stakeholders adapt their thought to take the most appropriate decision and apply the "goodness of fit" [9].

In mechanised agriculture, it offers a holistic approach that can help to understand the influences of processing (use of equipment), elements of sustainability by focusing on problems as the result of root causes and not the symptoms, while mediating it as a composition of interconnected events [8]. A process in agriculture is a transformation of materials and energy by labour and equipment in a social environment. In complex management situations of interacting of labour and equipment "the components and their interactions are changing and can never be quite pinned down" according to Snowden [10]. It is considered that social learning is needed to engage various stakeholders in problem formulation allowing purposeful action. A framework of the systems thinking approach fosters an analysis of the complexity of economic environmental and social benefit conflicts encountered in the agricultural ormanufacturing systems in terms of costs and benefits.

Indeed, once a system or policy is implemented and becomes part of the culture, it suggests the vision of the world, it conveys that interacts with local perception and interests leading to a translation of the cultural or system's goals [10]. The translation results in and results from social economic learning processes, a focus in adaptive management [11 bordieu] and which could be defined as knowledge co-production among stakeholders brought together [12]. As Mayo suggested in 1945, team work and spontaneous cooperation among all employees are possible if thoroughly planned or made possible in industry and society to meet and grow on a on a level more commensurate with the development and utilisation of technical skills. The suggested growth equation in the mixed open and closed systems environmentin the mathematical model by von Bertalanffy $L(t) - L_\infty - (L_\infty - L_0)e^{-kt}$, the founding father of Integrated Systems Thinking concept[2], The multi-disciplinary approach by von Bertalanffy is touched on in this paper, although the mathematical model itself is outside the material of this paper.

3. Research Focus

Researchers acknowledging the integrated systems thinking concept can identify not only the relations of certain human behaviour, but the dynamics of sustainable performance measures in manufacturingor agrarian economic-social systems. The economic motivations often hesitate to invest in environmental, safety, cultural-social and health efforts, as the long lead-time outcome is not clear and provides uncertainty to the returns on investment. The cause lies in the challenge of related environmental and social practices of the physical engineered project to meet the potential economic benefits envisaged by the stakeholders. Managers in agriculture want to measure human behaviour and other inputs on a cost basis, the usual yardstick in business. Key questions should be asked to provide clarity on the decisions to be made. For example, how much attention should be incorporated for cultural or religious dedicated activities or can be gained by having members in a skill upgrade? How much can be saved to appeal to staff or community members to contribute with ideas on process setting, such as making modification in the process, energy consumption of equipment and safety aspects? What are the economic performances contributing of these efforts over time? With a current focus on the sustainable practices in any environment, these variables can vary from location or country. This paper will include the artisan farmers, and commercial agriculturalists. This second group has greater access to capital, skills and have commercial visions reaching out into the future. From an engineering perspective, it will range from basic traditional tools to sophisticated cutting edge equipment, encountered in countries where labour costs are high and labour-saving devices are very important for survival of the farm.

4. Data and Methodology

The framework of data collection was developed in two stages. The first stage comprised of long separated investigations of communities with minimalistic technological interventions and agriculturalists with access to technological equipment. Empirical field notes were collected through direct observation of the activities and conversations with stakeholders in different locations. The data were gathered over an approximate period from February 2012 till April 2017 and spread over a geographical area from Australia to Japan and the tropical environments of Malaysia and Indonesia.

Rice cultivation in different social and economic environments

Knowledge skills and cooperation varies enormously, like the cost structure and cultural factors or the availability of seeds, soil fertilisers, infrastructure and capital means. Also, the marketability of rice as a commodity has a cultural and religious significance. For instance, in the Dayak groups surveyed in Borneo, an informant explained that till relatively recently rice was not available for sale in the market. From this recognition can be deducted that rice has values, hardto measure such as nutritional food value, human exertion, or in economic terms of money.

The images below are from four locations; two in Indonesia, one in Japan and one in Australia. The first two can perhaps be best described as artisan agriculture as tradition and religious aspects are finely interwoven in the rice cultivation. In Japan, because of its large number of smallnear hobby size rice farms, tradition is interwoven with the highly developed equipment for those rice plots.The Australian rice growing model is a commercial agro-operation led by a manager, who are up to date with the latest technology and news.

The purpose of the country comparison is to illustrate how local socio-cultural and economic driven aspects have created different settings of the agriculturalist's interaction in their environment. Figure one to four below provides notions of human and equipment involvement in the planting stage, five to eight are impressions at the harvesting stageof the rice cultivar.

4.1. Various planting techniques to establish rice fields

Dayak shifting cultivators in Borneo, Indonesia, apply the traditional technique of community rice seed broadcasting in unirrigated fields. Shallow holes are poked into the soil with a stick or tugal, by a group of community planters. A second group, as seen in the image, follow and broadcast the rice seeds in the holes. The"nugal" planting technique is being performed by Dayak groups. This cultivation pattern yields one crop a year, without the input of industrial fertilisers.

Figure 1. Nugal tradition, photo by rmolsumsel.com.

Traditional aquatic rice planting in Central Java, Indonesia. This seedlingplanting technique is carried out in a small group or by a singular planter. With improved rice varieties and irrigation, crops are harvested five times every two years in Central Java. In some parts of West Sumatra, it is not unusual to grow up to three crops a year with special cultivated rice seeds and input of industrial fertilisers.

Figure 2. Manual aquatic planting, photo by Gunawan Kartapranata

Agriculturalists in Japan grow rice for commercial and personal purposes. With limited space for cultivation due to the many mountains the average farm size is under one hectare. Rice is perceived as a culturally precious commodity. Because of the many part time farmers,it is the norm to apply a mechanised seedling planting technique.

Figure 3. Mechanised planting, photo by Jerry Anderson

Aerial sowing of rice seeds by hi-tech equipped crop planes is one of the options in cost effective farming in Australia. Engineered facilities guided by satellite navigation systems for accurate broadcasting of the seeds. In this image, a plane swoops low over a recently irrigated rice field that will be drained at the precise moment to conserve water and maximised crop quality and yield atharvest time.Usual two crops are produced per year.

Figure 4. Aerial seed broadcasting (Photo by V Bucello)

4.2. Various techniques in crop harvesting

Figure 5. Traditional panicle harvesting, photo upperkapuas.blogspot.com.au

The ladang or field are harvested by the Dayak groups about four to six monthsafter planting, depending on weather andseed variety.Individual growersusually sow three to six different varieties each year.The genetic rice pool might be forty to sixty species in any given village. According to Dayak magico-religious tradition the rice goddess, in the Iban Dayak culture Pulang Gana, is honoured by harvesting only the rice panicles. It means that each rice stalk is harvested individually by hand.Yields vary from 200 kg up to 800 kg a hectare (kg/h) in recordings taken in the period from 1985 and 1997 [14].The variation are mainly caused by soil fertility, drought, floods, pests, fungal attacts and animal intrustion (birds and rats).

Figure 6. Manual harvesting, photo by author

In West Sumatera, Java Bali and many other locations in Indonesia or for that matter in Asia, the rice crop is harvested by cutting the whole plant. The padi or whole rice grain with husk is mechanical or manual released by a firm bashing movement. The padi, is transported in bags from the field to storage places. When the farmer or trader sells the rice, the padi is taken to a mechanical rice huller to remove the husk and polish the grain.Yield on the main islands of Indonesia varies between 4.8 – 5.6 tonnes a hectare (t/h) in 2009 [15]
.

Figure 7. Mechanised harvesting,photofactsanddetails.com

Harvesting in Japan is being carried out on small rice fields with compact mechanised equipment. This is to meet the local conditions of the mainly part-time or "Sunday" farmers who derive their main income external from rice farming. Although food consumption patterns have shifted to include bread, cereals and pasta, the historical cultural significance of rice growing and consumption remains strong in Japan. To encourage the continuation the government is subsidising rice cultivation. Because of the staple food control act (1942) the government oversees distribution and sales. Average yearly crop yield is 5t/h [16].

Figure 8. Commercial harvester, photo by V Bucello

Harvesting in Australia is by means of special satellite navigated harvesters, to reduce damage and maximise rice yields. Monitoring equipment in the harvester records the precise rice yield in every section of the field. This data collection allows precise future application of fertilisers for optimal dosage. Agriculturalists are professional managers and work together with specialised technical rice production equipped contractors.The average 5 year yieldwas 10 tonnes a hectare in 2016[17].

5. Results and Discussion

Technological development supported by advanced engineering and IT has touched the agricultural sector in an uneven pattern. It is noted that agriculture is significantly influenced by local culture and the connecting worldviews in the community, as well as resources, education, infrastructure, financial capital and a whole range of interlocked variables, including government support. Farmers could be assisted in their rice cultivation by long established tradition and new modern sign posts. From traditional experience, to rice production needing high technically skills with the aid of electronic guided devices and software. Those advancements firstly enable farmers to be part of the participatory dynamics and suggest fostering skill ownership in the community which provides meaningfulness and self-confidence. Secondly, applied technology support the development of technological literacy in communities of non- or semi-skilled members. These utilisations provided engagements of opportunities on local and commercial matters of concern. Thirdly, they enabled teamwork and support the development of champions' in the community, able to orchestrate complex collaborations.

6. Conclusion

Locations have their emotional and commercial connectivity inrice cultivation, thus not a simple equation of people, growing, enhancements by engineering and capital investment. Von Bertalanffy allowed to foster a relative relationship of engineering, culture, commercial behaviour and human interaction. In Australia highly engineered assisted rice yields are 10 t/h while in higly cultural traditional Dayak cultivation, field yields between 200 to 800 kg/h. It is merely from the advantaged position in global rice cultivation that differences have come to our attention. The main findings in this paper suggest that an understanding of the different perspectives of the stakeholders in agriculture and open discussion could lead to better cooperation in engineering and understanding of cultural expectations. Future research in engineered solusions and cultural affinity could alleviate tension in crop procrument and food security.

References

[1] Czerniewicz L and Brown L 2014 *South Africa Journal of Education* 34 (1) art. 789
[2] von Bertalanffy L 1968 *General System Theory*, (New York: George Braziller)
[3] ns General System Theoryhttp://panarchy.org/vonbertalanffy/systems.1968.html Accessed 27 June 2017
[4] Cook S D N and Brown J S1999 *Organisation Science* 10(4):381-400
[5] Funtowicz S and Ravetz J 1999 *Futures* 31 (7):641-646
[6] Habermas J 1987 *The Philosophical Discourse of Modernity*(Cambridge: Polity Press)
[7] Checkland P 1981 *Systems Thinking, Systems Practice* (Chichester: John Wiley)
[8] Steyaert P and Ollivier G 2007 *Ecology and Society*12 (1): art. 25
[9] Friedman B D and Neuman Allen K 2014 Chptr1: *Systems Theory*In: *Theory & Practice of Clinical Social Work*by BrandellJ R(Sage Publication) DOI: http://dx.doi.org/10.4135/9781483398266.n1
[10] Zhang H 2014 *A Framework for Integrated System Thinking into Sustainable Manufacturing*, dissertation (Corvallis,Oregon State University)
[11] Snowden D 2002 *Journal of Knowledge Management* 6 (2):100-111
[12] Röling N G and Wagemakers M A 1998 *Facilitating Sustainable Agriculture: Participatory Learning and Adaptive Management in Times of Environmental Uncertainty*, (Cambridge: Cambridge University Press)
[13] Mayo E 1945 *The Social Problems of an Industrial Civilisation* (Boston: Harvard University)
[14] Barton H 2012 Quaternary International 249 96 -104
[15] Panuju D et al 2013 *Journal of the Saudi Society of Agricultural Sciences*12 27–37
[16] OECD 2017, *Crop Production (indicator)*. doi: 10.1787/49a4e677-en Accessed on 29 June 2017
[17] Field 2016 Rice grower breaks record achieves yield of 16.2 tonnes/ha http://www.weeklytimesnow.com.au/agribusiness/cropping/griffith-rice-grower-breaks-record-achieves-yield-of-162-tonnesha/news-story/a643e3034d93f5f9148131ef92c2917d Accessed 27 June 2017

IOP Conf. Series: Materials Science and Engineering **237** (2017) 012034 doi:10.1088/1757-899X/237/1/012034

E-precision agriculture for small scale cash crops in Tobasa regency

Panca Putra Simanjuntak[1], Pangeran Tiurniari Napitupulu[1], Soni Pratama Silalahi[1], Kisno[1], Norlina Pasaribu[2], and Libuše Valešová[3]

[1]Institut Teknologi Del, Faculty of Informatics and Electrical Engineering, Computer Network Management Study Program,Toba Samosir, 22381.

[2]Institut Teknologi Del, Faculty of Informatics and Electrical Engineering, Informatics Engineering Study Program, Jl. Sisingamangaraja, Sitoluama, Toba Samosir, Sumatera Utara, 22381, Indonesia

[3]Czech University of Life Sciences Prague, Department of Economics and Development, Kamýcká 129, Prague 6, CZ 16521, Czech Republic

* pancasimanjuntak47@gmail.com

Abstract. Cash crop is a promising sector in Tobasa regency; however, the trend showed a negative change of the cash crop production in. This research aims to develop an application which is based on Arduino for watering and fertilizing corn land. The result of using e-precision agriculture based on embedded system is 100% higher than the conventional one and the risk of harvesting failure using the embedded system decreased to 50%. Embedded system in this study acquired critical environment measurements which at last affected the yield raising and risk reduction. As the result, the use of e-precision agriculture provided a framework to be used by different stakeholders to implement e-agriculture platform that supports marketing of agricultural production since the system is proven to save the material and time which finally reduces the risk of harvesting failure and increases the yield. In other words, the system is able to economize the use of water and fertilizer on a small corn land. The system will be developed for more efficiency in material loss and the mobile-based application development to reach sustainable rural development particularly for cash-crop farmers.

1. Introduction

Agriculture is a never-ending effort since human needs food for living every day; however, human population trend shows that the number of people in Indonesia is getting increased. This becomes a big opportunity for everyone interested in cultivating agriculture land particularly cash crops. Cash crop, as is called, is the income resource for farmers since it has a direct relation with the market and the product can earn “cash” for them [1]. Cash crop is an agricultural sector that is potentially developed in Indonesia. Fertile land, tropical climate, high rainfall, and abundant sunlight are positive factors for the country which is 252 million populated. From the human resource point of view, Indonesia is now entering the phase of demographical bonus which is related to higher productive age compared to the non-productive one. North Sumatera Province is the part of Indonesia which has the potential for cash crop such as corn, cassava, soya beans, and peanuts; ironically, the total production of corn and peanuts decreased dramatically from 2012-2014. Corn production fell from 1.35 million tons to 1.15 million tons in only two years period while peanuts experienced 1.5 tons production reduction at the same time [2]. This is caused by the declination of the harvested land for human residence. Toba Samosir (Tobasa) is a regency located in the northern part of North Sumatera, is known as a center of paddy and corn supplier for North Sumatera. Other cash crops such as cassava, green nuts, and peanuts also grow in this area. Almost all districts in Tobasa grow cash crops as the alternative crops after wetland paddy. Despite its capacity to produce those crops, Tobasa also experienced the depletion of corn production for the last three years in which the production decreased sharply from 24.957 tons in 2011 to 14.247 tons in 2013 [3]. This situation is in line with a census carried out by Statistic Bureau of Indonesia which showed the negative trend of the number of cash crops farmers in Indonesia. In 2013, there were 10.5 million of them left compared to 16.5 million in 2003 [4]. These farmers tend to change their activities to other occupation since the number of harvested land was not promising anymore. One the problem faced by the farmers is about technological issue where all activities are still carried out manually such as in Tobasa regency. For example, with a farmer, it needs more than two hours for spreading water and fertilizer on a 2.500 m^2 corn land. In this paper, a technological solution is introduced to cash crop farmers in order to stimulate them to rework the land in order to increase corn production. The technology is a smart application based on Arduino for watering and fertilizing corn land. This application gives an effective solution for farmers since they do not have to visit their land all the time for monitoring [5]. The objective of this study is to create an application which helps the farmers use embedded system for their agricultural activities to actualize precision agriculture in Tobasa regency. The significance of this study gives benefit to the farmers for monitoring their corn land and implemented the basis of precision agriculture which finally influences the farmers' economy for sustainable rural livelihood.

2. Method

2.1 System design

This application uses Arduino open-source microcontroller which is compatible with any platform and it can be connected via Universal Serial Bus (USB) [6]. Arduino is then connected with ESP8266, a wifi module functions as an extra microcontroller to create TCP/IP. This module needs 3.3v and has three wifi modes such as station, access point and both. It is also equipped with processor, memory, and GPIO. pH sensor is deployed to provide the accuracy for reading the soil acidity and it has serial asynchronous interface connection. Water level sensor is an instrument used to give a signal that water is reaching a certain level. It sends dry contact signal (NO/NC) to the panel. This detector gives alert to stimulate other automatic items to move [7]. Other materials used for this application are water valve, water pipes, and water pump. Software which is the main focus of this research is built with a framework. Programming of a communication between the information system and automatic watering and fertilizing involves wifi module which is placed in Arduino. When the information system sends a message or commands carried out by Arduino and the message will be executed. After execution, Arduino will send back the message to the information system stating that the message has been completely executed.

Figure 1 illustrates the system design where the application is connected to Arduino. When a user accesses information system and sends "watering" command, Arduino receives the command and processes it. After that, Arduino sends the command to servo and servo sends the command to open the water valve and conducts the watering. In this web-based application, the user can see the soil's pH and is able to see the need of the crop based on the data stored in database. When the crop is in need of water or fertilizer, user can automatically monitor, control, and carry out watering or fertilizing. When the water in the tank gets lower, water level sensor will detect the water volume and sends the command to relay to refill the water and fertilizer.

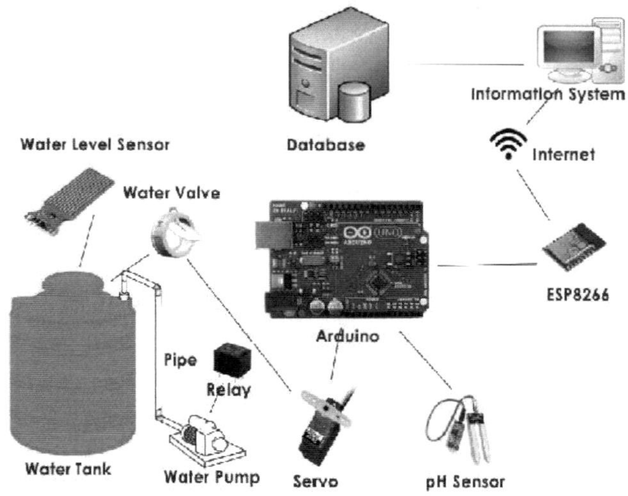

Figure 1. System Design

2.2 System flowchart

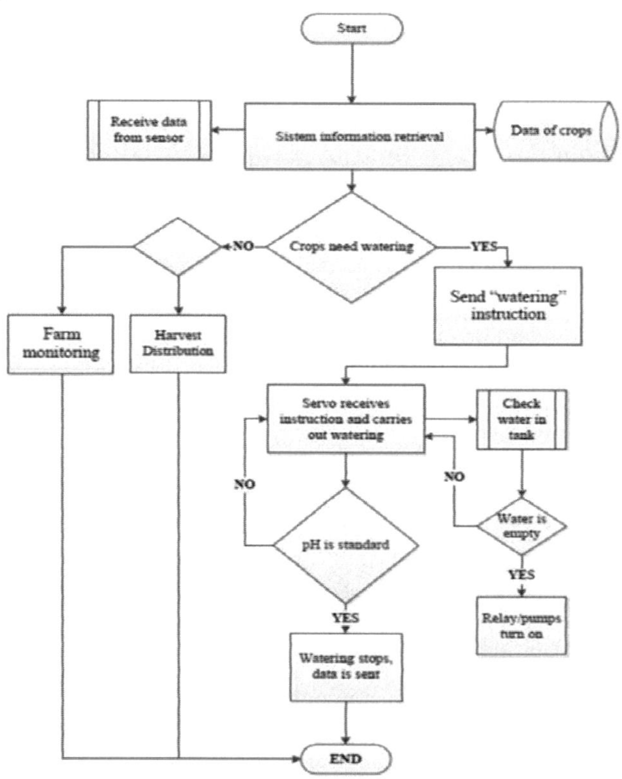

Figure 2. System flowchart

The flowchart of the system is illustrated in Figure 2. When the system is accessed, a user is able to check whether the crop needs watering based on pH sensor. If the crop needs watering, the user can click "water button" in the application. Then, a command is sent to Arduino and Arduino sends the command to servo while carrying out watering. Water level sensor also works simultaneously to check the water availability in the tank. When soil's pH meets the standard, watering and fertilizing is stopped and the user can execute other feature such as selling and crops management which are displayed by the application.

2.3 Graphical User Interface

The graphical user interface is designed to be user friendly so that the novice user can easily use it without any complicated training.

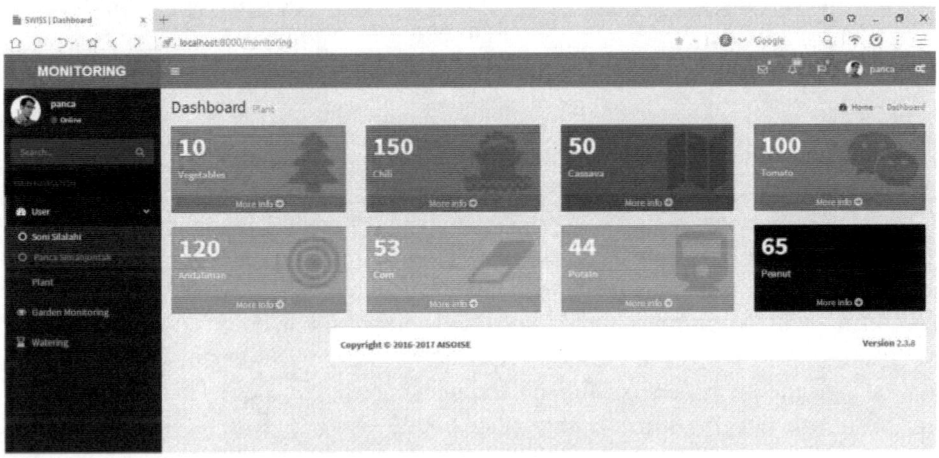

Figure 3. GUI of general control system

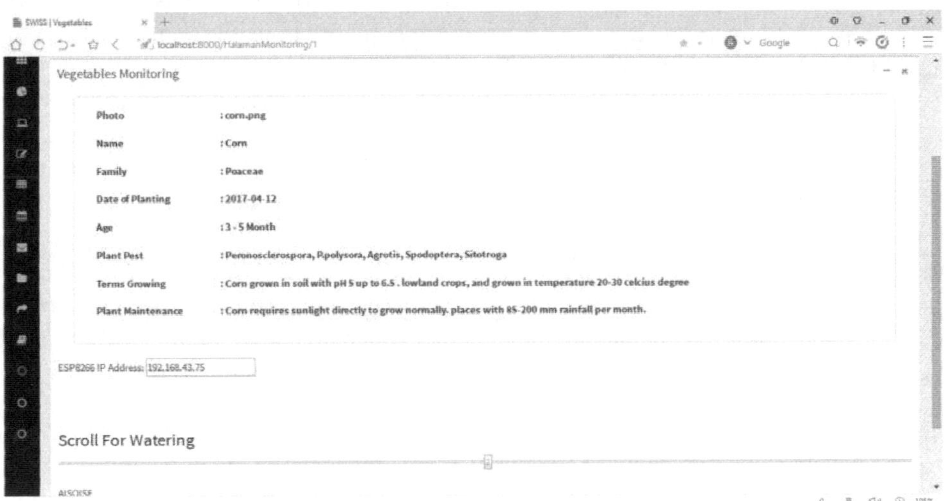

Figure 4. GUI of crop control system

Figure 3 displays the crops available in the land and when a specific crop is clicked, it displays the new page containing important information of the crop and carries out watering or fertilizing. For example in Figure 4, "corn" option is clicked and then the interface shows the family of the crop, growth information, and maintenance. There are two buttons below the data, "water button" and "back button". Watering and fertilizing are carried out after reading the necessity data of the crop e.g. if the pH is 6 or 7 based on the data, watering or fertilizing is not executed.

3. Result and discussion

The experiment was carried out on a 500 m^2 manual and e-precision corn land respectively. The comparison of both methods yield is shown in Table 1.

Table 1. Monitoring result on corn land (per 500 m^2)

Evaluation aspect	Manual	E-precision agriculture
Watering time	24 minutes	5 minutes
Fertilizing time	30 minutes	10 minutes
Water loss	2-2.25 l	0.5 l
Fertilizer loss	1.2 -1.5 kg	0.2 kg
Yield	1 ton	2 tons
Harvest fail risk	10%	5%

Table 1 shows the result of monitoring in six aspects using conventional method compared to e-precision agriculture on a corn land with 500 m^2 wide. With manual method, the loss of time for watering was five times higher and fertilizing time was three times higher than e-precision agriculture. The manual method required the farmers to carry water buckets and the watering was carried out based on the preferences of the farmer. This result indicates the increasing yield of e-agriculture using distributed system for monitoring temperature, moisture, and light intensity in the greenhouse using Arduino Uno. This system consisted of two sensor-actuator nodes and a controller node connected to internet using Ethernet Shield board. The system's ability to monitor indoor environment of greenhouse, and environment data access via web enabled the monitoring and plant management to be performed automatically [8]. Moisture on soil content is determined by the water content where the proper amount of water in the soil in this study was considered standard to grow the crop properly. The similar issue took place when fertilizing the crops where the farmers spread the fertilizer using the backpack-tank. With the help of the application, the pipes spread the water and fertilizer based on the sensor working on the soil and the time for executing the instruction. This means that the application implemented in this study is suitable for agriculture environment [9] since the result that is gained is more efficient and user-friendly. This also indicates the success of a design of a complete application with an electronic device that can help landlord agriculture to start out a dependable quality product in the marketplace. In terms of material loss using manual method, the amount of water that lost was four times higher and the fertilizer loss was six times higher. Since there was no standard of watering and fertilizing of the manual method, water fell out from the bucket during the period of bringing it from the water source to the land and the farmers had to carry the buckets many time. In addition, fertilizing process did not have the same standard from one to another corn. Smart agriculture irrigation control using wireless sensor networks can monitor and control the irrigation automatically with the information from the web server node [10]. These material and time loss indeed influences the result of harvesting where the result of using e-precision agriculture was 100% higher than the conventional one. The risk of harvest fail using the embedded system decreased to 50% since the precision in watering and fertilizing made the growth of the crop in good condition where all corn seed have the same nutrition from watering and fertilizing. Embedded system in this study acquired critical environment measurements which at last affected the yield raising and risk reduction [11]. In summary, the use of e-precision agriculture provided a framework to be used by different stakeholders to implement e-agriculture platform that supports marketing of agricultural production [12] since the system is proven to save the material and time which finally reduces the risk of harvesting failure and increases the yield.

4. Conclusion

The implementation of e-precision agriculture in this study gave evidence that time, material, and risk had been reduced effectively. The accuracy of the system was proven in the term of the harvesting yield which increased up to two folds. In other words, the system was able to economize the use of water and fertilizer on a small corn land. For future work, the system will be developed for more efficiency in material loss since there was still found little loss during the implementation. The system will also be implemented for other cash crops such as peanuts and cassava. Other parameter aspects such as temperature, light intensity, and pests will be included for the next development for each kind of cash crop. The data collection and analysis for each crop will be used for developing large-scale cash crop in Tobasa regency. Finally, the mobile-based application development is highly recommended for better result so as to raise the awareness of the farmers to work again on their land and to reach sustainable rural development for the people around.

Acknowledgments

A special gratitude is addressed to HIMATEK IT Del (Himpunan Mahasiswa Program Studi Teknik Komputer Institut Teknologi Del) for the time and efforts in completing this study.

References

[1] Sadjad S 2000 Agribisnis Berorintasi Ekspor: Kasus Tanaman Pangan: Mungkinkah? *Agrimedia* (5) 2 p 1-4.
[2] BPS Sumatera Utara 2014 *Pertanian: Statistik Tanaman Pangan Sumatera Utara 2012-2014* (Medan: BPS Sumatera Utara)
[3] BPS Tobasa 2014 *Toba Samosir dalam Angka* (Balige: BPS Tobasa)
[4] Mandiri T 2017 Angka petani hortikultura turun, siapkah AEC 2015? *Hortikultura Indonesia.* p 1-3.
[5] Siththtikumar N, Maduranga MWP 2016 Designing and implementing an Arduino based low cost automated water irrigation system for home gardens *IRSEA.* p 141-143.
[6] Djuandi F 2011 *Pengenalan Arduino* (Jakarta: www.tokobuku.com)
[7] Iyappan S, Ramaprabha R 2017 Design and Implementation of Brushless DC Motor based Solar Water Pumping System for Agriculture using Arduino UNO *International Journal of Engineering and Technology* IJET **(9)** 1 p. 224-232.
[8] Pamungkas BA, Rochim AF, Widianto ED 2017 *Perancangan jaringan sensor terdistribusi untuk pengaturan suhu, kelembaban dan intensitas cahaya.* (Semarang: Universitas Diponegoro)
[9] Hashim NMZ, Mazlan SR, Abd Aziz MZA, Salleh A, Ja'far AS, Mohammad NR 2015 Agriculture monitoring system: a study *Jurnal Teknologi* **(77)** 1 p.53-59.
[10] Gettha S, Priya Sathya R 2016 Smart agriculture irrigation control using wireless sensor networks, GSM and Android phone *Medwell Journals* **(19)** 15 p 2780-2786.

[11] Zachariadis S, Kaskalis TH 2012 An embedded system for smart vineyard agriculture *2nd Pan-Hellenic Conference on Electronics and Telecommunications PACET 12* p 1-4.

[12] Namisiko P, Aballo M 2013 Current status of e-agriculture and global trends: a survey conducted in TransNzoia County, Kenya *International Journal of Science and Research* IJSR **(2)** 7 p 18-21.

IOP Conf. Series: Materials Science and Engineering **237** (2017) 012035 doi:10.1088/1757-899X/237/1/012035

Performance of coffee origin and genotype in organoleptic and physical quality of arabica coffee in North Sumatra Province of Indonesia

Sabam Malau[1,*], Albiner Siagian[2], Bilter Sirait[3], Samse Pandiangan[1]
[1]Agroecotechlogy Department, Faculty of Agriculture, Nommensen HKBP University, Jalan Sutomo 4-A Medan 20234, Indonesia.
[2]Faculty of Public Health, North Sumatra University, Medan, Indonesia
[3]Agriculture Faculty, Darma Agung University, Medan, Indonesia

* drsabammalau@hotmail.com

Abstract. The objective of this research was to determine effect of coffee origin and genotype on organoleptic and physical quality of *Arabica coffea* L. growing in North Sumatra. Seven districts treated as origins and 28 genotypes were chosen. The research was conducted with nested design with 3 factors. Organoleptic parameters were fragrance/aroma, flavor, aftertaste, acidity, body, uniformity, balance, clean cup, sweetness, overall and total score. Physical quality was green bean weight. The results revealed that origins affected significantly organoleptic quality. Coffee from Dairi showed the highest total score (90,82). Genotypes were significantly different in organoleptic quality. Genotype Da17, Da18, Da19, Da20 and Hu4 had the best total score (89,85 - 91,68). Total score did not correlate with green bean weight but had positive correlation with altitude. Among organoleptic parameters, acidity was more significant for total score (r^2 = 0,836). Altitude had more effect on acidity (r^2 = 0,486).

1. Introduction

Coffee is one of the most important commodity in North Sumatra. This province contributed 55.600 ton green bean to total production of arabica coffee bean of Indonesia around 148.000 ton in year 2010 [1]. In North Sumatera, arabica coffee are growing in districts Dairi, Tapanuli Utara, Karo, Simalungun, Humbang Hasundutan, Samosir, Toba Samosir, Pakpak Bharat, Mandailing Natal and Deli Serdang. These districts vary in altitude and climate. In these areas, coffee plants are already cultured since approximately more than one hundred years ago so that the variation among genotypes may exsist. Plant growth and bean production of coffee are affected by environment such as drought [2], rain [3] and temperature [4]. Besides the environment, those two parameters are also influenced by genotype of coffee plant [5]. Taste of coffee is affected by altitude acording to [6] and [7]. Hence, the variation in enviroment and genotype of coffee may affect growth, taste and physical quality of coffee. The goal of this research was to determine variation of origin and genotype on organoleptic and physical quality of arabica coffee growing in North Sumatra.

2. Materials and Method

Seven districs (as coffee origins) in North Sumatra were selected. In each district, 2 subdistricts were chosen, and in each subdistrict 2 coffee farms (as genotypes) were selected. In each farm grew 200-400 coffee plants with age of 6-7 years. Each farm were divided into 4 equal size of plots which were treated as samples (4 samples). Altitude was recorded. The harvested ripe red fruits were processed with full washed method to produce green beans [8].

Organoleptic quality was tested by using SCAA Cupping Protocols [9] in the laboratory belongs to Goldenways Coffee company in Medan. A Licensed Q Grader (Dr. Ir. Sabam Malau) and his asistant run the cupping tests. The taste tests were done two times with a distance of one week. The average value were then used for statistical analysis. The research was conducted in July-August 2012. The nested design with three random factors were used.The factors were district (Z), subdistrict (L) nested in Z, and genotype (G) nested in L nested in Z [10]. Between significant means were compared by LSD method as well as analysis of simple correlation and determination coefficients were also conducted [11].

3. Result

The districts were highly significant different ($\alpha = 0{,}01$) in fragrance/aroma, flavor, aftertaste, acidity, body, balance, overall and total score, and significantly different ($\alpha = 0{,}05$) in sweetness and 100 green bean weight (Table 1). The genotypes were highly significant different in fragrance/aroma, flavor, body, balance, overall, total score, and 100 green bean weight significantly different in after taste, acidity and sweetness but not different in uniformity and clean cup.

Coffee from Dairi had significantly ($\alpha = 0{,}05$) the highest total score (90,82) (Table 2) Coffee from Dairi had the same 100 green bean weight (15,18 g) with Simalungun and Tapanuli Utara.

Genotype Da20 displayed the highest total score (91,68) but was similar with Da17, Da18, Da19 and Hu4 (Table 3). Da20 had the heaviest 100 green bean weight (18,32 g).

Total score of organoleptic quality correlated highly significant with flavor, aftertaste, acidity, body, balance, sweetness, and overall whereby coefficient determination was at rate 0,510 to 0,836 (Table 4). Most of organoleptic parameters correlated significantly one another. However, sweetness did not correlate with balance and overall. Total score and each parameter of organoleptic quality did not correlate with green bean weight. Most of organoleptic parameters had significant correlation with altitude. Sweetness and overall did not correlate with altitude.

Table 1. Mean square of district, subdistrict, genotype and error of organoleptic and physical quality.

	District (df 6)	Subdistrict (df 7)	Genotype (df 14)	Error (df 84)
Parameter	MS	MS	MS	MS
Organoleptic quality				
Fragrance/Aroma	4,913**	0,033ns	0,254**	0,092
Flavor	4,029**	0,096ns	0,271**	0,088
Aftertaste	2,070**	0,107ns	0,233*	0,106
Acidity	2,170**	0,098ns	0,116*	0,058
Body	1,781**	0,020ns	0,179**	0,064
Uniformity	7,67ns	2,00ns	0,79ns	0,50
Balance	1,635**	0,039ns	0,177**	0,068
Clean cup	1,536ns	1,929ns	1,000ns	0,690
Sweetness	5,036*	1,179ns	1,179*	0,607
Overall	2,669**	0,152ns	0,244**	0,097
Total score	268,694**	9,333ns	11,775**	4,108
Physical quality				
100 bean weight (g)	24,401*	4,041ns	3,218**	0,413

	$F_{0,05}$	3,87	2,77	1,82	
	$F_{0,01}$	7,19	4,28	2,32	

df = degree of freedom, MS = mean of square, ns = not significant, * = significant at α = 0,05; ** = significant at α = 0,01;. $F_{0,05}$ = value of F tabular at α = 0,05; $F_{0,01}$ = value of F tabular at α = 0,01.

Table 2. District performance in organoleptic and physical quality.

Nr	District	Fr/Ar	Fl	At	Ac	Bo	Un	Ba	Cl	Sw	Ov	To	Gb (g)
1	Humbang Hasundutan	8,94	8,68	8,13	8,19	8,08	9,50	8,32	9,50	9,50	8,86	87,69	13,23
2	Simalungun	8,04	7,78	8,03	7,87	7,92	8,25	8,09	9,13	8,00	8,18	81,29	16,32
3	Pakpak Bharat	7,82	7,57	7,84	7,81	7,74	8,13	7,95	9,50	8,38	8,04	80,76	14,39
4	Samosir	7,84	7,71	7,89	7,98	8,02	8,75	7,99	9,13	8,75	7,94	81,99	13,45
5	Dairi	8,93	8,64	8,77	8,83	8,74	9,88	8,83	10,0	9,50	8,73	90,82	15,18
6	Tapanuli Utara	8,16	7,93	8,16	7,97	8,03	8,63	8,10	9,25	8,75	8,34	83,31	14,56
7	Toba Samosir	7,52	7,42	7,64	7,75	7,76	8,13	7,91	9,25	8,50	7,74	79,61	12,76
	$LSD_{0,05}$	0,16	0,28	0,29	0,28	0,13	ns	0,18	ns	0,97	0,35	2,73	1,80
	$LSD_{0,01}$	0,24	0,41	0,43	0,41	0,19	ns	0,26	ns	ns	0,51	4,04	ns

$LSD_{0,05}$ = least significant difference at α = 0,05; $LSD_{0,01}$ = least significant difference at α = 0,01; ns = not significant; Nr = number, Fr/Ar = fragrance/aroma, At = aftertaste, Ac = acidity, Bo = body, Un = uniformity, Ba = balance, Cl = clean cup. Sw = sweetness, Ov = overall, To = total score. Gb = 100 green bean weight

Table 3. Genotypic performance in organoleptic and physical quality.

G	Fr/Ar	Fl	At	Ac	Bo	Un	Ba	Cl	Sw	Ov	To	Gw (g)
Hu1	9,03	8,75	8,25	8,15	8,13	10,00	8,85	10,00	8,50	8,50	88,15	13,02
Hu2	8,93	8,58	7,90	8,03	8,00	10,00	7,98	10,00	10,00	9,03	88,43	13,76
Hu3	8,88	8,80	7,95	8,08	7,95	8,00	8,28	8,00	9,50	8,90	84,33	13,71
Hu4	8,95	8,58	8,43	8,50	8,23	10,00	8,18	10,00	10,00	9,00	89,85	12,42
Si5	7,75	7,53	7,93	7,80	7,90	8,00	8,00	9,00	8,00	8,50	80,40	16,51
Si6	8,18	8,05	8,25	8,10	8,00	8,50	8,13	9,00	8,00	8,00	82,20	16,43
Si7	7,83	7,55	7,95	7,68	7,80	8,50	7,90	9,00	8,00	8,05	80,25	16,60
Si8	8,43	8,00	8,00	7,90	7,98	8,00	8,35	9,50	8,00	8,18	82,33	15,73
PB9	7,53	7,13	7,73	7,63	7,63	7,50	7,85	10,00	8,00	8,00	78,98	14,55
PB10	8,13	7,70	7,88	7,85	7,90	8,00	8,05	10,00	8,00	8,05	81,55	14,08
PB11	7,78	7,58	7,78	7,75	7,55	8,50	7,85	9,00	8,50	8,10	80,38	14,63
PB12	7,85	7,88	7,98	8,00	7,90	8,50	8,05	9,00	9,00	8,00	82,15	14,30
Sa13	7,85	7,73	7,78	7,98	8,03	8,50	7,93	8,50	8,50	7,90	80,68	13,20
Sa14	7,93	7,80	7,90	8,08	7,95	8,50	8,05	9,50	9,00	8,03	82,73	14,40
Sa15	7,63	7,40	7,73	7,95	8,10	9,00	8,00	9,50	9,00	7,80	82,10	12,93
Sa16	7,95	7,93	8,15	7,90	8,00	9,00	7,98	9,00	8,50	8,05	82,45	13,27
Da17	9,03	8,78	8,88	8,90	8,88	10,00	8,95	10,00	8,00	8,55	89,95	13,19
Da18	8,73	8,40	8,53	8,85	8,70	10,00	8,78	10,00	10,00	8,50	90,48	14,74
Da19	8,93	8,65	8,75	8,65	8,50	10,00	8,75	10,00	10,00	8,95	91,18	14,47
Da20	9,03	8,73	8,93	8,90	8,88	9,50	8,83	10,00	10,00	8,90	91,68	18,32
Ta21	8,30	8,13	8,60	8,10	8,05	9,50	7,90	10,00	9,00	8,75	86,33	14,44

Ta22	8,03	7,90	8,08	8,03	8,10	9,00	8,15	9,50	9,00	8,00	83,78	15,05
Ta23	8,33	8,13	8,33	8,10	8,20	8,00	8,08	9,00	8,50	8,63	83,28	14,33
Ta24	8,00	7,55	7,65	7,65	7,75	8,00	8,28	8,50	8,50	8,00	79,88	14,42
To25	7,43	7,38	7,53	7,85	7,90	8,00	7,93	9,50	8,50	7,85	79,85	12,44
To26	7,53	7,33	7,68	7,83	7,70	8,50	7,85	8,50	8,50	7,85	79,25	12,55
To27	7,23	7,33	7,63	7,53	7,75	8,00	7,95	9,50	8,00	7,48	78,00	13,14
To28	7,90	7,65	7,73	7,80	8,08	8,00	7,90	9,50	9,00	7,80	81,35	12,94
$LSD_{0,05}$	0,30	0,30	0,32	0,24	0,25	ns	0,26	ns	0,78	0,31	2,02	0,64
$LSD_{0,01}$	0,40	0,39	ns	ns	0,33	ns	0,34	ns	ns	0,41	2,67	0,85

$LSD_{0,05}$ = least significant difference at $\alpha = 0,05$; $LSD_{0,01}$ = least significant difference at $\alpha = 0,01$; ns = not significant; Nr = number, G = genotype, Fr/Ar = fragrance/aroma, Fl = flavor, At = aftertaste, Ac = acidity, Bo = body, Un = uniformity, Ba = balance, Cl = clean cup. Sw = sweetness, Ov = overall, To = total score. Gw = 100 green bean weight

Table 4. Simple coefficient of correlation (r, above) and coefficient of determination (r^2, below) among organoleptic quality, physical quality and altitude.

	Fl	At	Ac	Bo	Ba	Sw	Ov	To	Gw	Al
Fr/ Ar	0,968 ** 0,937	0,775 ** 0,597	0,788 ** 0,621	0,744 ** 0,553	0,775 ** 0,601	0,573 ** 0,329	0,864 ** 0,746	0,911 ** 0,830	0,142 ns 0,020	0,463 * 0,215
Fl	1	0,786 ** 0,618	0,800 ** 0,640	0,734 ** 0,538	0,756 ** 0,571	0,589 ** 0,347	0,837 ** 0,701	0,902 ** 0,814	0,095 ns 0,009	0,466 * 0,218
At		1	0,884 ** 0,782	0,847 ** 0,718	0,739 ** 0,546	0,417 * 0,174	0,707 ** 0,500	0,864 ** 0,747	0,338 ns 0,114	0,679 ** 0,460
Ac			1	0,943 ** 0,889	0,811 ** 0,658	0,599 * 0,359	0,662 ** 0,438	0,914 ** 0,836	0,159 ns 0,025	0,697 ** 0,486
Bo				1	0,795 ** 0,633	0,510 ** 0,260	0,589 ** 0,347	0,856 ** 0,732	0,199 ns 0,040	0,771 ** 0,594
Ba					1	0,337 ns 0,114	0,514 ** 0,264	0,780 ** 0,609	0,197 ns 0,039	0,633 ** 0,401
Sw						1	0,630 ns 0,397	0,714 ** 0,510	-0,031 ns 0,001	0,307 ns 0,094
Ov							1	0,819 ** 0,670	0,201 ns 0,041	0,317 ns 0,100
To								1	0,116 ns 0,013	0,618 ** 0,381
Gw									1	0,273 ns 0,075

n = 28, * = significant at α 0,05 = 0,374, ** significant at α 0,01 = 0,478; ns = not significant; Fr/Ar = fragrance/aroma, Fl = flavor, At = aftertaste, Ac = acidity, Bo = body, Un = uniformity, Ba = balance, Cl = clean cup. Sw = sweetness, Ov = overall, To = total score. Gb = 100 green bean weight, Al = altitude

4. Discussion

This research revealed that the geographical origins of the coffee affected significantly the organoleptic and physical quality of coffee (Table 2). The genotypes varied significantly in total score of organoleptic quality (Table 3). The same results was found by [6], [12], [13], [14], [15], [7] and [16]. Total score of the organoleptic quality correlated significantly with all parameters of of the organoleptic quality (Table 4). The similar results in some parameters was revealed by [12], [14], [15], [17] and [16]. However, [15] found that total score did not correlate with body. Based on the determination coefficient (r^2 = 0,836), rather than other parameters, acidity was more important for total score. This research showed total score of organoleptic quality did not correlate with green bean weight (Table 4). The same result was reported by [15]. However, [12] found that green bean weight affected organoleptic quality. In contrary to [12] and [15], this research showed that body did not correlate with green bean weight. Total score of organoleptic quality correlated significantly with altitude (Table 4). This result was the same with the research result conducted by [7]. Due to coefficient of determination (r^2 = 0,486), altitude affected more to acidity than to others. The higher elevation of coffee farm, the higher acidity.

5. Conclusion

The origins of Arabica coffee affected the organoleptic quality whereby Arabica coffee from Dairi had the highest total score (90,82). Genotypes were significantly different in organoleptic quality whereby the best total score (89,85 - 91,68) was showed by Da17, Da18, Da19, Da20 and Hu4. Total score of organoleptic quality did not correlate with green bean weight but positively correlated with altitude. Rather than other parameters, acidity was more important for total score (r^2 = 0,836). Altitude affected more to acidity (r^2 = 0,486) than to others. In the future research, Da17, Da18, Da19, Da20 and Hu4 can be used as seed source for coffee production as well as for coffee breeding.

Acknowledment

The authors are thankful to the owner of Goldenways Coffee company for the equipments needed for this research.

References

[1] Ditjenbun 2012 Kebijakan perkebunan dalam pengelolaan perkebunan yang lestari, berpotensi ekonomi dan berkontribusi pada REDD+. Ditjenbun. Jakarta. (Indonesian Language)

[2] Silveira HRDO, Santos MDO, Alves JD, Souza KRDD, Andrade CA and Alves RGGM 2014 Growth effects of water excess on coffee seedlings (Coffea arabica L.). Acta Scientiarum. Agronomy Maringá, v. 36, n. 2, p. 211-218. http://dx.doi.org/10.4025/actasciagron.v36i2.17557

[3] Leonel LE and Philippe V 2007 Effect of altitude, shade, yield and fertilization on coffee quality (Coffea Arabica L. var. Caturra) produced in agroforestry systems of the Northern Central Zones of Nicaragua. In: 2nd International Symposium on Multi-Strata Agroforestry Systems with Perrennial Crops. Costa Rica. DOI: 10.13140/RG.2.1.4689.1289.

[4] Gichimu BM 2013 Arabica coffee breeding : challenges posed by climate change. In: 10th AFCA Conference, Kampal, Uganda. https://www.researchgate.net/publication/281626916_Coffee_Growing_in_Kenya_Climate_Change_Challenges_and_their_Mitigation

[5] Kufa T and Burkhard J 2013 Studies on Root Growth on Coffea Arabica Populations and its Implication for Sustainable Management of Natural Forests. J. Agric. Crop Sci., Vol 1(1), pp.1-9.

[6] Yusianto, Hulupi R, Sulistyowati, Mawardi S and Ismayadi C 2005 Sifat Fisiko-Kimia dan Cita Rasa Beberapa Varietas Kopi Arabika *Pelita Perkebunan*, 21(3):200-222 http://iccri.net/download/Pelita%20Perkebunan/vol_21_no_3_desember_2005/Sifat%20FisikoKimia%20dan%20Cita%20Rasa%20Beberapa%20Varietas%20Kopi%20Arabika.pdf (Indonesian Language)

[7 Wahyuni E, Karim A and Anhar A 2013 Analisis Citarasa Kopi Arabika Organik pada Beberapa Ketinggian Tempat dan Cara Pengolahannya di Datararan Tinggi *Gayo Jurnal Manajemen Sumberdaya Lahan* http://www.jurnal.unsyiah.ac.id/MSDL/article/view/2199/2153 (Indonesian Language)

[8] Yusianto 2008 Pengolahan Basah Penggerbusan Kering (Model Pengolahan Hindia Barat/WIB) in Mawardi S, Hulupi R, Wibawa A, Wiryadiputra S and Yusianto (eds.) Panduan Budidaya dan Pengolahan Kopi Arabica Gayo, p142-147, (PPKKI, Jember) (Indonesian Language)

[9] SCAA 2009 Specialty Coffee Association of America Cupping Protocols (Revised: November 21, 2009) . https://www.scaa.org/PDF/PR%20%20CUPPING%20PROTOCOLS%20V.21NOV2009A.pdf

[10] Quinn GP and Keough MJ 2002 Experimental design and data analysis for biologists (Cambridge University Press NY) http://www2.ib.unicamp.br/profs/fsantos/apostilas/Quinn%20&%20Keough.pdf

[11] Gomez AK and Gomez AA 1984 Statistical procedures for agricultural research (John Wiley & Sons NY) http://pdf.usaid.gov/pdf_docs/PNAAR208.pdf.

[12] Salla M H 2009 Influence of Genotype, Location, Processing Method on the Quality Coffee (Coffea arabica L.) (MSc Thesis Hassawa University, Hassawa, Ethiophia) https://cgspace.cgiar.org/bitstream/handle/10568/669/Thesis_HailemichaelInflurence.pdf?sequence=1

[13] Abubakar Y , Karim A and Fahlufi F 2011 Flavor of arabica coffee grown in Gayo Palteau as affected by varieties and processing techniques *Proceedings of The Annual International Conference Syiah Kuala University 2011 Banda Aceh, Indonesia. November 29-30* http://www.jurnal.unsyiah.ac.id/AICS-SciEng/article/view/1915/1843

[14] Kitila O, Alamerew S, Kufa T and Garedew W 2011 Organoleptic Characterization of Some Limu Coffee (Coffea arabica L.) Germplasm at Agaro, Southwestern Ethiopia. Int. J. Agric. Res., 6(7): 537-549. DOI: 10.3923/ijar.2011.537.549

[15] Tessema A, Alamerew S, Kufa T and Garedew W 2011 Variability and association of quality and biochemical attributes in some promising Coffea arabica germplasm collection in Southwestern Ethiopia *Int. J. Plant Breed. Genet.* 5(4): 302-316. DOI: 10.3923/ijpbg.2011.302.316.

[16] Kathurima CW, Gichimu BM, Kenji GM, Muhoho SM and Boulanger R 2015 Evaluation of beverage quality and green bean physical characteristics of selected Arabica coffee genotypes in Kenya *African Journal of Food Science Vol 3.(11) :* 365-371 https://agritrop.cirad.fr/552843/1/document_552843.pdf

[17] Odeny D, Chemining'wa G, Shibairo S and Kathurima C 2015 Sensory Attributes of Coffee under Different Shade Regimes and Levels of Management *Food Science and Quality Management* Vol.46 : 19-26 http://www.iiste.org/Journals/index.php/FSQM/article/view/27544

The role of oil palm companies in Indonesia as a nation's competitive advantage

N Tampubolon[1] and M Pasaribu[2]
[1]Bina Nusantara University (Binus), Jakarta, Indonesia.
[2]Indonesia University, Jakarta, Indonesia.

nurdin.tamp@gmail.com, manerep_kupang@yahoo.co.id

Abstract. Indonesia is the largest world Crude Palm Oil (CPO) producer with Malaysia in second place. This agricultural commodity has become a chief Indonesian foreign exchange earner behind fossil fuel exports. In 2016, the export value of this commodity reached USD 17.8 billion. Historically, Malaysia has been more advanced in the CPO delivery, which can be explained by the general companies'environment of management, technological advancement and engineering, human resource skills and superior external support, such as road infrastructure, regulations and research& development by the industry and government. It is clear from data that the Indonesian production is disadvantaged by a wide range of inefficiencies. They range from limited technology and production management skills to limitedcultivation advancement. Applications of technical improvements are desired to enhance the national competitive advantage to the next level. This paper is an exploration of the current management culture and to consider a strategic management model that would be the most appropriate for Indonesia and would enourage high end technology and plantation management. A gradual level improvement would enable Indonesia to compete on a global scale as an industry leader in the palm oil market.

1. Introduction

Palm oil is produced from the fruits of the African oil palm (*Elaeis guineensis*) which were introduced in plantationsearly last century. As many Indonesians from individual farmers to companies are depending on the well-being of the industry, it is important to the engineering industry to keep abreast of the latest development to keep the industry efficient in the competitive vegetable oil world market. The sustainability of the palm oil industry can be viewed from several perspectives. The technical engineered perspective is touching all of the elements that are often integrated in the business strategic plan of the oil palm industry. The economic, environmental and social elements are the long term strategic management plan emphasizes sustainability of the firm and revenue flow. Advanced engineering can increase the economic output, assist in the environmental responsibility of the company and be part of the social undertaking of the industry. An implementation of a strategic competitiveness plan is currently jointly undertaken by the oil palm industry and the government of Malaysia. The Indonesian government has not been as advanced in this perspective, leaving the development largely to the private sector. The future challenge of individual palm oil producers remains that their production cost is controlled in order to compete effectively with the competitors in the industry. The focus on reducing costs is based on the idea that the business of CPO is very dependent on the market situation and can fluctuate greatly from year to year. Low production cost will ensure the company remains economically viable in a low global market price environment without being threatened by possible economic failure. This strategy would strongly support the company in competing on a global scale which market is predicted to grow and be more competitive in the future [1].

2. Palm oil development

The expansion of palm oil cultivation has been very rapid in the last few decades in Indonesia. This is partially due to the geographical location and weather pattern near the equator, which is very suitable for palm oil cultivation. The Government of Indonesia has also viewed the development of palm oil plantations as a solution to overcome rural poverty by creating job opportunities and develop rural infrastructure in remote areas. Since the 1980s, palm oil cultivation has grown exponentially, either in plantation estates under corporate management or independent planters' small holding management.

According to Bastion [2] palm oil productivity could normally reach a yield of approximately 3.3 tons/ha/year. However, in Indonesia the reality is that palm oil companies only produce 2.89 tons/ha/year. Despite Indonesia has become the world's leading palm oil producer and exporter, Indonesia struggles in applied technology to compete. The Palm Oil Board of Malaysia which is the main government research body, owns 87 US patents. The Indonesian counterpart, the Indonesia Oil Palm Research Institute has not one patent registered in its name [3].

In addition, the current global market has demanded that the CPO has to be produced in a sustainable manner by the industry, it means that cultivation and processing has to improve in Indonesia. The processing requires many engineering skill inputs with a focus on reliableequipment, low maintenance, energy efficiency, easy to use and to make the industry sustainable in the future. This directs the palm oil industry to meet the global cultivation and processing benchmarks to be competitive in the global export market. The palm oil industry in Indonesia ought to include the application of sustainable development and environmentally friendly palm oil agricultural management that complies with Indonesian Sustainable Palm Oil (ISPO) and Round Table Sustainable Palm Oil (RSPO) standards [4].

3. Competition and engineering

The government ought to create an environment in which firms can reach out to the competitive advantages of their established industries by introducing advanced technologies, improve methods and by penetrating more advanced qualities in their production segments [1].

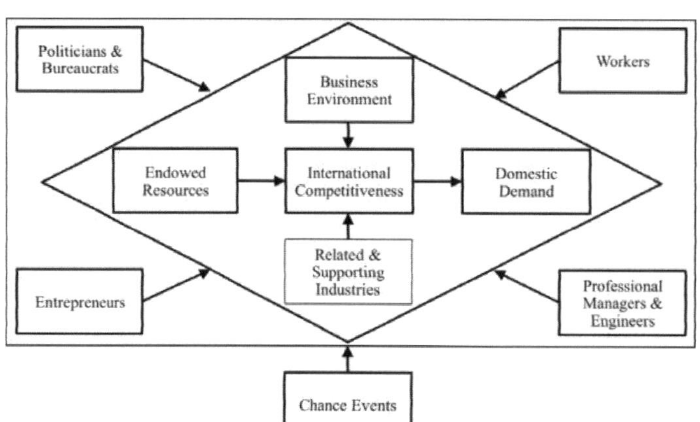

Figure 1. A New Paradigm of International Competitiveness (The Nine Factor Model)

What is competitiveness? A nation or standard of living (wealth) is determined by the productivity in which it uses its human resources, financial capital and natural resources [1]. The appropriate definition of competitiveness is productivity. Productivity depends both on the value of products and services (e.g. Uniqueness, quality) as well as the mode of efficiency in which they are produced [5]. It requires the development of a strong supporting engineering industry to create a high level of efficiency. From a managementperspective, productivity should be measured in terms of the value (revenue) produced, per unit of labour or capital, not just the volume. It is not, what industries or nation or region it competes in, to measure prosperity, but how firms compete in those industries. Productivity in a nation or region is a reflection of what both domestic and foreign firms choose to do in that particular location. The productivity of local industries is of fundamental importance to competitiveness, not just the trading sectors of the businesses. Nations and regions compete in offering the most productive environment for businesses and industries [6].

What are the Forms of Competitive Advantage? Absolute advantage is the ability to produce something more efficiently than any other country [1]. Furthermore, comparative advantage is the ability to produce some products more efficiently or better than other products. The national's competitive advantage is an international competitive advantage stemming from a combination of factor conditions; demand conditions, related and supporting industries, firm strategies, structures and rivalries [6].

In order to assess Indonesia's international competitiveness, two major considerations should be addressed; e.g. the business and the government services utility function. A new model of international competitiveness for less developed countries, named "The Nine Factor Model" is specified. This is used to provide a new model to analyse the palm oil industry in Indonesia. Government and businesses have to utilise introduced capital and employ cutting edge technology which are developed in other countries if not available in Indonesia or create resources and other factors which influence economic growth from their initial stages. Till now the principal key engine of Indonesia's economic growth has been the exploitation of natural resources and the deployment of a low skilled work force. It has been lacking to tackle high skill advanced production.

Indonesia's population can be grouped into four sectors of deployment. Firstly, many unskilled, semiskilled and a limited number of members in the highly skilled workforce. Secondly, local and national politicians and thirdly, bureaucrats, who formulate and implement economic plans. The fourth group is a diverse number of entrepreneurs who make investment decisions despite high risks, highly skilled professional managers who are in charge of operations and skilled engineers who implement new technologies [6]. There are four physical determinants of international competitiveness. Namely; endowed resources, the business environment, related and supporting industries and a domestic demand. In another skill classification there are four human factors, namely the general work force, politicians and bureaucrats, entrepreneurs and professional managers and engineers. External chance events should be noted as the ninth factor (See Figure 1).

The different between the new model and Porter's diamond model is to be found as much in the division of factors as well as in the addition of new variables. The diamond model included both natural resources and labour in factor conditions, but the nine factors model places natural resources under endowed resources, while labour is included within the category of workers. A detailed investigation of the nine factors of international competitiveness is needed. The physical factor consists of endowed resources, business environment, related and supporting industries, and domestic demand. While the human factors be mobilised in the above-mentioned physical factors. People combined and arranged with the physical factors can aim to gain an international competitiveness. Workers, politicians and bureaucrats, entrepreneurs, and professional managers and engineers have to be considered in this remodelled structure [6].

The business and engineering environment should be viewed at the level of the nation's provision of favourable conditions to industry and company formation. At the national's level, there are visible and invisible components. They firstly include physical provisions such as roads, ports, telecommunication and other forms of provided infrastructure. The second component is concerned with the people's psychological acceptance of competitive values and market mechanisms. A commitment of producers, merchants, consumers, managers, engineers and other participants in the economy to the legitimacy and obligations in commercial based arrangements and notion of credit provision is essential. At an industrial level, the business environment is determined by the number and size of competitors, the type and height of entry barriers, the degree of product differentiation, and other factors shaping the nature of rivalry and the level of economic activities. At a company level, the strategy and organisation of businesses and the attitudes and behaviour of individuals and groups ethics within enterprises are a major consideration.

4. Stages of processing with high engineering input

Research and development in engineering are needed in many areas from chemical to mechanical engineering. Large scale fully mechanised processing is demanded in a competitive environment resulting from a sequence of processing steps designed to extract from a harvested oil palm bunch to a high yield of a product of acceptable quality for the international edible oil trade. The palm oil processing takes place in the oil palm mill which are located a short distance from the plantations. The processing needs to take place shortly after the fruit bunches are harvested as the fruits deterioraterapidly. Figure 2 is a visual simplified sequence of the stages of required processing.

The palm oil extraction process involves the harvesting of fresh fruit bunches from the plantation. First of all the oil palm bunches requires to be sterilised and to be softened. Next step is to thresh the fruit bunches to free the palm fruit, which is being mashed before the crude palm oil is pressed. The crude oil requires further treated to purify and dry it for storage and transportation to the nearest harbour. Most of the CPO is exported. Below is a table of digital data on production, export quantity and value in US dollars.

Table 1. Indonesian Palm Oil Production and Export Statistics

	2011	2012	2013	2014	2015	2016
Production (million tons)	23.5	26.5	30.0	31.5	32.5	
Export (million tons)	17.6	18.2	22.4	21.7	26.2	25.7
Export (in USD billion)	20.2	21.6	20.6	21.1	16.5	17.8

Source Indonesian Palm Oil Producers Association (GAPKI) and Ministry of Agriculture

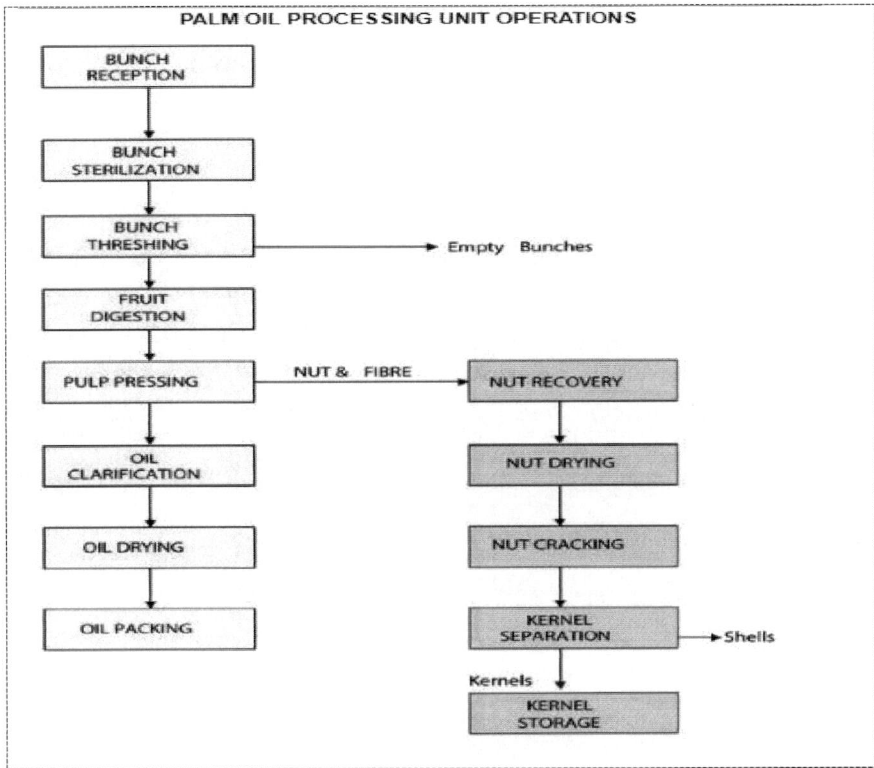

Figure 2. Processing stages requiring highly engineered equipment.
Source: http://www.fao.org

5. Conclusion

Indonesia is the largest producer, while Malaysia is the second largest producer of CPO. The oil palm industry is a highly competitive industry where the commodity prices are set by the vegetable oil world market price. Individual CPO producers are aware that they have to protect their ventures by controlling the cost of production as they unable to influence the market price.The industry's challenge is to become a market leader by following the worlds best practices and to make use of Indonesia's comparative advantage in CPO production. The New Paradigm of International Competitiveness (The Nine Factor Model) indicates the engineering input is vital in the overall management plan for CPO production. Currently it shows the Palm Oil Board of Malaysia owns 87 US patents, versus the Indonesia Oil Palm Research Institute which has no patents registered in its name. It indicates that research, including engineering research at government level is disappointing in Indonesia, depending mainly on the private sector. It is up to the individual enterprise to perform best with the given environment. Engineering will play an important role to keep abreast of the latest developments in technical and chemical advances in the their field.

Acknowledgments

The author wishes to acknowledge the generous assistance and encouragement from colleagues and the Indonesian Palm Oil Producers Association (GAPKI).

IOP Conf. Series: Materials Science and Engineering **237** (2017) 012036 doi:10.1088/1757-899X/237/1/012036

References

[1]. Porter M.(1990). *The Competitive Advantage of Nations.* New York: The Free Press.
[2]. Basiron Y and Chan-Cook W (2004). The Oil Palm and Its Sustainability, *Journal of Palm Oil Research*, **16** (1), 1-10
[3]. Bruno L (2016). http://www.ehs.org.uk/press/the-palm-oil-industry-technological-change-global-competition-social-and-environmental-impact. accessed 27 5 2017
[4]. BasironY(2007)*Eur. JLipid Sci. Technol.* **109** p 289–295DOI 10.1002/ejlt.200600223
[5]. Tampubolon N (2017) The role of oil palm companies in Indonesia as a nation's competitive advantage, *3rd IIFAS Proceedings*, UGM, Jogjakarta
[6]. Cho, Dong-Sung and Moon, Hwy-Chang. 2002. *Asia-Pacific Business Series* – Vol. 2. *From Adam Smith to Michael Porter, Evolution of Competitiveness Theory*. Singapore: World Scientific

IOP Conf. Series: Materials Science and Engineering **237** (2017) 012037 doi:10.1088/1757-899X/237/1/012037

A field survey on coffee beans drying methods of Indonesian small holder farmers

Parulian Siagian[1,2], Eko Y. Setyawan[1,3], Tumiur Gultom[4], Farel H. Napitupulu[1], HimsarAmbarita[1,*]
[1]Department of Mechanical Engineering, Faculty of Engineering, Universitas Sumatera Utara, Jl. Almamater Kampus USU Medan 20155, Indonesia
[2]Mechanical Engineering, Universitas HKBP Nommensen, Medan, Indonesia
[3]Mechanical Engineering, National Institute of Technology Malang, Indonesia
[4]Biology Department, Universitas Negeri Medan, Jl. William Iskandar Medan, 20222, Indonesia

[*] himsar@usu.ac.id

Abstract. Drying agricultural product is a post-harvest process that consumes significant energy. It can affect the quality of the product. This paper deals with literature review and field survey of drying methods of coffee beans of Indonesia farmers. The objective is to supply the necessary information on developing continuous solar drier. The results show that intermittent characteristic of sun drying results in a better quality of coffee beans in comparison with constant convective drying. In order to use energy efficiently, the drying process should be divided into several stages. In the first stage when the moist content is high, higher drying air temperature is more effective. After this step, where the moist content is low, lower drying air temperature is better. The field survey of drying coffee beans in Sumatera Utara province reveals that the used drying process is very traditional. It can be divided into two modes and depend on the coffee beans type. The Arabica coffee is firstly fermented and dried to moisture content of 80% using sun drying method, then followed by Green House model of drying up to moisture content about 12%. The latter typically spends 3 days of drying time. On the other hand, The Robusta coffee is dried by exposing to the sun directly without any treatment. After the coffee beans dried follow by peeled process. These findings can be considered to develop a continuous solar drying that suitable for coffee beans drying.

1. Introduction

Coffee (*CoffeaSp*) is one of the famous agricultural products in the world. It is easily find in our daily life. According to International Coffee Organization [1], in 2016 the total coffee production in the world is about 9.09 million of tons. Indonesia is known as the fourth largest coffee producer in the world after Brazil, Vietnam, and Colombia. According to Ministry of Agricultural Republic of Indonesia, in 2016 Indonesia produces 667,655 tons of coffee [2]. Indonesia people consumed around 270,000 tons of coffee. The Indonesian coffee is mainly exported to USA, Germany, Japan, Italy, etc. Coffee production has a significant effect for economy growth of Indonesia. The main player of the coffee production in this country is small holder farmer. The small holder farmers produce up to about 95% of Indonesia coffee. Sumatera Utara province is known as a coffee producer in Indonesia. Not only in quantity but it also knew for the good quality of coffee. The arabica coffee from Mandailing, Lintong, Sidikalang regencies are known as arabica coffee of Indonesia with special quality.

Producing coffee from raw beans into roasted beans or coffee powder can be divided into several steep. In general, small holder coffee farmers practice the following coffee processing. The coffee cherry is piled-up usually for 2 to 10 days and followed by drying for 5 days. After this, it is heaped and dried again. The last drying process usually takes 10 days to meet moisture content of 11% to 20%. These steps show that drying process is a significant step in coffee process. Even though Indonesia is one of the biggest of coffee producers in the world, however, study on the drying process of coffee beans in Indonesia is very limited.

In this study, we perform a literature review on drying process of coffee beans and field survey of drying process practiced by small holder coffee farmers in Sumatera Utara province of Indonesia. As a note, our final goal is to develop continuous solar drier with integrated hybrid solar collector and desiccant as thermal energy storage. The developed drier will be tested for drying coffee. Thus, the present study is a first step for this study. The results are expected to supply the necessary information on developing the continuous solar drier for coffee beans.

2. Method

As stated above this studies consist of two types of review. The first review is a literature review on the reported study on drying coffee. The second review is a field visit to review the drying methods of Indonesian small holder coffee farmers. In the first review, peer reviewed reported studies related to the drying coffee will be reviewed. The review includes drying method, quality, heat and mass transfer mechanism. In the second review, a field survey is carried out in several regencies in Sumatera Utara province. The survey includes interviews to explore the drying time and barriers.

3. Results and Discussions

3.1. Drying coffee beans

Drying is a process to reduce the moist content of an object dried. This process is known and has been used by mankind since ancient era. Drying process of dried grape using solar energy can be tracked has been done by Greek in 1490 BC [3]. Drying is a simultaneous heat and mass transfer problems between drying media and the object as shown in Figure 1. In brief, the moist inside the object diffuses to the surface due to concentration difference. In the fluid-solid interface, the water evaporates into vapour and it will flow out with drying air. In addition, evaporation needs an amount of energy to supply latent heat of phase change of water into vapour. The need of energy for evaporation will be a driving force for heat transfer between the object and drying air. This fact shows that there are three different mechanisms occur at the same time, they are heat transfer, mass transfer and phase change. This process is also known as a significant energy consuming process [4].

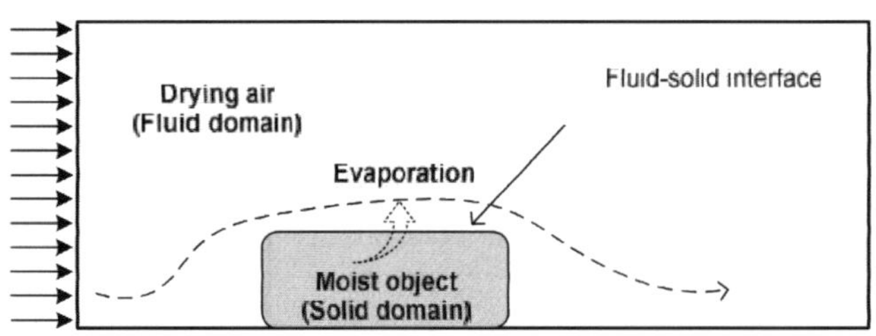

Figure 1. Heat and mass transfer mechanism in drying

Study on drying coffee had been reported by several researchers. Livramento et al [5] reported a study on the proteomic analysis of coffee grains exposed to different drying process. The objective of the study is to analyse the proteomic profile and identify differentially abundant proteins in coffee arabica L. grains subjected to two different drying conditions. The samples of coffee grains were collected from the municipality of Bom Sucesso, state of Minas Gerais, Brazil, from a plot at 1100 m altitude. Two different drying methods are natural coffee and pulped coffee dried with mechanical hot air drying at 60°C. It is observed that fruits dried in a dryer at 60°C showed an altered proteomic profile, with a reduction in the most abundant proteins compared to those yard-dried grains. The results show that post-harvest processes of the coffee quality are related to changes in protein abundance, indicating that proteomic analysis may be effective in the identification of biochemical changes in coffee grains subjected to different post-harvest processes. The study concluded that drying has been considered as one of the factors that cause impact on the final beverage quality. Ishwarya and Anandharamakrishnan [6] reported a study on spray-freeze-drying (SFD) approach for soluble coffee processing and its effect on quality characteristics. The SFD method is compared with spray-dried (SD) and freeze-dried (FD) methods. The analysed qualities are aroma, volatile, solubility, morphology, flow properties and colour measurement. The results show that different drying technique can affect the quality of the dried coffee. In their study, SFD technique can be potentially employed for the production of soluble coffee with improved product characteristics. Keinwachter and Selmar [7] investigated the influence of drying on the content of sugars in wet processed green Arabica coffees. It is well known that sun drying yields much better green coffee quality than machine drying. However, this conclusion was made based only on traditional knowledge and experience. Thus, the objective of their research is to elucidate the reasons for the putative quality differences between sun and machine dried green coffees. Two different drying methods are compared. The first method is continuous drying and the second method is pause method to mimic day-and-night-rhythm. The comparison quality is carbohydrates content in dried coffee. Dong et al [8] reported their study on the effect of different drying techniques on bioactive components, fatty acid composition, and volatile profile of Robusta coffee beans. There tested five different drying methods. They are room-temperature drying (RTD), solar-drying (SD), heat-pump drying (HPD), hot-air drying (HAD), and freeze drying (FD). The compared qualities are bioactive components, fatty acid composition, and volatile compound profile. The results showed that drying methods affect the quality of the dried coffee beans. All drying methods show significant differences in the contents of organic acids, total UNSFAs, and amino acids. However, the other parameters such as SFA, caffeine, and trigonelline content are not affected by drying methods. The HAD and HPD drying methods resulted in a significant alteration in the dried coffee bean volatile compound profile likely. This is because of the high temperatures and intensive ventilation treatments. Overall, the different metabolic responses to the drying methods resulted in the multiple metabolic association and clustering on the planes of PCA and heat map. The FD method is a good drying method that can be utilized to preserve fat, organic acids, and monounsaturated fatty acids. The HAD method is ideal for retaining polyunsaturated fatty acids and amino acids, whereas the HPD method produced the largest number of volatiles as well as the highest content of saturated fatty acids and ratio of unsaturated fatty acids to saturated fatty acids. The study suggested that the conventional drying methods can be effectively replaced by HAD or HPD methods in the future. Considering its drying efficiency, good quality retention and production cost, the HPD has the broadest market prospect for potential application.

Suzihaque and Driscoll [9] investigated the effect of solar radiation, buoyancy of air flow and optimization study of coffee drying in heat recovery drier. The objective was to develop an efficient heat recovery dryer and to investigate its performance so as to optimize design parameters.

Hernandez-Diaz et al [10] reported a study on modelling heat and mass transfer during drying of green coffee beans using prolate spheroidal geometry. Three-dimensional governing equation in prolate spheroidal coordinate system was solved numerically. The objective was to describe the drying of green coffee beans. The results showed that the method can provide valuable information about the moisture distribution profiles inside the grain during drying. It is suggested to use the model to design the optimal design of coffee driers. Burmester and Eggers [11] performed experimental work to investigate heat and mass transfer during the coffee drying process. The objective is to determine transport coefficient of the coffee beans during drying process. The experiments were performed using a hot air drying where the temperature and humidity were controlled. The results show that drying rate is mainly influenced by the temperature of the drying gas. Increasing drying temperature can be used to enhance efficiency of the process. However, sensitivity of the coffee beans should be taken into account. An increase of drying medium velocity does have an influence on the process, as long as drying medium becomes saturated. Lower relative air humidity has small impact on drying rate, but strongly affect the equilibrium humidity of the coffee. In order to optimize the drying process, the application of higher gas temperature at the beginning of the process and lower temperature at the end of the drying process is recommended. Putranto et al [12] proposed reaction engineering (REA) approach to develop mathematical modelling of intermittent and convective drying of coffee. Results show that the REA models the drying kinetics of coffee with well agreement between predicted and experimental data. The comparison with diffusion-based model, it is claimed the REA perform comparably or even better. Nilnont et al [13] performed finite element simulation for coffee drying to describe moisture diffusivity, shrinkage, equilibrium moisture content. It was shown that the mean diffusivity value of the coffee bean is 1.173×10^{-10} m^2/s. The shrinkage of coffee bean is a linear function of moisture removal.

Those literatures reveal that the focus of the researchers can be divided into two groups. The first group focuses on the investigation on the effect of drying method to the quality of the dried coffee bean. The qualities of dried coffee beans can be expressed in several parameters such as qualities bioactive components, fatty acid composition, volatile compound profile, sugar content, etc. It is clear that drying method strongly affect the quality of coffee beans. Another valuable information can be drawn here is that the intermittent characteristic of sun drying results in a better quality in comparison with constant convective drying. This information can be used to develop an optimum sun drier. However, there is a very limited study on the effect of solar drier and the quality of the coffee. The second group of literatures focuses on the numerical solution to the heat and mass transfer during drying process. As a note, drying process consume a huge amount of energy to overcome the latent heat during evaporation. Thus, optimization of the drying process is extremely needed to develop an energy efficient drier. This is the main background to the second focus of those studies. The valuable information can be drawn from the second group is that drying process can be divided into several stages. In the first stage when the moist content is high, higher drying air temperature is more effective. For the steps where, the moist content is low, lower drying air temperature is better.

3.2 Drying methods of Indonesia smallholder coffee farmers

In this study, a field survey of drying methods that are used by Indonesian small holder farmers has been carried out. The survey is conducted in three regencies in Sumatera Utara province which are known as coffee producers. They are Dairi regency known as centre for "Kopi Sidikalang", Humbang Hasundutan regency known for "Kopi-Lintong", and Toba Samosir regency.

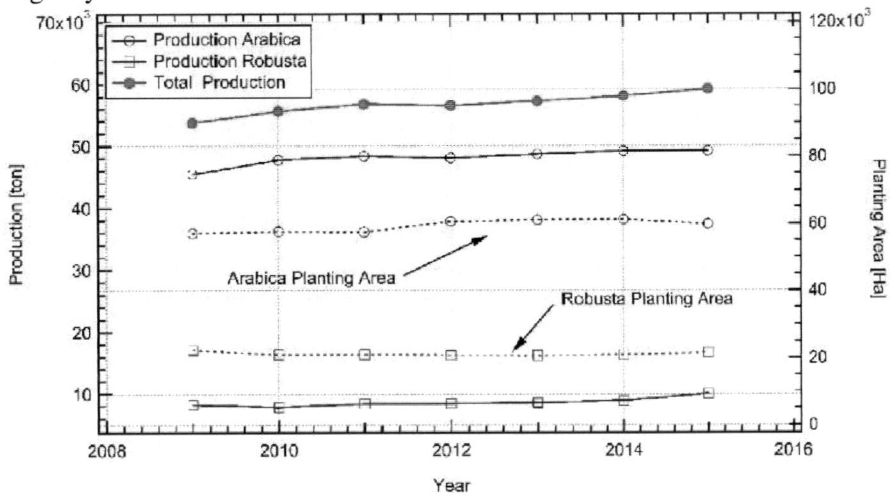

Figure 2. Production and planting area of small holder coffee farmers in Sumatera Utara province

Figure 2 shows production and planting area of small holder coffee farmers in Sumatera Utara province. It can be seen that the production and planting area of coffee in this province are almost flat. In other word, there is no significant change in production and planting. There are two types of coffee plants are found in Sumatera Utara province, Arabica and Robusta coffees. The production of Arabica coffee is the higher than Robusta coffee. In year 2015, the production of Arabica coffee and Robusta coffee are 49,176 ton and 10,022 ton, respectively. The productivity of Arabica coffee is higher than Robusta coffee. During the year 2009 to 2015, the productivity of Arabica coffee varies from 0.79 ton/Ha to 0.84 ton/Ha. On the other hand, the productivity of Robusta coffee varies from 0.37 ton/Ha to 0.46 ton/Ha. This productivity strongly affected by weather conditions during the production year. These values reveal that Arabica coffee is more productive than Robusta coffee. The ratio of Arabica coffee and Robusta coffee productivity is about 50%.

In the field survey in those three regencies, there is a clear different of drying method found in the farmers. The drying method and treatment depend on the type of the coffee. The data was collected from 20 farmers which are located in 5 villages (or Desa). They are Desa Sosor Lontung Kec.Sidikalang (Dairi regency), DesaLintong and Desa Sigumpar (Humbang Hasundutan regency), Desa Tampahan and Desa Parsoburan Tengah (Toba Samosir regency). The survey reveals that drying methods of coffee in this regency is categorized as a traditional method. It is sun drying method. There are several level drying methods are found. In the farmer level, for the Arabica coffee the used method is fully washed and the drying is performed after fermentation. In this step, the target of moisture content is only 80%. The farmers then sell the coffee beans to the dealer. In dealers, Green House drier is used to dry the coffee from moisture content of 80% to about 12%. The drying time usually takes 3 days

IOP Conf. Series: Materials Science and Engineering **237** (2017) 012037 doi:10.1088/1757-899X/237/1/012037

drying time. On the other hand, the Robusta coffee is treated differently as found in Sidikalang regency. The Robusta coffee is dried by exposing to the sun directly without any treatment. After the coffee beans dried followed by peeled process. The typical of drying process of coffee in Sumatera Utara province is shown in Figure 3.

(a) Direct sun drying (b) Green House drying

Figure 3. Coffee beans drying method found in Sumatera Utara province

4. Conclusions

Literature review and field review of drying method of coffee in Indonesia small holder farmers have been performed. In the literature review it was shown that drying method influences the quality of cocoa beans. The intermittent characteristic of sun drying results in a better quality in comparison with constant convective drying. However, there is a very limited study on the effect of solar drier and the quality of the coffee. The numerical solution review shows that in order to use energy efficiently, the drying process should be divided into several stages. In the first stage when the moist content is high, higher drying air temperature is more effective. For the steps where, the moist content is low, lower drying air temperature is better. The field survey of drying coffee in Sumatera Utara province reveals that drying process is very traditional. It can be divided into two types depend on the coffee beans. The Arabica coffee is firstly fermented and dried to moisture content of 80% using sun drying method. Then followed by Green House model of drying up to moisture content about 12% and it typically spends 3 days of drying time. On the other hand, The Robusta coffee is dried by exposing to the sun directly without any treatment. After the coffee beans dried followed by peeled process.

References

[1] ICO, *Internationa Coffee Organization* (www.ico.org) Accessed on June 26, 2017.
[2] Ministry of Agricultural of Indonesia, *The crop estate statistics of Indonesia*, Jakarta, 2015.
[3] SharmaAK, PG Adulse 2007*Raisin production in India*. NRC for Grapes, Pune.
[4] Dina SF, Ambarita H, Napitupulu FH, Kawai H 2015 *Cases Studies in Thermal Engineering*5 32-40.
[5] Kalinka Gabriella do Livramento, Flavio Meira Borem, Anderson Cleiton Jose, Agenor Valadares Santos, Derlan Einstein do Livramento, Jose Donizeti Alves, Lucianno Vilela Paiva, 2017 *Food Chemistry* 221. 1874-1882.
[6] S. Padma Iswarya, C Anandharamakrishan, 2015. *Journals of Food Engineering* .149. 171-180
[7] Maik Kleinwacter, Dirk Selmar.2010. *Food Chemistry*. 119.500-504

[8] Weinjiang Dong, Rongsuo Hu, Zhong Chu, Jian Ping Zhan, Lehe Tan.2017. *Food Chemistry* 234. 121-130
[9] M.U.H. Suzihaque, Robert Driscoll. 2016. *4th International Conference on Process Engineering and Advance Materials.* 148. 812-822
[10] W.N. Hernandez-Diaz, I.I. Ruiz-Lovez, M.A. Salgado-Cervantes, G.C. Rodriguez-Jimenes M.A , Garcia-Alvarado.2008 *Journal of Food Engineering* .86. 1-9
[11] Katrin Burmester, Rudolf Eggers. 2010 *Journal Food Engineering*. 99. 430-436
[12] Aditya Putranto, Xiao Dong Chen, Zongyuan Xiao, Paula. Webley. 2011. *Journal of Food Engineering* .105.638-646
[13] W. Nilnont, S. Thepa, S. Jinjai, N. Kasayapanand, C. Thamrongmas, B.K. Bala.2011. *Food and Bioproducts Processing.*

Influence of heating rate and temperature on austenite grain size during reheating steel

Richard A.M. Napitupulu
Mechanical Engineering Department, Universitas HKBP Nommensen, Medan, 20234, Indonesia

richard_alf@yahoo.com

Abstract. Controlling the final microstructure is one effective way to get HSLA steel with good mechanical properties. The structure of the desired item on the final microstructure depends on the initial grain size formed during the initial heating process, where to get super fine ferrite grains, it should form the initial austenite grain smooth during the heating process. Austenite grain size at the beginning of the heating process is important in order to obtain the size of the final microstructure that provides maximum mechanical properties. In this study, HSLA steel reheated to a temperature of 960°C, 1060°C and 1120°C with holding time variation of 10, 30 and 60 minutes at a heating rate of 5°C/minute, 7.5°C/minute and 10°C/minute, then water quenching. The austenite saw by using optic microscope and count by ASTM E112 method. From the results it is concluded that there is a relationship between temperatures interrelated heating, heating rate and holding time on the growth of austenite grain. The higher the temperature, the heating occur austenite grain size. While the most optimal results obtained for reheated temperature 1060°C with a heating rate 7.5°C/minute and the heating temperature 1120°C with a heating rate 5°C/minute.

1. Introduction

Controlling of microstructure is one of effective ways to get the alloyed steels with a good mechanical properties[1]. Grain structure desired in the end depend on grain microstructure formed during the heating process and the beginning of its deformation, where to get a ultrafine ferrite grain, it must be formed small austenite under reheating process [1, 2].

The research that originated from prior austenite where austenite grain size was reheating before deformed, has an important role where the prior austenite grain size will determine the grain size after recrystallization and grain growth after the steel is deformed [3,4]. If we examine first the heating of alloyed steel, the austenite grain coarsening has to be considered an important process for grain size development. The austenite grain size represents the initial grain size for the subsequent rolling process. According to A Kern [5] the influence factors for grain size evolution during heating steel are chemical composition, solution behavior of carbonitride, heating time and temperature. According to T.A. Kop, et. all, the ferrite-start temperature depends both on the manganese content and on the amount of niobium in solution [6]. The effect of deoxidants on size distribution of particles and austenite grains has been discussed by Karasev based on the relationship between surface area of particles and that of grains per unit volume of metal [7]. Most of the research based on the investigation of grain growth kinetic under isothermal annealed condition based on Beck formula according to the empirical equation $D = kt^m$, where D is the average grain size at time t, t is holding time, and m (the time exponent) and k are temperature-dependent parameters [8].

During reheating the steel, the effects of heating rate on the precipitate distribution at the onset of grain growth will depend on the initial precipitate size, amount of solute in the matrix, and solute diffusivities, along with the time–temperature history controlled by the heating rate [9]. Further, grain growth in non-isothermal systems can be a very complex process, as recently summarized by Mishra and DebRoy [10]. For example, even in the classical grain growth experiments austenite grain growth kinetics at the holding temperature may depend on the heating rate employed to reach this temperature.

Currently in industrial applications, the grain size of the prior austenite as the basis for calculation is only obtained through a trial-and-error by heating the steel to a reheating temperature and hold in a several time under isothermal condition according to the kinetic grain growth based on Beck formula. To obtain the real condition, it is important to simulate the non-isothermal condition by changing the heating rate during annealed the steel. Therefore, the present work was undertaken to experimentally investigate the influence of temperature, holding time and heating rate on austenite grain size, designed for use in reheating a HSLA steel.

2. Experimental Procedure

The composition of the HSLA steel used was, as expressed in wt pct on table 1. A section of industrial steel plate with 6 mm thick has been cut for reheating specimens that were used to this study. The specimens were reheating at varied temperature (960°C, 1060°C and 1120°C) by varied reheating rate (5°C/minute, 7.5°C/minute and 10°C/minute) and hold for varied holding time (10, 30 and 60 minutes), while the use of heating rate with three different interval rates; during reheating has been performed in non isothermal condition.

Table 1. HSLA Steel Composition (% wt)

C	Si	Mn	P	S	Al	Nb	V	N	Fe
0,12	0,266	0,645	0,009	0,005	0,034	0,028	0,011	0,0037	Balance

On examining the final microstructure, the specimens were quenched until room temperature, which allowed the identification of the austenite grains. It is well known that revealing the austenite grain boundaries always be a challenge for some researchers. In this work, using a modification of Villella's reagent [11] which is diluted with aquadest, the austenite grain boundary was successively revealed. The resulting grain size was measured with the linear intercept method [12].

To determine the condition of the microstructure occur, it must be observed, especially the phase austenitic microstructure. In observing the austenite grain boundary phase, there was an obstacle in the observation using an existing etching substance. By etching solution of 4 g of picric acid in 100 ml of distilled water added with 8 drops Teepol and 8 drops of HCL, with the technique of etching solution in hot conditions and samples are also in hot conditions, which for the overall sample indicates the grain boundaries are fairly obvious.

3. Result and Discussion

Austenite Evolution

From the results of the evaluation of the etching substance sightings austenite grain boundaries, and the obtained mixture etching method that is appropriate for this study. For examples of the appearance of grain boundaries for each treatment process experienced by the sample can be seen in the picture below.

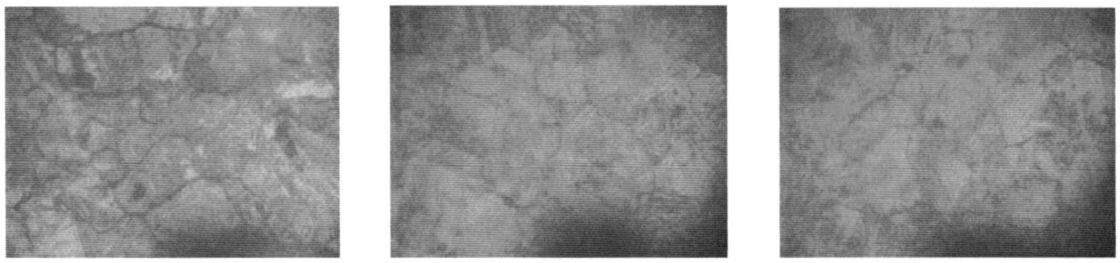

Figure 1. Austenite Grain formed during reheating 960°C with heating rate 5°C/minute and holding time 10, 30, 60 minute.

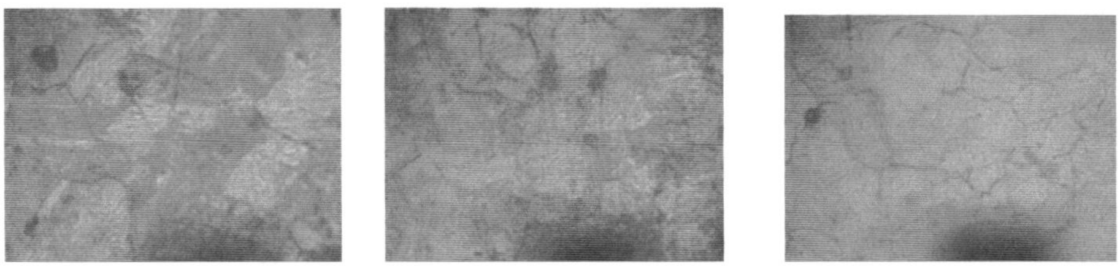

Figure 2. Austenite Grain formed during reheating 1060°C with heating rate 5°C/minute and holding time 10, 30, 60 minute.

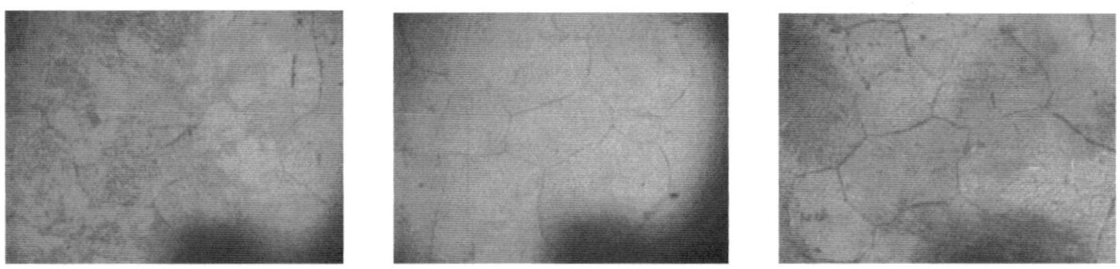

Figure 3. Austenite Grain formed during reheating 1120°C with heating rate 5°C/minute and holding time 10, 30, 60 minute.

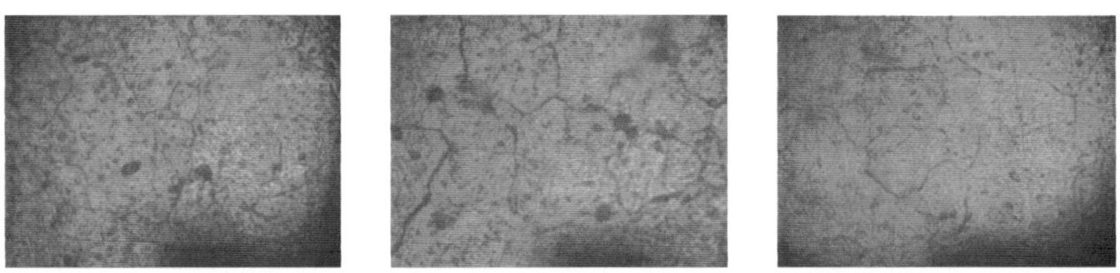

Figure 4. Austenite Grain formed during reheating 1060°C with heating rate 7.5°C/minute and holding time 10,30, 60 minute.

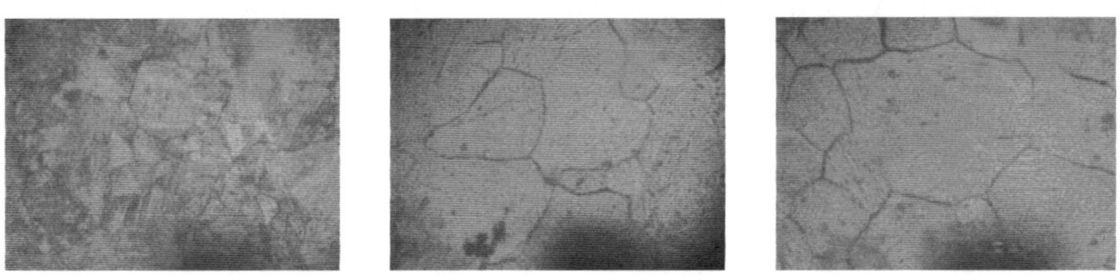

Figure 5. Austenite Grain formed during reheating 1120°C with heating rate 7.5°C/minute and holding time 10,30, 60 minute..

Figure 6. Austenite Grain formed during reheating 1060°C with heating rate 10°C/minute and holding time 10, 30, 60 minute.

Austenite Grain Size

After the solution and etching method to show austenite grain boundaries, the grain size was calculated by using the intercept method, in accordance with ASTM E112 standard metallographic calculation. From the calculations, the average diameter of the austenite grain for each temperature, the rate of heating and holding time, such as shown in figure 7, 8 and 9.

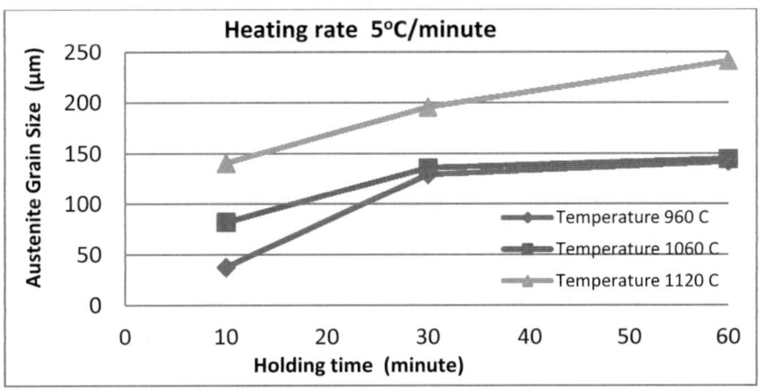

Figure 7. The relation graphs between grain size, reheating temperature and holding time at the heating rate 5°C/minute.

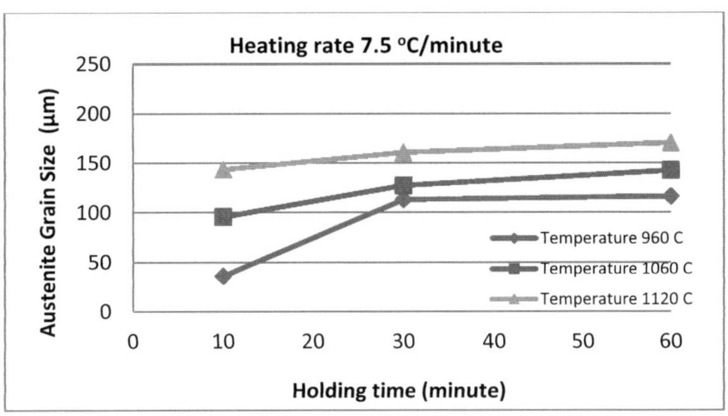

Figure 8. The relation graphs between grain size, reheating temperature and holding time at the heating rate 7.5°C/minute.

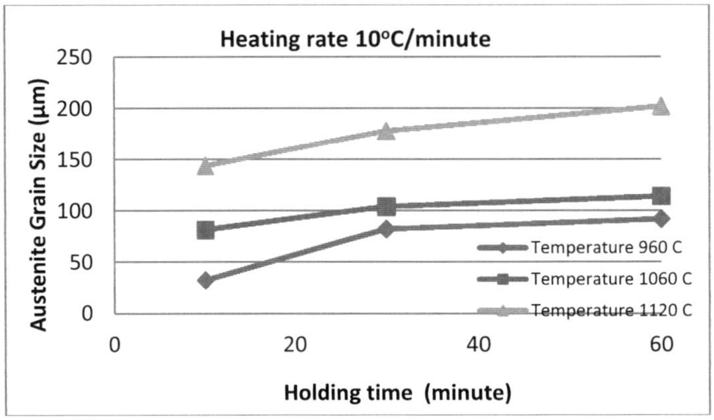

Figure 9. The relation graphs between grain size, reheating temperature and holding time at the heating rate 10°C/minute.

Figure 7, 8 and 9 shown that a heating rate of 5°C/min, 7.5°C/min and 10°C/min for each of the heating temperature by increased the holding time will increase the austenite grains size. When the heating temperature increased, the grains size formed increasing. This is due to the increase of holding time and heating temperature that will increase the activation energy of growth which will negate previous dislocations that exists on the grain boundaries so the growth of grain boundaries is not hindered by dislocation for growing.

In addition when we compared the austenite grains formed, the heating rate of 7.5°C/min will formed smaller. From this comparison it can be seen that the heating rate also plays a role in grain growth. From the pictures it can be presumed that on the low heating rate, the disintegration homogeneity of the dislocations more bigger so it will lowering obstacles that inhibit grain growth especially on the grain boundaries. In addition, the minimum energy needed for atoms to diffuse to obtain a more stable state (Q) will be smaller which will increase the grain boundaries migration to continue to grow even greater [13]. With a lower heating rate, homogeneity disintegration of sediment contained in the matrix is higher so it does not hinder the grain boundaries to continue to grow. However this needs to be investigated further by using Electron Microscope.

Figures 7, 8 and 9 shown that increased temperature will increase austenite grain that occurs. In addition to great affect grain growth activation energy, temperature rise will also increase the segregation of the solute contained in the matrix. Please also allege that at high temperatures will cause pressure maturation "Zener" which is a factor that can increase the growth of the grain size [14].

4. Conclusions

From the observation and discussion of the data obtained the following conclusions:

1. Large grain austenite formed with increasing heating temperature.
2. Austenite grain size increased with increasing holding time at the same heating temperature.
3. Austenite grain size will be smaller by increasing heating rate at the same heating temperature.
4. There is a relationship that is intertwined between the heating temperature, heating rate and holding time on the austenite grains growth.

References

1. R. Bengochea. B. Lopez and l. Gutierrez, 1999, "Influence of the Prior Austenite Microstructure on the Transformation Products Obtained for C-Mn-Nb Steels after Continuous Cooling", ISIJ International, Vol. 39, No. 6, pp. 583-591.
2. C.M Sellars, 1980, " The Physical Metallurgy of Hot Working", *Proc. Int. Conf.on Hot Working and Forming Processes , Sheffield, England*, pp. 3-15.
3. I.V. Samarasekera and E. B. Hawbolt, 1995, "Overview of Modelling the Microstructural State of Steel Strip During Hot Rolling" The Journal of The South African Institute of Mining and Metallurgy, pp. 157-165.
4. C.M Sellars, 1982,"Static Recrystallization and Precipitation During Hot Rolling of Microalloyed Steels" , *Mat Science Seminar, Sheffield, England.*
5. Andreas Kern. Joachim Degen kolbe, Bruno Musgen, Udo Schriever, 1992, "Computer Modelling for the Prediction of Microstructure Development and Mechanical Properties of HSLA Steel Plate", ISIJ International, Vol. 32 , No. 3, pp, 387-394
6. T. A. Kop, P. G. W. Remun, V. Svetchnikov, J. Sietsma, S. Van Der Zwaag, 2001, "Some observations on the effect of austenitisation conditions on the transformation kinetics in an HSLA steel and related C-Mn steels", Journal of Materials Science **36**, pp. 1863 – 1871
7. Andrey V. Karasev and Hideaki Suito, 2006, "Effect of Particle Size Distribution on Austenite Grain Growth in Fe–0.05 mass % C Alloy Deoxidized with Mn–Si, Ti, Mg, Zr and Ce", ISIJ International, Vol. 46, No. 5, pp. 718–727
8. Manohar P.A, D. P Dunne, T Chandra and C. R Killmore, 1995, "Grain Growth Prediction in Microalloyed Steel", *ISIJ International,* Vol 36 , 1996, pp. 194 – 200,.
9. K. A. Alogab, D. K. Matlock, J. G. Speer and H. J. Kleebe, 2007, The Effects of Heating Rate on Austenite Grain Growth in a Ti-modified SAE 8620 Steel with Controlled Niobium Additions, ISIJ International, Vol. 47, No. 7, pp. 1034–1041.
10. S. Mishra, T. Debroy , 2006, "Non Isothermal Grain growth in metals and alloys, *Materials Science and Technology;* **22**, 3; pg. 253 .
11. Anon., Metallography, ASM Metals Handbook (1991).
12. Anon., 19-th Annual Book of ASTM Standards Part E112, ASTM, Philadelphia (1996), p. 636.
13. D. Priadi, R.A.M. Napitupulu and E.S. Siradj, 2011, *"Austenite grain growth calculation of 0.028% Nb steel"*, Journal of Mining and Metalurgy B, ISSN 1450-5339, University of Belgrade, Serbia, 15, hal. 199 – 209.
14. K. C. Russel, Precipitate Coarsening and Grain Growth in Steels, Department of Materials Science and Engineering, Massachusetts Institute of Technology.

Preliminary study on aluminum-air battery applying disposable soft drink cans and Arabic gum polymer

S Alva[1], R Sundari[1], H F Wijaya[1], E H Majlan[2], Sudaryanto[3], I G A Arwati[1,2] and D Sebayang[1]
[1]Mechanical Engineering Department, Faculty of Engineering, Mercu Buana University, Meruya Selatan, Kembangan, West Jakarta, 11650, Indonesia.
[2]Fuel Cell Institute, The National University of Malaysia, Bangi, Selangor DE, 43600, Malaysia.
[3]Center for Sains and Technology Advanced Materials – BATAN Kawasan Puspiptek Serpong, Tangerang Selatan, Banten 15314.

* sagir.alva@mercubuana.ac.id

Abstract. This study is in relation to preliminary investigation of aluminium-air battery using disposable soft drink cans as aluminium source for anode. The cathode uses commercial porous carbon sheet to trap oxygen from air. This work applies a commercial cashing to place carbon cathode, electrolyte, Arabic gum polymer, and aluminium anode in a sandwich-like arrangement to form the aluminium-air battery. The Arabic gum as electrolyte polymer membrane protects anode surface from corrosion due to aluminium oxide formation. The study result shows that the battery discharge test using constant current loading of 0.25 mA yields battery capacity of 0.437 mAh with over 100 minute battery life times at 4M NaOH electrolyte and 20 % Arabic gum polymer as the best performance in this investigation. This study gives significant advantage in association with beneficiation of disposable soft drink cans from municipal solid waste as aluminium source for battery anode.

1. Introduction

Issues on friendly environmental energy have been growing remarkable interest in many disciplines. Serious issues on carbon dioxide since Kyoto Protocol (1997) increasing global temperature due to rapid technology development and vigorous industrial activities have drawn significant attention corresponding with limitations of application of fossil fuel energy from oil and gas resources. Alternative energy sources starting from hydropower, solar, wind, geothermal, electricity until nuclear energy have developed through serial work and thorough investigations expecting to yield low polluted energy [1].

In the context of electricity as energy source, currently, many applications use electric current based on galvanic reaction, for instance, part of automobiles have started to replace petrol energy with electrical battery. Nowadays batteries have been used in large electronic devices including notebooks, tablet computers, pocket calculators, clocks and hand watches, until children toys due to mobility, lightweight, low cost and safety. Having special interest on metal batteries such as lithium – aluminium batteries combined with other components with regard to its electrolyte and cathode material contribute to large industrial applications, however, limitation in their life time have been still a serious obstacle that needs further improvements up to date [2]. Nevertheless, aluminium galvanic based batteries have still attracted many scientists to do more researches and investigations due to aluminium low cost, low weight, and nontoxic material [3].

With regard to aluminium ion batteries, several previous investigations in relation to efforts yielding long lasting and safe battery have been reported [4-10]. The energy of an aluminium ion battery is dependent on several factors like battery cell voltage, capacity and chemical composition. According to Armand & Tarascon [4], an aluminium battery yielded elevated output by increasing chemical potential difference between two electrodes, reducing mass of reactants, and avoiding electrolyte from by-side chemical reactions. By addressing specifically to aluminium-air batteries, an electronic industry produced electricity through chemical reaction of oxygen in atmospheric air and aluminium, however, it is reported that by-product reaction may impede the surface of aluminium anode and as a result, it may result shorter battery survival [4]. Modesto & Yulie reported that simple aluminium-air battery could produce energy of 1.0 V and 100 mA [8]. Earlier investigation as reported by Rao et al. [11], an aluminium-air battery produced a 400 W energy with its replaceable electrolyte. Nevertheless, neglected from its restriction with regard to by-product reaction resisting to further oxidation of aluminium anode, application of aluminium-air batteries have placed the first priority in electronic industries due to economics reason and low weight material [11]. According to Modesto &Yulie [8] and Mohamad [9], the basic concept of electrochemical reaction in aluminium-air batteries was illustrated as follows:

Anode: $Al_{(s)} + 3OH^-_{(aq)} \rightarrow Al(OH)_{3(s)} + 3e^-$ $E^o = -2{,}35$ V (1)
Cathode: $O_{2(g)} + 2H_2O_{(l)} + 4e^- \rightarrow 4OH^-_{(aq)}$ $E^o = +0{,}40$ V (2)
Overall: $4Al_{(s)} + 3O_{2(g)} + 6H_2O_{(l)} \rightarrow 4Al(OH)_{3(s)}$ $E^o = 2{,}75$ V (3)

With regard to the electrochemical reaction, eq. (1) describes the oxidation of aluminium by releasing electrons, while eq. (2) describes the reduction of oxygen from air to form hydroxyl ion (OH-) by capturing electrons, and eq. (3) shows the reduction oxidation reaction of aluminium oxygen in the battery.

Apart from the advantage-disadvantage issues on the application of aluminium-air batteries, this study currently investigates the role of aluminium-air battery using some modification. The application of Arabic gum polymer as hydro gel polymer in this investigation assumes to reduce by-product reaction interfering aluminium anode surface in association with the role of heteroatomic organic compounds (N, O, and S). This study applies porous carbon cathode and sodium base electrolyte. On account of that reason, the main objective of this study is to design aluminium-air battery applying aluminium anode obtained from disposable soft drink cans in municipal solid waste.

2. Material and Experimental

2.1 Materials

Materials used in this paper include aluminium metal obtained from disposable soft drink cans as an anode. Other materials used are air-cathodes and coin cell casing purchased from Magna value Sdn BhD Malaysia. The Arabic gum purchased from Elnasr Ltd., and both electrolytes NaOH and NaCl purchased from Merck.

2.2 Experimental

The fabrication of aluminium-air battery was conducted in the laboratory of production, Mechanical Engineering Department, University of Mercu Buana, Jakarta. For the preparation of aluminium anode, a plate of disposable soft drink cans were cut to adjust the size of proposed diameter and height, respectively, i.e. 16 mm and 2 mm, as shown in Fig. 1a,

IOP Conf. Series: Materials Science and Engineering **237** (2017) 012039 doi:10.1088/1757-899X/237/1/012039

followed by cleaning with rubbing paper. The preparation of carbon cathode, porous carbon commercial has cut to fit cashing size with its diameter and height. i.e. 16 mm and 2 mm, respectively, and placed at cashing bottom as shown in Fig. 1b. A sandwich-like arrangement of carbon cathode, Arabic gum with drop wise of NaOH or NaCl, and aluminium anode designed from bottom to surface to form the proposed battery. Different concentrations of Arabic gum in aqueous NaOH solution were prepared to produce optimum result. The Arabic gum stirred mixing with NaOH solution to form homogeneous solution. The homogeneous solution of Arabic gum and electrolyte solution (NaOH or NaCl) placed in a container installed by porous carbon cathode and aluminium anode, and then dried at room temperature for 2 days. After forming gel solution, battery assay was ready for electrical measurements in terms of electrical voltage and current using a battery tester. Prior to assembling the aluminium-air battery,variations of some brands of disposal soft drink cans were characterized by AAS. The aluminium-air battery assay was includes several examinations: (i) electrical voltage and current measurements of variable electrolyte concentrations (NaOH and NaCl); (ii) electrical voltage and current measurements of variable Arabic gum concentrations; (iii) electrical measurements of aluminium-air battery using the highest aluminium content among branded disposable soft drink cans using optimum electrical performance of electrolyte and Arabic gum; and (iv) measurements of aluminium-air battery capacity.

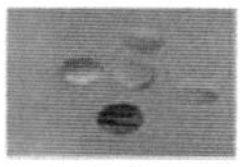
(a)

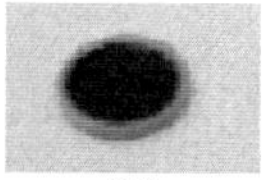
(b)

Figure 1 Fabrication of aluminium anode (a) and porous carbon cathode with 20% Arabic gum (b) placed in commercial cashing.

3. Results and Discussions

3.1 AAS Analysis

With regard to aluminium contents of selected four branded disposable soft drink cans (Pocari Sweet, Cap Kaki Tiga, Greensand, and Coca-Cola), this investigation selected Pocari Sweet as the branded disposable soft drink can for aluminium anode in aluminium-air battery assay since this given brand disposable drink can contains the highest aluminium content (Table 1).

Having compared to previous investigation as reported by George DS [12], the results of element content shown in Table 1 is generally not much different from previous work. Previous investigation reported 92.5-97.5% Al, 1.0% Mg, 1% Mn, 0.4% Fe, 0.2% Si, and 0.15% Cu obtained from soft drink cans commonly used in Asia [12].Furthermore, the disposal soft drink cans with pocari sweet branded could be select as an anode material in the assembly of Al-air batteries.

3.2 Battery assay

Variable concentrations of NaCl (3%, 3.60%, 4%, and 2M) and NaOH 4M were prepared to examine the average of electrical voltage (V) and current (mA) based on triplicate measurements. The highest result of electrical measurement of electrolyte solution in terms of voltage and current is taken as the best electrolyte solution to be used for further battery assay and this electrolyte solution will be mixed with Arabic gum until homogeneous solution obtained. The results of electrical measurements using variable concentrations of NaCl and

NaOH shows that 4M NaOH yielded the highest voltage and current (Table 2) and therefore, 4M NaOH electrolyte is selected as the best electrolyte (1.63V and 24.57 mA) for further measurements.

In addition, the reason of using 4M NaOH electrolyte for further study is strengthened by data of electrolyte performance reported in previous investigation. Marliyana et al. [7] reported that 2M NaCl using aluminium (99.8%) anode gave a value of – 0.87 V, while 4M NaOH using pure aluminium (99.999%) anode yielded a value of 1.531 V at room temperature, and seawater using aluminium mixed with minor element (0.1% thalium and 0.1% indium) as anode produced a value of 0.9 V.

Table 1. AAS results of elements for given type of disposable soft drink cans obtained from municipal inorganic waste.

	Element content (%) in given type of disposable drink cans			
Element	**Pocari Sweet**	**Cap Kaki Tiga**	**Greensands**	**Coca-Cola**
Aluminum (Al)	96.38	89.74	90.87	93.28
Magnesium (Mg)	1.14	3.28	2.25	1.17
Mangan (Mn)	0.75	1.93	1.21	1.04
Iron (Fe)	0.51	1.79	1.52	1.72
Silicon (Si)	0.19	0.88	1.33	0.68
Copper (Cu)	0.19	2.36	1.92	1.26

Table 2. Results of electrical voltage and current measurements applying variable concentrations of NaCl and NaOH

Electrolyte	**Concentration**	**Measurements**						**Average**	
		Sample 1		**Sample 2**		**Sample 3**		**Volt (V)**	**Current (mA)**
NaCl	3%	0.705	7.55	0.701	7.77	0.704	7.51	0.70	7.61
	3.60%	0.716	9.21	0.714	9.35	0.71	9.53	0.71	9.36
	4%	0.715	10.59	0.714	10.28	0.712	10.42	0.71	10.43
	2 M	0.73	10.7	0.728	10.78	0.725	10.85	0.73	10.78
NaOH	4 M	1.64	25.4	1.63	24.4	1.62	23.9	1.63	24.57

The selection of NaOH over NaCl as electrolyte in this study was also supported by Hofmeister series that the ability of OH- from NaOH to form hydrogen bonding is stronger than that of Cl- causing faster mobility of OH- in electrolyte solution yielding higher electrical voltage and current. [13].

Similar treatment as done for NaOH and NaCl electrolyte solution has applied for Arabic gum solution. Variable concentrations of Arabic gum solution (10%, 20%, and 30%) were prepared and the highest voltage and current of Arabic gum solution has taken as reference after triplicate measurements. The optimum results in terms of electrical voltage and current for a given electrolyte solution (NaOH or NaCl) and Arabic gum solution were mixed together expecting to obtain the optimum condition of aluminium-air battery. In order to examine the corrosion effect of aluminium anode in relation to the existence of Arabic gum in aqueous electrolyte solution of NaOH, this study uses different concentrations of Arabic gum (0%, 20%, and 30%). The results of corrosion effect were shown visually in Fig. 2.

IOP Conf. Series: Materials Science and Engineering **237** (2017) 012039 doi:10.1088/1757-899X/237/1/012039

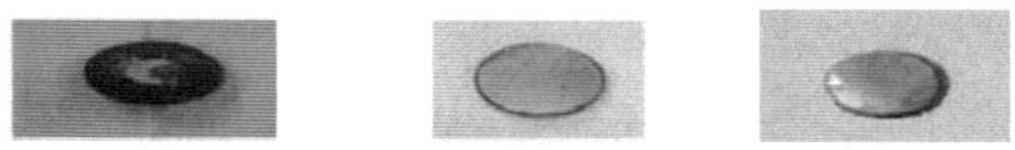

(a) (b) (c)

Figure 2 Corrosion effect on Al anode surface with Arabic gum 0% (a); 20% (b); 30% (c). 4M NaOH electrolyte; Carbon cathode; Pocari Sweet disposable cans.

Based on Figure 2 it appears that battery without Arabic gum experienced corrosion, while the other one using Arabic gum looks much cleaner. This is because the Arabic gum membrane has an OH^-functional group capable of slowing the movement of ions through hydrogen interactions [14]. Thus, the corrosion process caused by the electrolyte used can be inhibited.

This study used discharge assay applying a battery tester with constant current loading of 0.25 mA to examine the performance of proposed aluminium-air battery. Table 3 shows the discharge data of aluminium-air battery in terms of electrical voltage, current, duration time, and battery capacity for different concentrations of Arabic gum (0%, 20%, and 30%). At first, variable concentrations (0%, 20%, 30%, 40%, and 50%) of Arabic gum were planned to be examined, however, finally only three concentrations (0%, 20%, and 30%) of Arabic gum were examined due to dramatically dropped of battery performance after using 30% Arabic gum (Fig.3 and Table 3). Finally, the results showed that 20% Arabic gum performed the best discharge test with achieved battery capacity of 0.437 mAh.

Higher solution viscosity may hinder ions migration in electrolyte solution resulting lowered battery performance and therefore, the higher Arabic gum concentration (30%) showed lower battery performance rather than that one of Arabic gum 20%.

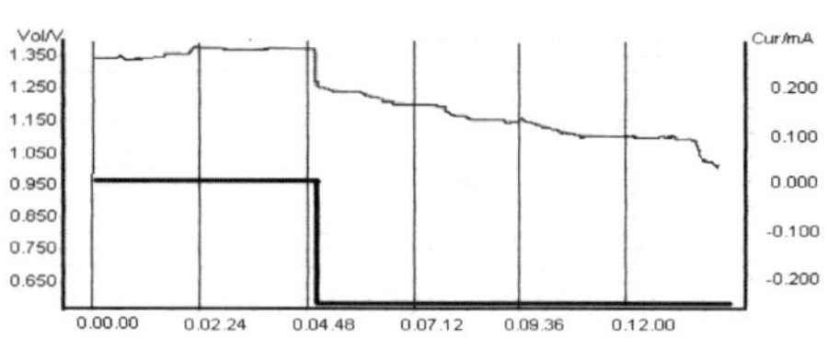

(a)

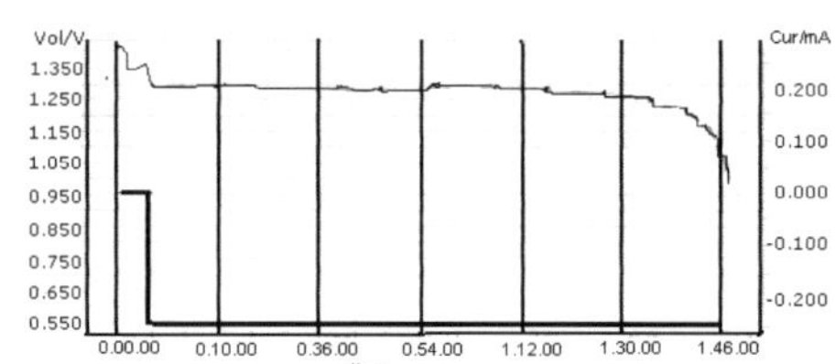

(b)

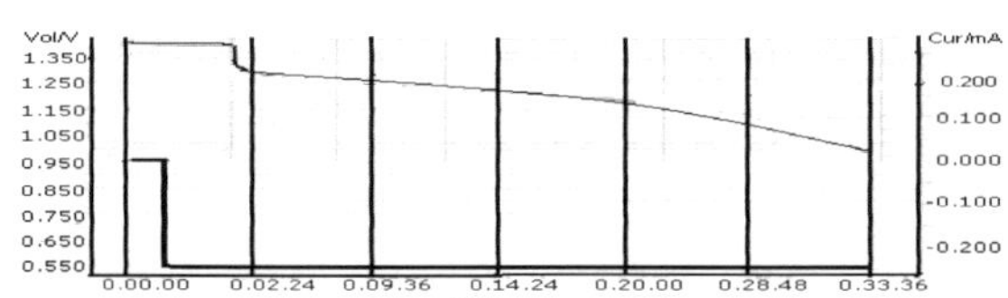

(c)

Figure 3 Discharge test with battery tester applying Arabic gum 0% (a); 20% (b); 30% (c). 0.25 mA current loading. 4M NaOH. Al anode. Porous carbon cathode. Pocari Sweet disposable cans.

Table 3. Battery performance related to electrical voltage, current, and capacity applying Arabic concentrations (0%, 20%, and 30%). 0.25 mA current loading.

specification	Time (h:m:s)	Voltage (mV)	Current (mA)	Duration (min)	Capacity	Total capacity*)
0% Arabic gum;	0:00:00	1262	-0.2496	4	-0.01664	
4M NaOH;	0:04:45	1147	-0.2496	4	-0.01664	-0.037
carbon cathode;	0:08:45	1030	-0.2496	1	-0.00416	
pocari sweet cans	0:09:03	1000	-0.2496		0	
(Al anode)						
20% Arabic gum;	0:00:00	1316	-0.2496	90	-0.3745	
4M NaOH;	1:29:15	1249	-0.2496	12	-0.04992	
carbon cathode;	1:41:00	1150	-0.2496	2	-0.00832	-0.437
pocari sweet cans	1:43:50	1047	-0.2496	1	-0.00416	
(Al anode)	1:44:28	999	-0.2496			
30% Arabic gum;	0:00:00	1348	-0.2495	11	-0.04574	
4M NaOH;	0:11:40	1249	-0.2495	12	-0.0499	
carbon cathode;	0:23:25	1144	-0.2495	4	-0.01663	-0.116
pocari sweet cans	0:27:10	1047	-0.2495	1	-0.00415	
(Al anode)	0:28:04	999	-0.2495			

*) round-off data.

The data of battery discharge graphs as presented in Fig. 3 was obviously observed in Table 3. Regarding to battery with 0% Arabic gum (Table 3) only after 9 minutes the capacity dropped to zero, however, battery with 20% Arabic gum (Table 3) was still survived over 100 minutes duration, while battery with 30% Arabic gum survived about 20 minutes. Moreover, with respect to total battery capacity, the 20% Arabic gum battery posed the highest value, i.e. 0.437 mAh, and the reason of this phenomenon has already elucidated above.

4. Conclusions

The preliminary battery discharge test of Al anode and porous carbon cathode shows moderate satisfaction result with regard to achieved battery capacity of 0.437 mAh applying 4M NaOH and 20% Arabic gum polymer. The 20% Arabic gum polymer successfully achieved over 100 minute battery life time. The disposable soft drink cans as Al source for battery anode looks promising for future broader application.

Acknowledgement

The researchers give many thanks to Mercu Buana University for the financial support through UMB Internal Funding. Reseachers also say thank you to the National University of Malaysia and the Government of Malaysia for the financial support they provided through research grants GGPM-2013-044, FRGS/2/2013/TKO4/UKM/02/3.

References

[1] Farahiyah AR, Maniruzzaman A A M, Saidur R, Abu Bakar W A W, Hainin M R, Putrajaya R, and Norhidayah Abdul Hassan N A 2017 *Renewable and Sustainable Energy Reviews***71** 112.

[2] Li Y, Yang J and Song J 2017 *Renewable and Sustainable Energy Reviews***71** 645.

[3] Kazazi M, Abdollahi P and Moghadam M M 2017 *Solid State Ionics***300** 32.

[4] Armand M and Tarascon J M 2008*Nature*, **451** 652.

[5] Egan DR, Ponce CL, Wood RJK, Jones RL, Stokes K R and Walsh F C 2013*Journal of Power Sources***236** 293.
[6] Gelman D, Shvartsev B and EinE Y 2014 *Journal of Materials Chemistry A***2** 20237.
[7] Marliyana M, Meor ZMT, Edy HM, Siti M, Wan MFWR, Wan RW D and Jaafar S 2015*Journal of Industrial and Engineering Chemistry***32** 1.
[8] Modesto T and Julie H Y 2007 *Journal of Chemical Education***84** 1936A.
[9] Mohamad AA 2008 *Corrosion Science***50** 3475.
[10] Mori R 2015 *Journal of The Electrochemical Society***162** A288.
[11] Rao BML, Cook R, Kobasz W and Deuchars G D 1992 *International on Power Sources Symposium, 35th IEEE*, USA.
[12] George D S 1988 From Monopoly to Competition: The Transformation of Alcoa. Cambridge University Press.
[13] Nathaniel V N and Jane M V 2008 *Journal of Molecular Liquids***143** 160
[14] Abu-Dalo M A and Othman AA 2012 *International Journal of Electrochemical Science***7**9303.

IOP Conf. Series: Materials Science and Engineering **237** (2017) 012040 doi:10.1088/1757-899X/237/1/012040

The effect of microstructure on abrasive wear of steel

A Kešner[1,*], R Chotěborský[1] and M Linda[2]
[1]Department of Material Science and Manufacturing Technology, Czech University of Life Sciences Prague, Kamýcká 129, 165 21, Czech Republic
[2]Department of Electrical Engineering and Automation, Czech University of Life Sciences Prague, Kamýcká 129, 165 21, Czech Republic

* kesner@tf.czu.cz

Abstract. Abrasive wear of agricultural tools is one of the biggest problems in currently being. The amount of abrasive wear, depending on the microstructure, has been investigated in this work. Steels 25CrMo4 and 51CrV4 were used in this work to determine the effect of the microstructure on the abrasive wear. These steels are commonly used for components that have to withstand abrasive wear.SEM analysis was used to detect the microstructure. The standardized ASTM G65 method was used to compare the abrasive wear of steels. The results show that the abrasive wear depends on the microstructure of steels.

1. Introduction

Abrasive wear is the important problem for agricultural tools. Many producers of agricultural tools solves a way to increase the abrasion resistance of steel at the same or lower cost of manufacturing.[1, 2].Abrasive wear causes removal of material from a surface agricultural tools and thus reduces the ability to perform the required function and it increases service life for component replacement[3, 4]. Resistance to abrasion wear can be increased by a suitable selection of material [5–7]. Material of agricultural tools, however must be chosen with regard to other mechanical properties such as strength and toughness[8]. Another option to increasing abrasion wear resistance is heat treatment[6, 9, 10]. The greatest resistance to abrasive wear is the combination of bainite and martensite microstructures[11, 12]. Isothermal quenching allows the formation of this combination of microstructures.

Soil is the main factor causing abrasive wear of agricultural tools. Other factors affecting the agricultural instrument are various barriers such as stone or large soil consolidation. For this reason, the agricultural tool must not only be resistant to abrasive wear, but must also have the required hardness and toughness. Each microstructure (in this case, bainit and martensit) has different properties[13–18].

In the works[5, 9, 11]are stated that with increasing hardness the resistance to abrasive wear increases.

The aim of this work is to evaluate the effect of the bainitic - martensitic structure on the resistance to abrasive wear of the selected steels. There are many contradictions in literature on the influence of hardness on abrasion resistance. Hardness and its depending on abrasive wear and it has been investigated in this work.

2. Material and method

Samples with dimension of 25 x 10 x 50 mm from selected steels 25CrMo4 a 51CrV4 have been used in this work. The chemical composition of steels is shown in Table 1.

IOP Conf. Series: Materials Science and Engineering **237** (2017) 012040 doi:10.1088/1757-899X/237/1/012040

Table1. Chemical composition of steels used in this work (wt. %)

material	C	Mn	Si	P	S	Cr	Ni	Cu	Al	Mo	Sn	V	Ti
25CrMo4	0.25	0.71	0.23	0.018	0.022	1.03	0.09	0.23	0.023	0.21	0.011	0.004	0.015
51CrV4	0.53	0.89	0.26	0.012	0.025	1.02	0.08	0.13	0.028	0.02	—	0.12	—

The heat treatment procedure include: heating temperature 800 °C in air for 1200 seconds. The cooling parameters are shown in Table 2.The heat treatment parameters were taken from the TTT diagrams for the selected steels.The cooling parameters were chosen to create a combination of bainite and martensite. Steel 25CrMo4 is marked with a letter C, and 51CrV4 is marked by letter V.

Table 2. Overview of sample cooling

	cooling 1			**cooling 2**			**cooling 3**		
	temp.	**medium**	**time [s]**	**temp.**	**medium**	**time [s]**	**temp.**	**medium**	**time [s]**
C40	400	salt bath	37	400	air	163	20	air	to 20°C
C41	400	salt bath	37	20	water	to 20°C	X	X	X
V40	300	salt bath	40	300	air	1000	20	air	to 20°C
V41	300	salt bath	40	20	water	to 20°C	X	X	X
V42	400	salt bath	40	400	air	1200	20	air	to 20°C
V43	400	salt bath	40	20	water	to 20°C	X	X	X

Abrasive wear tests were performed by the ASTM G65 standardized test[20, 21]. The disk was pressed to the sample with a force of 100 N and the fraction of abrasive particle size from 0.2 mm to 0.315 mm was applied between the disc and the sample. The diameter of the rubber disc was 210 mm and the wheel width was 12.5 mm. During one measurement, the disc ran a distance of 210 meters. After each measurement, mass loss were measured on analytical balance of 0.1 mg accuracy. The mass losses were recalculated to volume loss which is indicated in $mm^3 \cdot m^{-1}$ distance. Ten measure procedures were performed for each sample.

Metallographic samples were cut in cross sectional area after full wear test. Samples were prepared for image analysis in the cut surface by grinding, polishing and etching. The grinding was done under water, followed by a diamond suspension of 9 and 3 μm. Polishing was carried out with the addition of a chemical suspension of OP-S. For etching, 2% nital (1 ml HNO3 + 50 ml ethanol) was used.

Image analysis was performed on an electron microscope SEM (Tescan Mira 3 GXM) equipped with an energy dispersive X-ray (Oxford X-MaxN) detector. Parameters which were set on the electron microscope are shown in Figure 1. Images area (104 x 104 μm) were taken over the entire cross-section of the sample at a distance of 0.2 mm.

The image analysis was performed in Quick PHOTO INDUSTRIAL 3.1. The proportions of the phases were found - see Figure 1.

Hardness was measured on the analyzed area for abrasive wear test at a distance of 2 mm. Micro hardness was measured on the PMT 3 hardness machine using the HRC method.

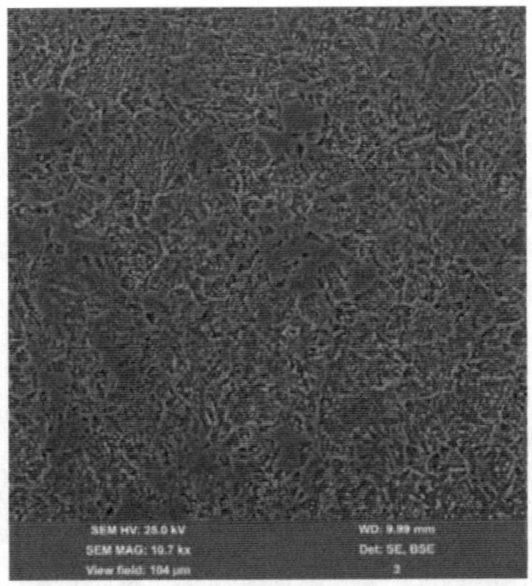

Figure1. Image analysis microstructure for V43 in Quick PHOTO INDUSTRIAL 3.1. The boundaries between the phases of bainite and martensite are indicated in red colour.

3. Results and discussion

Standardized ASTM G65 abrasive tests were evaluated. Relationships between weight losses on distances (wear traces)are shown in Figure 2. The weight loss was greatest for material 25CrMo4 (designated C). The material 51CrV4 (designated V) was significantly smaller in weight loss at the same length. The differences were minimal between 51CrV4. However, these differences are noticeably large in the overall life of the material. Differences in weight loss were statistically processed by the F-test method. There was no statistically significant difference between samples V40 and V42, samples V40 and V43, samples V42 and V43. Among the other samples, statistically significant differences were found, that is the heat treatment of these samples had a significant effect on abrasion resistance.

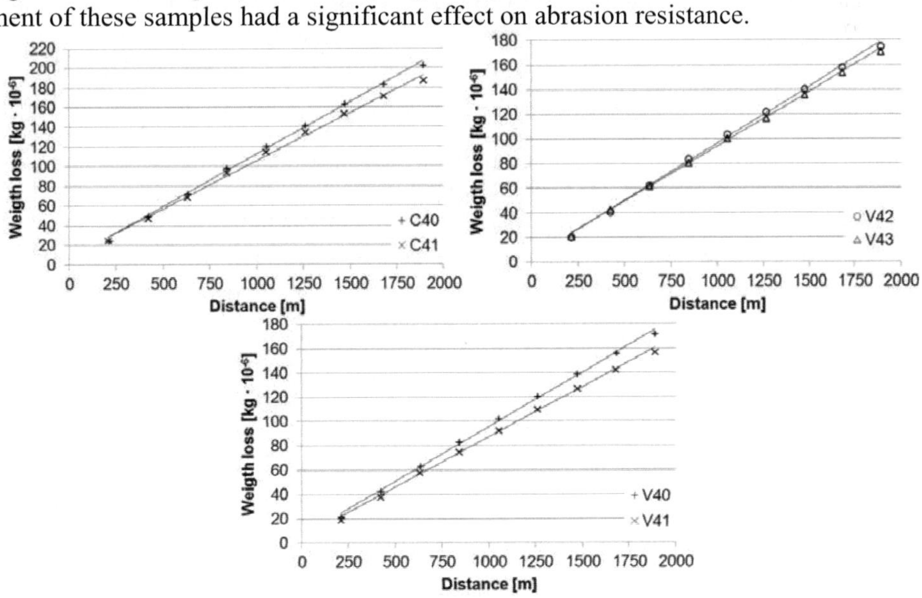

Figure 2. Dependence of weight loss on the distance traveled by the disc

The effect of microstructure on mass losses of steel was investigated. The structure of bainite and martensite was determined from the image analysis for all samples – shown in Figure 3.

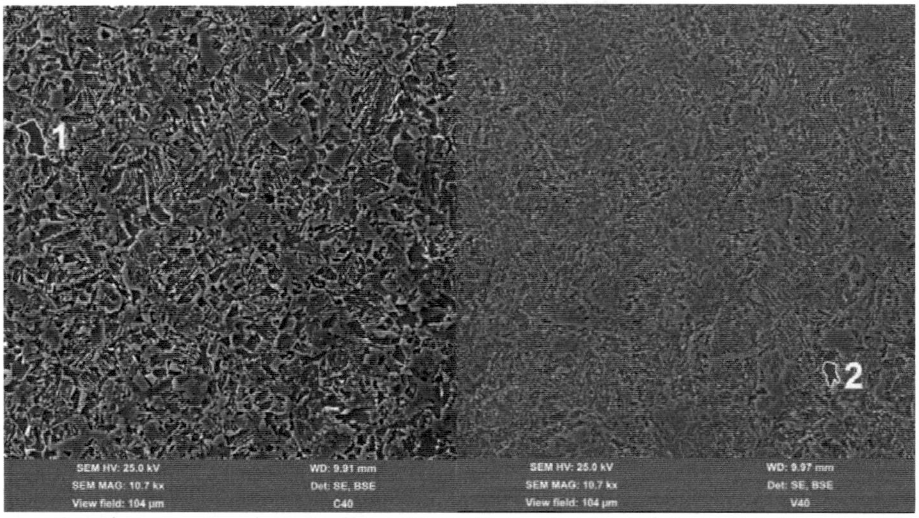

Figure 3. Microstructure images for steel. Left: Steel C40 with 12% bainite (example of bainite microstructure is marked number 1), 88% martensite. Right: SteelV40 with 99% bainite and 1% martensite (example of martensite microstructure is marked number 2).

The dependence of weight loss on the bainite structure is shown in Figure 4.The results show that the ratio volume of phases of bainite (60 to 80%) and martensite (20 to 40%) has the smallest weight loss. Increased weight loss was found in the ratio of volume phases bainite (95 to 100%) and martensite (0 to 5%) but less than the weight loss in bainite (20-30%) and martensite (70-80%) volume of bainite in martensite phase does not cover the entire distribution ratio of volume phases bainite and martensite. For this reason, it is good to determine the ratio of volume phases with 0 to 10% and 45 to 50% bainite and martensite.

The effect of microstructure on hardness is shown in Figure 4. The highest hardness of 56HRC was found highest in case if microstructure content 60% bainite and rest volume of martensite. If microstructure included 73% bainite and rest volume of martensite, a hardness of 43HRC was measured, which is less than 99% bainite where the hardness value was 45HRC. Ratio of volume phases with the difference between the hardness in dependence on the microstructure can be seen in Figure 5 on the left. A microstructure of similar composition shows great differences in hardness (for example bainite 98%).

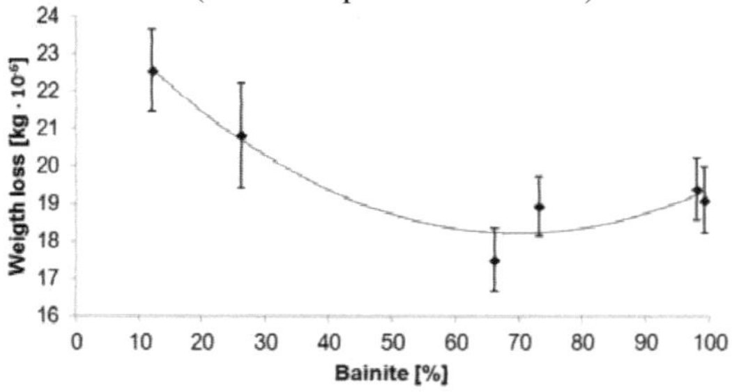

Figure4. Dependence of weight loss on bainite structure

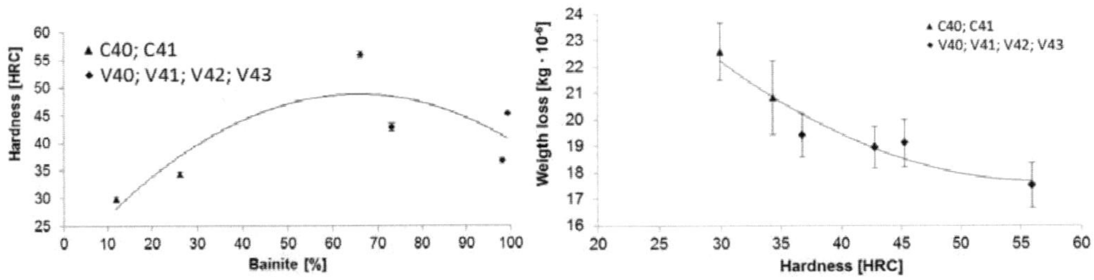

Figure 5. The dependence of hardness on the volume of bainitic microstructure

Authors AK Bhakat[8], M Vite-Torres and etc. [19], H Sabet and etc. [22] report that the greatest impact on abrasive wear has hardness.In fact, the microstructure is the most important, as the authors suggest V Jankauskas a R Skirkus [1], J J Coronado a kol. [23] and the results of this work. The results of this work it is necessary to add further representation bainite in martensite because some representation% bainite in martensite in this work was not done yet.

4. Conlusion

The results of this work show that:

- Heat treatment affects resistance to abrasion wear due to the formation of a different microstructure. Microstructure affects abrasive wear.
- Experimental measuring did show the effect of hardness on abrasive wear. With increasing hardness increases abrasion resistance.
- The 51CrV4 steel, due to its chemical composition, has greater abrasive wear resistance than 25CrMo4 steel.
- The highest abrasion resistance was measured for 51CrV4 steel containing 60-80% bainite and rest of volume martensite microstructure.

ACKNOWLEDGEMENT. Supported by Internal grant 31140/1312/3114 agency of Faculty of Engineering, Czech University of Life Sciences in Prague

References

[1] Jankauskas V and Skirkus R 2013 Steel abrasive wear forecasting by wearing surfaces microgeometric parameters. Mechanika.Vol. 19, no. 4.

[2] Chotěborský R and Linda M 2015 FEM based numerical simulation for heat treatment of the agricultural tools. Agronomy Research. Vol. 13, no. 3, p. 629–638.

[3] Baburaj A, Chaudhary S S K B, Khatirkar R K and Sapate S G 2013 Abrasive Wear Behaviour of Heat Treated En31 Steel. ISIJ International. Vol. 53, no. 8, p. 1471–1478.

[4] Chacon-Nava J G, Martinez-Villafañe A, Almeraya-Calderon F, Cabral-Miramontes J A and Stack M M 2010 Some remarks on particle size effects on the abrasion of a range of Fe based alloys. Tribology International [online]. Vol. 43, no. 8, p. 1307–1317.

[5] Herian J, Aniolek K, Cieśla M and Skotnicky G 2014 Shaping the structure during rolling and isothermal annealing, and its influence on the mechanical characteristics of high-carbon steel. Materials Science and Engineering: A. Vol. 608, p. 149–154.

[6] Chotěborský R, Hrabě P, Müller M, Savková J, Jirka M and Navrátilová 2009 M Effect of abrasive particle size on abrasive wear of hardfacing alloys. Research in Agricultural Engineering. 2009. Vol. 55, no. 3.

[7] Müller M, Chotěborský R and Hrabě P 2015 Application of overlay material on soil processing tools for purpose of increasing their abrasive wear resistance.ListyCukrovarnicke a Reparske. 2015. Vol. 131, no. 9–10.
[8] Bhakat A K, Mishra A K and and Mishra N S 2007 Characterization of wear and metallurgical properties for development of agricultural grade steel suitable in specific soil conditions. Wear. Vol. 263, no. 1–6, p. 228–233.
[9] Jankauskas V, Katinas E, Skirkus R and Alekneviciene V 2014 The method of hardening soil rippers by surfacing and technical-economic assessment. Journal of Friction and Wear. Vol. 35, no. 4, p. 270–277.
[10] Chotěborský R and Linda M 2016 Determination of chemical content of soil particle for abrasive wear test. Agronomy Research. Vol. 14.EestiPollumajandusulikool.
[11] Das Bakshi S, Shipway P H and Bhadeshia H K D H 2013 Three-body abrasive wear of fine pearlite, nanostructured bainite and martensite. Wear. Vol. 308, no. 1–2, p. 46–53.
[12] Zhou H, Wu Z W, and Fu H A study of self-hardening bainitic cast steel. Materialwissenschaft und Werkstofftechnik. Vol. 44, no. 9, p. 753–759.
[13] Richadson RCD (1967) The wear of metallic materials by soil—Practical phenomena. Journal of Agricultural Engineering Research. Vol. 12, no. 1, p. 22–39.
[14] Tagar AA, JiCh, Ding Q, Adamwoski J, Chandio FA and Mari I A (2014) Soil failure patterns and draft as influenced by consistency limits: An evaluation of the remolded soil cutting test. Soil and Tillage Research. Vol. 137, p. 58–66.
[15] Mclaughlin N B, Drury C F, Reynolds W D, Yang X M, Li Y X, Welacky T W and Stewart G (2008) Energy Inputs for Conservation and Conventional Primary Tillage Implements in a Clay Loam Soil. Transactions of the ASABE. Vol. 51, no. 4, p. 1153–1163.
[16] Mikolajczak P and Napiórkowski J (2016) Analysing the reliability of working parts operating in abrasive soil pulp taking into consideration confounding factors. Vol. 18, no. 4.
[17] Manuwa S I (2009) Performance evaluation of tillage tines operating under different depths in a sandy clay loam soil. Soil and Tillage Research. Vol. 103, no. 2, p. 399–405.
[18] Aluko OB and Chandler H W (2006) A Fracture Strength Parameter for Brittle Agricultural Soils. Biosystems Engineering. Vol. 93, no. 2, p. 245–252.
[19] Vite-Torres M, Vite J, Laguna-Camacho J R, Castillo M, and Marquina-Chávez A 2011 Abrasive wear on ceramic materials obtained from solid residuals coming from mines. Wear. Vol. 271, no. 9–10, p. 1231–1236.
[20] Kamdi Z, Shipway P H, Voisey K T and Sturgeon A J Abrasive wear behaviour of conventional and large-particle tungsten carbide-based cermet coatings as a function of abrasive size and type. Wear. Vol. 271, no. 9–10, p. 1264–1272.
[21] Chotěborský R and Linda M 2014 Evaluation of friction force using a rubber wheel instrument. Agronomy Research. Vol. 12, no. 1.
[22] Sabet H, KhierandishSh, MirdamadiSh and Goodarzi M 2011 The Microstructure and Abrasive Wear Resistance of Fe–Cr–C Hardfacing Alloys with the Composition of Hypoeutectic, Eutectic, and Hypereutectic at Cr /C6. Tribology Letters. Vol. 44, no. 2, p. 237–245.
[23] Coronado J J, Caicedo H F and Gómez A L 2009 The effects of welding processes on abrasive wear resistance for hardfacing deposits. Tribology International. Vol. 42, no. 5, p. 745–749.

IOP Conf. Series: Materials Science and Engineering **237** (2017) 012041 doi:10.1088/1757-899X/237/1/012041

Piezoreflectance study of Nb-doped MoS_2 single crystals

Kerista Sebayang[1] and MulaSigiro[2,*]
[1]Department of Physics, Faculty of Mathematics and Natural Sciences, University of Sumatera Utara, Medan 20155, Indonesia.
[2]Department of Physics Education, Faculty of Teacher Training and Education, Universitas HKBP Nommensen, Medan 20234, Indonesia.

* mulasigiro@gmail.com

Abstract. We have measured the temperature dependence of the spectral features in the vicinity of direct band-edge excitonic transition of the Nb-doped MoS_2 single crystals from 25 to 300 K using piezoreflectance (PzR). The energies and broadening parameters of the A and B excitons of the Nb-doped MoS_2 samples have been determined accurately via a detailed line shape fit of the PzR spectra. we can infer that the Nb ions are most likely intercalated between the van der Waals gap and stabilize the rhombohedral 3R phase of the MoS_2 crystal.

1. Introduction

Molybdenum disulphide MoS_2 belongs to the group IVA layer type transition metal dichalcogenides having C_7 type crystal structure which is formed of unit layers consisting of transition metal (Mo) atoms sandwiched by chalcogen (S) atoms [1]. The MoS_2 compound has been extensively investigated because of the possible practical application as a solid state lubricant and a catalyst for hydrodesulfurization (HDS) and hydrogen evolution reaction (HER) [2-4], photoelectrochemical solar cells [5-6] , and can also be synthesized in large scale by chemical vapor deposition [7-11]. The monolayer MoS_2 single crystal exhibits high luminous energy photoconversion efficiencies [12-13], excellent electrical [14-16] and optical performance [17, 18] compared to its bulky counterpart. The successful application of this semiconductor compound originates largely from the sandwich interlayer structure, loosely bound by the weak van der Waals forces, as evidenced by easy cleavage in the c-direction along which the S-Mo-S layers are stacked to form the crystal [1, 19]. There are two known polytypes of MoS_2 [1, 19]; two-layer hexagonal and three-layer rhombohedral termed 2H and 3R, respectively. Both have regular layered structures with six-fold trigonal prismatic coordination of the Mo atoms by the sulfur atoms within the layers; 2H-MoS_2 has two layers per unit cell stacked in the hexagonal symmetry and belongs to the space group D_6^4 while 3R-MoS_2 has three layers in the c-direction but has rhombohedral symmetry and belongs to the space group C_3^5 . Naturally occurring 3R-MoS_2 has been found to be consistently rich in certain minor elements such as Re, Nb etc [20]. The incorporation of the impurity will essentially influence the structure symmetry of MoS_2, this is the adoption of the polytype 3R-MoS_2. For Nb-doped MoS_2, the niobium substitutions improve the photocurrent gain much more slowly than in the case of rhenium [21]. Niobium dopants seems to serve more than just as an impurity donor, it is most likely that niobium transition ions can either substitute for the Mo metal ions interstitially or intercalate between the Vander Waals gap resulting in a distortion of crystal structure [20, 21]. The sandwich Nb ions create stronger bonds than the original van der Waals forces and transform the two-layer hexagonal MoS_2 (2H-MoS_2) into three-layer rhombohedral MoS_2 (3R-MoS_2). The transformation of 2H-MoS_2 into 3R-MoS_2 by doping with niobium is perhaps not too surprising as it has been found that naturally

IOP Conf. Series: Materials Science and Engineering **237** (2017) 012041 doi:10.1088/1757-899X/237/1/012041

occurring 3R-MoS_2 is consistently rich in certain minor elements such as Nb, Re etc [21]. Despite the potentially attractive fundamental research offered by Nb-doped MoS_2, very few studies concerning the influence of niobium dopant on the electrical and optical properties of the material have been reported.

In this study, we present piezoreflectance (PzR) measurements which has been proven to be useful in the characterization of semiconductors [22, 23] of Nb-doped MoS_2 which grown by chemical vapor transport method with I_2 as transport agent. An elaborate temperature dependence of the spectral features near the direct band-edge excitonic transitions of the Nb-doped MoS_2 in the range 25 to 300 K has been obtained. The derivative nature of modulation spectra suppresses uninteresting background effects and greatly enhances the precision in the determination of inter band excitonic transition energies. The sharper line shape as compared to the conventional optical techniques has enabled us to achieve a greater resolution and hence to detect weaker features. The PzR spectra are fitted with a form of the Aspnes equation of the fist derivative Lorenzian lineshape [23, 24]. From a detailed lineshape fit we are able to determine accurately the energies and broadening parameters of the excitonic transitions. The parameters which describe the temperature behaviour of excitonic transitions indicate that A-B, caused by interlayer interaction and spin-orbit splitting, correspond to excitonic transitions with different origin. The origin of A, B excitons and the effects of dopant are discussed. The temperature variation of the transition energies has been analyses by the Varshni equation [25] and an expression containing the Bose-Einstein occupation factor for phonons [26,27]. The temperature dependence of the broadening function also has been studied in term of a Bose-Einstein equation that contains the electron (exciton)-longitudinal optical (LO) phonon coupling constant [26,27]. The addition of dopant impurities to a semiconductor is what enables the fabrication of a multitude of interesting devices. The physical role of niobium in influencing the electronic states of the MoS_2 crystal will also be discussed.

2. Experimental

Niobium-doped MoS_2 single crystals have been grown by the chemical vapour transport method with I_2 as a transport agent. The total charge used in each growth experiment was about 10 g. The stoichiometrically determined weight of the doping material was added in the hope that it would be transported at a rate similar to that of Mo. The quartz ampoule containing Br_2 (~5 mg cm^{-3}) and uniformly mixed elements (99.99% pure Mo, Nb and S) was sealed at 10^{-6} Torr. The ampoule was then placed in a three-zone furnace and the charge prereacted for 24 h at 800 ^{0}C with the growth zone at 950 ^{0}C, preventing the transport of the product. The temperature of the furnace was increased slowly to avoid any possibility of explosion due to the exothermic reaction between the elements. The furnace was then equilibrated to give a constant temperature across the reaction tube, and programmed over 24 h to produce the temperature gradient at which single-crystal growth took place. Optimal result were obtained with a temperature gradient of approximately 960 $\rightarrow$ 930 ^{0}C. Single crystalline platelets up to 10 x 10 mm^2 surface area and 2 mm in thickness were obtained. After 24 h, the furnace was allowed to cool down slowly (40^{0}C/h) to about 200^{0}C.

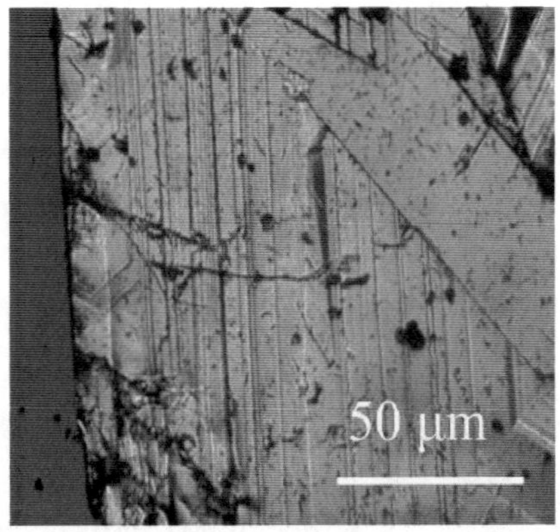

Figure. 1 Photograph of the as-grown niobium-doped MoS_2 single crystal with the surface normal to ***c***-axis

The ampoule was then removed and wet tissues applied rapidly to the end away from the crystals to condense the I_2 vapor. When the ampoule reached room temperature, it was opened and the crystals removed. The crystals were then rinsed with acetone and deionized water. Single crystalline platelets up to 10×10 mm^2 surface area and 2 mm in thickness were obtained. The as-grown niobium-doped MoS_2 single crystal is shown in Fig.1. MoS_2 crystallizes with 2H or 3R structure, while NbS_2 crystallizes in a distorted C6 structure, so that only a small solubility range is to be expected. We do not expect the two solid solutions to be miscible. It was found that a 5% nominal doping of MoS_2 prevented the growth of single crystals.

The PzR measurements were achieved by gluing the thin single crystal specimens on a 0.15 cm thick lead-zirconate-titanate (PZT) piezoelectric transducer driven by a 200 V_{rms} sinusoidal wave at 200 Hz. The alternating expansion and contraction of the transducer subjects the sample to an alternating strain with a typical rms $\Delta l/l$ value of ~ 10^{-5}. A 150 W tungsten-halogen lamp filtered by a model 270 McPherson 0.35 m monochromator provided the monochromatic light. The reflected light was detected by EG&G type HUV-2000B silicon photodiode. The DC output of the silicon photodiode was maintained constant by a servo mechanism of a variable neutral density filter. A dual-phase lock-in amplifier was used to measure the detected signal. Modulated spectra were normalized to the reflectance to obtain $\Delta R/R$. An RMC model 22 close-cycle cryogenic refrigerator equipped with a model 4075 digital thermometer controller was used to control the measurement temperature between 25 and 300 K with a temperature stability of 0.5 K or better.

3. Results and Discussions

Displayed by the dashed curve in Fig. 2 are the PzR spectra near the direct band edge over the range 1.7-2.25 eV for the Au-doped MoS_2 single crystals. The spectra are characterized by two prominent excitonic transitions, A and B excitons. In order to determine the position of the transitions accurately, we performed a theoretical line shape fitting. The energies of both excitonic transitions show the general trend of up-shifting as the temperature is lowered. The linewidths also become narrower in the process. The functional form used in the fitting procedure correspons to a first derivative Lorenzian line shape function of the form [23, 24]

$$\frac{\Delta R}{R} = \mathrm{Re}\left[\sum_{i=1}^{n} A_i^{ex} e^{j\varphi_i^{ex}} \left(E - E_i^{ex} + j\Gamma_i^{ex}\right)^{-n_i}\right] \tag{1}$$

where A_i^{ex} and φ_i^{ex} are the amplitude and phase of the line shape, E_i^{ex} and Γ_i^{ex} are the energy and broadening parameters of the inter band excitonic transitions and the value of n_i depend on the origin of the transition. For the first derivative functional form, $n = 2.0$ is appropriate for the bounded states, such as excitons or impurity transition, while $n = 0.5$ is applicable for three dimensional critical point inter band transitions [28]. The least-squares fits using equation (1) with $n = 2$ can be achieved and the fits are shown as solid curve in figure 1.The fits yield the parameters A_i, E_i and Γ_i. The obtained values of E_i are indicated as arrows and denoted as A and B. The nomenclature used here follows closely that of Wilson and Yoffe [1] and Beal *et al* [29]. The value of E_i obtained here show a general agreement with slight deviation from the corresponding low temperature transmission data of Beal *et al* [29]. We believed the derivative nature of the PzR spectra should offer better accuracy. The fitted values of E_i are displayed in table 1 together with the relevant works of [29-33] for comparison. The prominent A and B excitons are observed to be red-shifted. The energy position of A and B excitons are measured accurately with the PzR experiment. The splittings of excitons A and B ($\Delta_{BA} = E_B - E_A$) are determined to be 0.16 ± 0.013 eV. These number agreed well with the corresponding previously work [31,32], transmission data of Beal *et all* [29], Reflectance [33], Wavelength modulated reflectance (WMR) spectra of Fortin and Raga and Photoconductivity [30]. In the case of the WMR data of Fortin and Raga, their observed signature, B^* is almost certain to be due to the presence of 3R-MoS_2 in their 2H-MoS_2 sample. From the more recent theoretical and experimental studies [34-36], the A and B excitons are attributed to the smallest direct transitions at the K point of the Brillouin zone split by interlayer interaction and spin-orbit splitting [34, 35]. The A exciton belongs to K_4 to K_5 while the B exciton corresponds to K_1 to K_5 optical transitions. The K states have been shown by Coehoorn *et al* [34, 35] to be predominantly metal *d* states with a small contribution from the non-metal *p* states. According to the figure 2, during growth of Nb-doped MoS_2, the niobium transition ion can interfere between two layer, according to the weak van der Waals forces which link them together. This well known process involves a reduction in stoichiometry, distortion of the crystal structure and/or a variation in the position of the Fermi level. The sandwich Nb ions creates stronger bonds than the original van der Waals forces and converts the two-dimensional structure MoS_2 (2H-MoS_2) of the compound into a three-dimension structure MoS_2 (3R-MoS_2). It is very likely that Nb ions stabilize the 3R polytype of MoS_2. The resulting van der Waals forces of 3R-MoS_2 should be weaker than the 2H polytype and this is reflected by the measured reduction in Δ_{BA}.

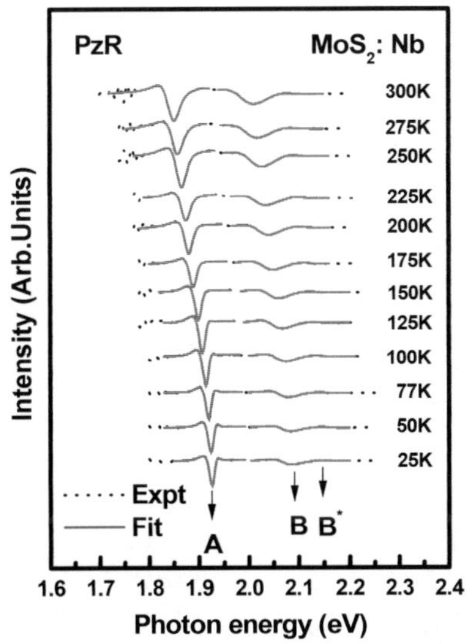

Figure.2. Piezoreflectance spectra of Nb-doped MoS_2 at several temperatures between 25 and 300 K. The dashed curves are the experimental results and the solid curves are least squares fits of Eq. (1).

Table.1 Energies of the excitons A, B and their splitting in MoS_2, MoS_2:Au using equation (1). Relevant values for previously works are included for comparison.

Material	A (eV)	B (eV)	E_{A-B} (eV)	Temperature (K)
MoS_2:Nb [a]	1.928 ± 0.005	2.167 ± 0.008	0.157 ± 0.013	25
	1.920 ± 0.005	2.160 ± 0.008	0.159 ± 0.005	77
	1.852 ± 0.008	2.086 ± 0.01	0.160 ± 0.018	300
MoS_2:Re [b]	1.915	2.066	0.151	15
Undoped MoS_2 [b]	1.928	2.136	0.208	15
MoS_2 [c]	1.88	2.06	0.18	300
MoS_2 [d]	1.9255	2.137	0.2115	4.2
MoS_2 [e]	1.92	2.124	0.204	4.2
MoS_2 [f]	1.929 ± 0.005	2.136 ± 0.008	0.207 ± 0.013	25
	1.845 ± 0.008	2.053 ± 0.01	0.208 ± 0.018	300
2H-MoS_2 [g]	1.910	2.112	0.202	5
3R-MoS_2 [g]	1.908	2.057	0.149	5

[a]Present work.
[b]PzR [31].
[c]Reflectance [33].
[d]WMR [30].
[e]Photoconductivity [30].
[f]PzR [32].
[g]Transmission [29].

4. Summary
In summary we have measured the temperature dependence of the energies and broadening parameters of the direct band-edge excitonic transitions of Nb-doped MoS_2 using PzR in the temperature range 25 to 300 K. From the experimental observations and detailed analysis of

the broadening parameters and energies of the excitonic features A and B, we can infer that the Nb ions are most likely intercalated between the van der Waals gap and stabilize the rhombohedral 3R phase of the MoS_2 crystal instead of forming the ternary $Mo_{1-x}Nb_xS_2$ system. As a result the electronic states of the MoS_2 crystals are modified with a reduction of energy splitting of A and B excitons.

Acknowledgements
The authors would like to acknowledge Professor Ying-Sheng Huang and Professor Ching-Hwa Ho for advising and providing materials and experimental equipment at Semiconductor Characterization Laboratory, National Taiwan University of Science and Technology (NTUST).

References

[1] Wilson J A and Yoffe A D (1969), *Adv. Phys.* **18**, 193
[2] Chang, Y. H. et al. (2013) Highly Efficient Electrocatalytic Hydrogen Production by MoSx Grown on Graphene-Protected 3D Ni Foams. Adv Mater 25, 756–760.
[3] Li, Y. et al. (2011) MoS2 Nanoparticles Grown on Graphene: An Advanced Catalyst for the Hydrogen Evolution Reaction. J Am Chem Soc 133, 7296–7299.
[4] Kong, D. et al. (2013) Synthesis of MoS_2 and $MoSe_2$ Films with Vertically Aligned Layers. Nano Lett 13, 1341–1347.
[5] Kam K K and Parkinson B A, (1982) *J. Phys. Cem.* **86**, 463.
[6] Li S J, Bernede J C, Pouzet J and Jamali M, (1996) *J. Phys.: Condens. Matter* **8**, 2291
[7] Shi, Y. et al. (2012) van der Waals Epitaxy of MoS_2 Layers Using Graphene As Growth Templates. Nano Lett 12, 2784–2791.
[8] Liu, K.-K. et al. (2012) Growth of Large-Area and Highly Crystalline MoS_2 Thin Layers on Insulating Substrates. Nano Lett 12, 1538–1544.
[9] Lin, Y.-C. et al. (2012) Wafer-scale MoS_2 thin layers prepared by MoO_3 sulfurization. Nanoscale 4, 6637-6641.
[10] Lee, Y.-H. et al. (2012) Synthesis of Large-Area MoS_2 Atomic Layers with Chemical Vapor Deposition. Adv Mater 24, 2320–2325.
[11] Zhan, Y., Liu, Z., Najmaei, S., Ajayan, P. M. & Lou, J. (2012) Large-Area Vapor-Phase Growth and Characterization of MoS_2 Atomic Layers on a SiO2 Substrate. Small 8, 966–971.
[12] D. S. Ginley, R. M. Biefeld, B. A. Parkinson and K. Keung-Kam, (1982) J. Electrochem. Soc., 129 145.
[13] A. Casalot, M. Chaouch and G. Vacquier, Ann. (1986) ('him.Fr., 11 509.
[14] Radisavljevic, B., Radenovic, A., Brivio, J., Giacometti, V. & Kis, A. (2011) Single-layer MoS2 transistors. Nat Nano 6, 147–150.
[15] Lembke, D. & Kis, A. (2012) Breakdown of High-Performance Monolayer MoS_2 Transistors. ACS Nano 6, 10070–10075.
[16] Wang, H. et al. (2012) Integrated Circuits Based on Bilayer MoS_2 Transistors. Nano Lett 12, 4674–4680.
[17] Eda, G. et al. (2011) Photoluminescence from Chemically Exfoliated MoS_2. Nano Lett 11, 5111–5116.
[18] Splendiani, A. et al. (2010) Emerging Photoluminescence in Monolayer MoS_2. Nano Lett 10, 1271–1275.
[19] R. Murray, B.L. Evans, (1979) J. Appl. Crystallogr. **12** 312.

[20] Clark A H 1970 N. Jahrbuchf. Mineral. Monatsheffe **3** 33
[21] J. B Legma, G. Vacquier, Traorè and A. casalot, 1991 Mater. Sci. Eng. B **8** 167.
[22] Mathieu H, Allegre J and Gil B, (1991) *Phys. Rev*. B **43**, 2218
[23] Pollak F H and Shen H, (1993) *Mater. Sci. Eng*. R **10**, 275
[24] Aspnes D E 1980 *Optical Properties of Semiconductors (Handbook on Semiconductors 2)* ed M Balkanski (Amsterdam: North-Holland) p 109
[25] Varnish Y P, (1967) *Physica* **34**, 149
[26] Lantenschlager P, Garriga M, Logothetidis S and Cardona M, (1987) *Phys. Rev*. B **35**, 9178
[27] Lantenschlager P, Garriga M, Vina L and Cardona M, (1987) *Phys. Rev*. B **36**, 4821
[28] Aspnes D E 1980 *Optical Properties of Semiconductors (Handbook on Semiconductors 2)* ed M Balkanski (Amsterdam: North-Holland) p 109
[29] Beal A R, Knights J C and Liang W Y, (1972) *J. Phys. C: Solid State Phys*. **5,** 3540
[30] Fortin E and Raga F, (1975) *Phys. Rev*. B **11**, 905
[31] Malikova L, Krystek W, Pollak F H, Dai N, Cavus A and Tamorgo M C, (1996) *Phys. Rev*. **B** 54, 1819
[32] Qiang H, Pollak F H, Sotomayor Torres C M, Leitch W, Kean A H, Stroscio M, Jafrate G J and Kim K W, (1992) *Appl. Phys. Lett*. **61**, 1411
[33] Beal A R, Liang W Y and Huges H P, (1976) *J. Phys. C: Solid State Phys*. **9,** 2449
[34] Coehoorn R, Haas C, Dickstra J, Flipse C J F, de Groot R A and Wold A, (1987) *Phys. Rev*. B **35**, 6195
[35] Coehoorn R, Haas C, and de Groot R A, (1987) *Phys. Rev*. B **35**, 6203
[36] Straub Th, Fauth K, Finteis Th, Hengsberger M, Claessen R, Steiner P, Hufner S and Blaha P, (1996) *Phys. Rev*. B **53**, 16152

Cellulose acetate layer effect toward aluminium corrosion rate in hydrochloric acid media

K S Andarany[1], A Sagir[1], A Ahmad[1], S K Deni[1,2] and W Gunawan[3]
[1] Mechanical Engineering, Mercubuana University, South Meruya No. 1 Kembangan West Jakarta 11650 Indonesia
[2] Physic Research Center, Indonesian Institute of Science, Building 440 Serpong South Tangerang, Indonesia
[3] Electrical and Electronic Engineering Department, University Technology Petronas 32610 Seri Iskandar, Perak Darul Ridzwan, Malaysia

* sagir.alva@mercubuana.ac.id

Abstract. Corrosion occurs due to the oxidation and reduction reactions between the material and its environment. The oxidation reaction defined as reactions that produce electrons and reduction is between two elements that bind the electrons. Corrosion cannot be inevitable in life both within the industry and household. Corrosion cannot eliminate but can be control. According to the voltaic table, Aluminum is a metal that easily corroded. This study attempts to characterize the type of corrosion by using a strong acid media (HCl). Experiment using a strong acid (HCl), at a low concentration that occurs is pitting corrosion, whereas at high concentrations that occurs is corrosion erosion. One of prevention method is by using a coating method. An efforts are made to slow the rate of corrosion is by coating the metal with "cellulose acetate" (CA). cellulose acetate consisted of cellulose powder dissolved in 99% acetic acid, and then applied to the aluminum metal. Soaking experiments using hydrochloric acid, cellulose acetate is able to slow down the corrosion rate of 47 479%.

1. Introduction

Aluminium known as the second largest attractive materials next to the iron and widely applied in daily life and industrial process such as in automotive, aerospace, construction and electrical application due to lightweight, ductility, and good corrosion resistance. The good corrosion resistance of aluminium attributed to a protective thin invisible oxide film that forms on aluminium surface. When exposed, acid or alkaline conditions (systems) destroy the protective coating and corrosion of aluminium occurs, yielding Al^{3+} (trivalent aluminium) ions in the acid and Al_2O_3 (aluminate) ions in the alkaline medium Hence, it is needed to add some inhibitor to increase corrosion resistance which has low or zero environmental impact[1].

Nowadays, many researches have conducted to developed environmentally safe corrosion inhibitor for aluminium in acidic media. Abdulwahab et al. (2011) studied the inhibitor of an aqueous extract of bitter leaf on aluminum alloy in 0.5 M HCl[2]. Chaubey et al. (2015) studied about extract papaya peel as potential corrosion inhibitor for Aluminium alloy in 1 M HCl[3]. And Khadraoui et al. (2015) studied extract aromatic plant Thymus algeriensis as a new eco-friendly corrosion inhibitor for 2024 aluminium alloy in 1 M HCl[4]. Recently, most of the study to observe corrosion behaviour conducted in low concentration acidic media (not more than 1M). in that case, this present study conducted not only in low concentration but also in higer concentration of acidic media (up to 3M).

In this study, we have attempted to prevent the corrosion of Al using cellulose acetate membrane. Cellulose acetate (CA) produced by the reaction of cellulose with acetic acid, the common material used for industrial acetylation of cellulose are wood and cotton. CA is widely applied in photographic film, textile, in cigarette filter and surface coating. But, the main application of CA is in production of membrane due to easy manufacture, cost effectiveness, environmental safety and strong mechanical properties [5-7]. The present paper is focused to evaluate CA inhibition effect due to corrosiveness behavior Al in HCl solutions which there is no study conducted on CA as corrosion inhibitor for Al in HCl solution.

2. Experimental

2.1. Material

Aluminum specimen with dimension 2cm x 2cm and thickness 0.1 cm was used for the study was obtained from beverage cans. It was sanded and cleaned with alcohol 70%.CA membrane were obtained by dissolving 5% cellulose powder with acetic acid.

2.2. Methods

Corrosion rate determined using weight loss method. It is the simplest and common conventional method for determining corrosion rate. The method involves the exposing of the specimen to a process environment for a given period, then removing the specimen for measurement. The basic measurement which determined from corrosion specimen, is weight loss, the weight loss taking place over the period of exposure being expressed as corrosion rate.

Specimen weight measured before and after specimen immersed in acidic media. The specimens immersed in HCl solution with concentration 0.5 M, 1M, 2M and 3M for 5 minutes in the present and absence of inhibitor. Corrosion rate are calculated from weight loss method with the formula :

$$CR = \frac{87.6 \ x \ W}{D.A.T} \tag{1}$$

Where: W = weight loss (g),
D = density (g/cm3)
A = area of specimen (cm2) and
T = time (hr)[8].

The inhibition efficiency (%IE) were calculated using formula:

$$\% \ IE = \frac{W1 - W2}{W1} . 100\% \tag{2}$$

Where: W1 and W2 are the corrosion rates in the absence and presence of the inhibitor[9].

2.3. Scanning electron microscopy (SEM)

Surface morphology of specimen immersed in HCl solution with presence and absence of inhibitor were analysed by recording SEM image using HITACHI SU3500.

3. Result and Discussion

Corrosion research on Al is very interesting to learn, considering Al is a material that is often used in various fields. In this study we studied the effect of CA membrane coating on Al on corrosion rate. Corrosion rate measurement was done by using 2 methods namely weight loss

IOP Conf. Series: Materials Science and Engineering **237** (2017) 012042 doi:10.1088/1757-899X/237/1/012042

method and using imaging with FE / SEM. Media test solution used is HCl with concentration 0.1M, 0.5M, 1M, 2M, and 3M. The corrosion rate of aluminium in HCl solution has shown in table 1. Corrosion rate were determined at room temperature after 5 minutes immersion period.

Table 1. Corrosion rate by weight loss method.

Blank		CA Inhibitor	E%
HCl (Molar)	Corrosion rate (mm/y)	Corrosion rate (mm/y)	
0.1	0.05324	0.02904	45.45
0.5	0.11132	0.10648	4.35
1	0.21781	0.19844	8.89
2	0.72119	0.31945	55.71
3	1.67956	0.80831	51.87

Based on tables 1 and 2 above, it is seen that the higher concentration of HCl solution then the corrosion rate will be faster [10] This is because HCl is a strong acid solution and so can damage the surface of Al metal more quickly. Through reaction:

$2Al + 6HCl \rightarrow 2Al^{3+} + 3H_2\uparrow + 6Cl^-$

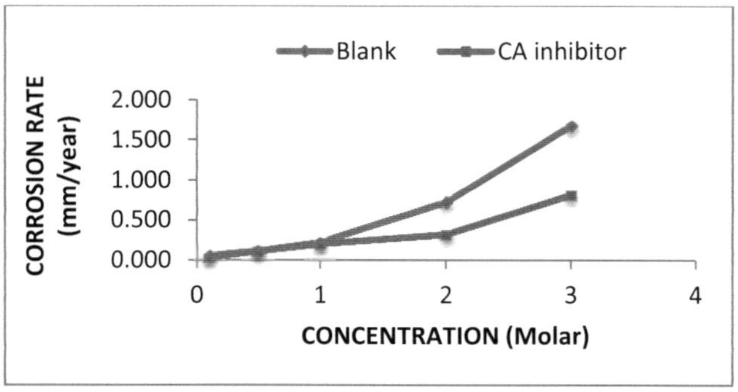

Figure 1. The corrosion rate of Aluminium in various concentration of HCl

The effect of the addition of CA membrane as inhibitor tested at different concentration of HCl solution studied by weight loss measurement. The value of percentage inhibition efficiency and corrosion rate obtained from weight loss method at different concentration of HCl solution summarized in Table 1. it was observed that the increasing corrosion rate associated with the increasing of HCl concentration solution shown in figure 1. By using CA as an inhibitor it can decreased corrosion rate with efficiency up to 50%. Although coating has performed the corrosion rate still occurs in proportion to the concentration. This is inseparable from the morphological changes of cellulose acetate on HCl acid media, as seen in FE / SEM, the higher the porosity of the CA the more enlarged due to the deacetylation in CA. (yamashita et al, 2003)[11].

IOP Conf. Series: Materials Science and Engineering **237** (2017) 012042 doi:10.1088/1757-899X/237/1/012042

Table 2. Corrosion thickness.

	Blank	CA Inhibitor
HCl (Molar)	Thickness	Thickness
0.1	6.41	0.038
0.5	9.615	5.769
1	22.435	6.538
2	74.358	7.435
3	91.025	8.974

Its observed that corrosion started occur in HCl solution with 1M represent from morphological aluminium surface recorded by SEM in figure 2

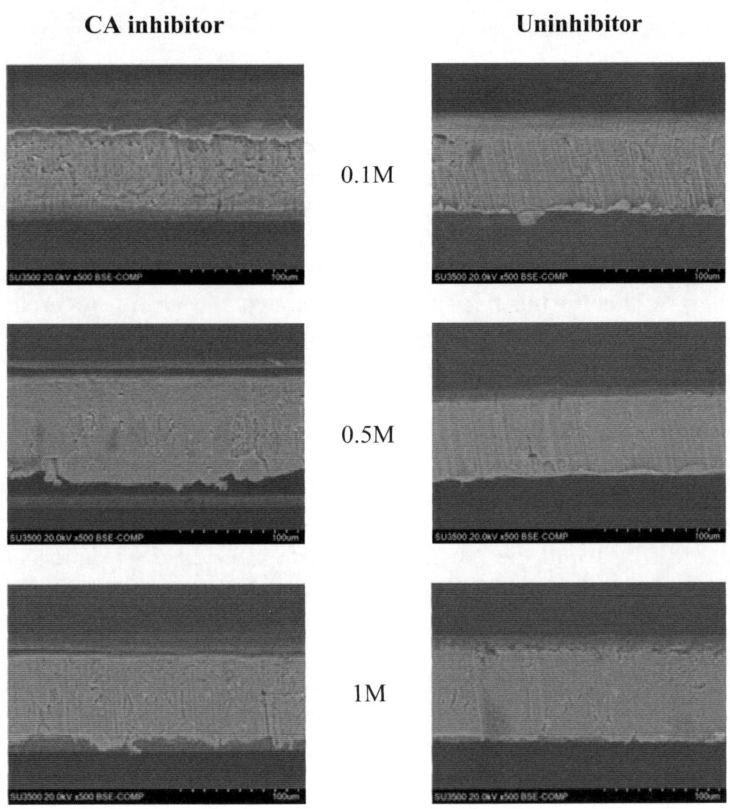

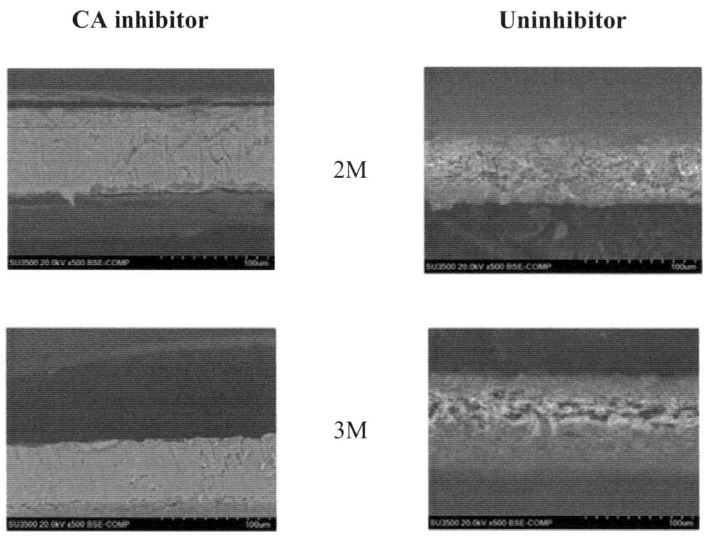

Figure 2. Microstructure image of Aluminium in various HCl by FE-SEM

The increase of the HCl immersion concentration on the CA inhibitor membrane can still be resolved well up to 3 M concentration in the sample by coating. This could be proven in SEM surface view morphology, where membrane shows relatively similar porous morphology in all conditions. Figure 2. Nevertheless, the increasing in HCl concentration affected to the adhesivity of the inhibitor to the Al substrate. Although the ability to protect the sample against acid corrosion can still be remedied even when under strong acid conditions with high concentrations (3 M).

As for the sample without the CA inhibitor, SEM morphological shows clearly differences when compared with the sample using CA inhibitor. Under conditions of 0.1 and 0.5 M, corrosion growth is not significant when compared to 1, 2 and 3 M., although there is some corrosion (oxide) on the surface under condition of 0.1 and 0.5 M. However, with increasing acid concentrations starting from 1 M to 3 M, the corrosion layer (oxide) begins to cover the metal surface and diffuses into the Al metal substrate affected degradation on Al substrate especially under condition of 2 and 3 M as evidenced by the SEM cross section [8,10].

4. Conclusion

Cellulose acetate acts as a good inhibitor for the corrosion of aluminum metal in HCl solution with concentration 0.1, 0.5, 1, 2 and 3M. The inhibition efficiency increases 55.71% under condition of 2M HCl solution and 51.87 % under condition of 3M HCl solution. CA can be considered as corrosion inhibitor due to easy manufacture, cost effectiveness and environmental.

Acknowledgement

Thank you to Mercu Buana University especially to Mechanical Engineering Department, Physical Research Centre LIPI and Universiti Petronas Malaysia for all the research support given.

References

[1] PrabhuDeepa., & Rao, Padmalatha. (2013). CoriandrumSativum L.—A Novel Green Inhibitor For The Corrosion Inhibition Of Aluminium In 1.0 M Phosphoric Acid Solution. Journal of Mechanical Engineering http://dx.doi.org/10.1016/j.jece.2013.07.004

[2] Abdulwahab M, Kasim A, Bello KA, Gaminana JO (2011). Corrosion inhibition of multi-component aluminium alloy in hydrochloric acid solution by aqueous extracts of bitter leaf (Verniniaamygdalina) powder. Adv. Mater. Res. 367:319-325.

[3] Chaubey, N., Singh, V. K., &Quraishi, M.A. (2016). Papaya Peel Extract As Potential Corrosion Inhibitor For Aluminium Alloy In 1M Hcl: Electrochemical And Quantum Chemical Study. Ain Shams Engineering Journal. http://dx.doi.org/10.1016/j.asej.2016.04.010.

[4] Khadraoui, A., Khelifa. A.,Hachama. K., &Mehdaoui. R. (2015). Thymus Algeriensis Extract As A New Eco-Friendly Corrosion Inhibitor For 2024 Aluminium Alloy In 1 M HCl Medium. Journal of Molecular Liquids. http://dx.doi.org/10.1016/j.molliq.2015.12.064.

[5] Cheng, H. N., Dowd, M. K., Selling, G. W., & Biswas, A. (2010). Synthesis Of Cellulose Acetate From Cotton By Products. *Carbohydrate Polymers, 80*, 449-452.

[6] Kamal, H., Abd-Elrahim, F. M., &Lotfy, S. (2014). Characterization And Some Properties Of Cellulose Acetate-Co-Polyethylene Oxide Blends Prepared By The Use Of Gamma Irradiation. *Journal of RadiationResearch and Applied Science, 7*, 146-153.

[7] Candido, R.G., Godoy, G.G., &Goncalves, Adilson R. (2017). Characterization And Application Of Cellulose Acetate Synthesized From Sugarcane Bagasse. *Carbohydrate Polymers,* http://dx.doi.org/10.1016/j.carbpol.2017.03.057.

[8] Kumar. P.S., Smart. D.S.R., Alexis. S.J. (2017). Corrosion BehaviorOf Aluminium Metal Matrix Reinforced With Multi-Wall Carbon Nanotube. Journal of Asian Ceramic Society. 5.71-75. http://dx.doi.org/10.1016/j.jascer.2017.01.004.

[9] Marques. M.S., Zepon. K.M., Petronilho. F.C., Valdir. S., &Kanis. L.A. (2017). Characterization Of Membranes Based On Cellulose Acetate Butyrate/Poly (Caprolactone) Triol/Doxycycline And Their Potential For Guided Bone Regeneration Application. Materials Science & Engineering C. 10.1016/j.msec.2017.03.095.

[10] Uwiringiyimana. Eugene., Sylvester. O'Donnell P., Joseph. Ifeoma V., & Adams. Feyisayo V., (2016). The Effect Of Corrosion Inhibitors On Stainless Steels And Aluminium Alloys: A review. African Journal of Pure and Applied Chemistry. 10(2). 23-32. 10.5897/AJPAC2016.0676

[11] Yamashita. Yoichiro ., Endo. Takeshi., (2003). Deterioration Behavior of Cellulose Acetate Films in Acidic or Basic Aqueous Solutions. Journal of Applied Polymer Science, Vol. 91, 3354–3361.

The analysis of effect of heat treatment temperature on micro structure, crystal structure and hardness material on alloy $Zr_{96,2}Sn_{2,3}Nb_{1.1}Fe_{0,4}$

Bungaran Saing[1] and Budiarto[2]
[1]Chemical Engineering, Faculty of Engineering, University of Bhayangkara Jakarta, Indonesia
[2]Mechanical Engineering, Faculty of Engineering, Christian University of Indonesia, Jakarta, Indonesia

* bungaran.saing@dsn.ubharajaya.ac.id, budidamaz@gmail.com

Abstract. The effect of heat treatment temperature (in 500 ^{0}C, 600 ^{0}C, and 700 ^{0}C) from $Zr_{96,2}Sn_{2,3}Nb_{1.1}Fe_{0,4}$ as fuel cladding material candidate's reactor nuclear power plant; for microstructure, crystal structure, and hardness has been carried out. Several characteristic was conducted by using an optical microscope, x-ray diffractometer and Vickers method. The result showed the crystalline characteristic peaks by a tendency to a single crystal formation and microstructure is getting better with less precipitation and the hardness of the alloy is 329.6 4.5 HVN after the homogenization process.

1. Introduction

The increasing of the global energy demand causes the need of more sustainable energy sources including the nuclear energy. In order to developed the nuclear technology, it required a new design and the construction of nuclear fuel cladding. As a current material, commercial zirconium alloy was expensive and unable to self produced (limited to patent right). Thus, the new material must have a characteristic such as strength, corrosion resistance, fabrication capabilities, neutron adsorption cross section, and cheap [1].

In recent years, the research of zirconium alloys with Nb, Sn, Fe, Mo and Ge has been developed. In order to make the new material cheaper, the cladding material must use zirconium element that weight at least 95% [3.7] and must made domestically. One of the properties of the material that needed is the resistance in high temperatures and corrosion. In accidental condition with a lot of loss of coolant (LOCA), the integrity of the nuclear plant need to be maintained. The good material should be forming a protective layer from continues oxidation process due to high temperature oxidation that occurred on a microscopic scale. The candidate of this nuclear fuel cladding was $Zr_{96,2}Sn_{2,3}Nb_{1.1}Fe_{0,4}$ alloy materials.

2. Theory

The heat treatment is a process that can control the material's properties, especially metal; it also can alter the material's structure. The effects of this process can make the material harder, malleable, increase toughness and to refine the metal grain size.

There are several processes of heat treatment which are annealing, sintering, calcination, tempering, normalization and homogenization. Annealing was a heating process that can reduce dislocation density and growth grain by recrystallization mechanisms. Sintering was a heating process of pellet's powder at its recrystallization temperature for material's strengthening and porosity elimination. Calcination is a process which powder's is heated at 1000 ^{0}C to remove the carbonates and water [4]. Tempering is a special heating process for softening the steel [5]. Homogenization is process heat treatment at high temperature for reducing unwanted chemical elements by diffusion and to get ordered crystal structure and grain. Normalization is a heat treatment at recrystallization temperature and cooled down at room temperature. Factors that can affect the process of heat treatment [4,5]: heating temperature, restraining temperature and cooling rate.

2.1 Atomic Diffusion

The atoms will be on a stable position at 0 K (-273 C). In this temperature, they are at the lowest energy. If the temperature is increased, the atom's energy would be increased too and affect the atom's movement. This process called diffusion. Diffusion mechanism was classified based on displacement and position atoms which is:

a) Empty mechanism
   There is always a blank space in crystal structure configuration. This happens by hopping of the atom to the vacant position. When one atom jumped to fill the blank space and the other will replaced it, there will be a continuity process.
b) Interstitials mechanism
   If there was a different size between 2 atoms, it can be happen interstitial mechanism. For example: the diffusion of atom nitrogen to metal surface.
c) Interchange mechanism
   Another name was ring mechanism, an atom movement such as cycle. It has a large distortion, so need a big energy.

2.2 Zirconium Alloys

Zirconium is a very hard metal grey which have melting point temperature 1860°C, and a low neutron absorption 0.18 barn. Zirconium can be easily corroded [6,7]. Therefore, if we wanted some good mechanical properties we must made alloy of Zirconium (Zr-Nb-Fe-Sn) as a nuclear fuel cladding. The development of power reactors are still on going by the goal energy efficiency and reducing of waste and cost. Power efficiency can easily achieved at 400-600^{0}C [9].

3. Research Methodology

3.1 Materials and Equipment

<u>Material</u> : alloy material was made by smelting techniques in arc melting furnace, the composition were 96.2% Zr, 2.3% Sn,1.1% Nb, and 0.4% Fe. Metallographic material and complete etching materials.

<u>Equipment</u>: Arc furnace melting, grinding, polishing, cutting machine, optical microscope, Vickers hardness tester method, and an X-ray diffractometer.

3.2 Procedures

- Weigh Zr, Sn, Nb, and Fe and placed them in the crucible to be made $Zr_{96,2}$ $Sn_{2,3}Nb_{1.1}$ $Fe_{0,4}$ alloy material. Furnaced them by arc melting furnace in argon gas condition to avoid them oxidation.
- Melted and cooled them then cut into 2x2 by diamond blade cutter type JMQ-12 at low speed to ensure a neat piece surface and clean of impurities.
- Normalized them by heated at 1000 ° C.
- Homogenized them by heat treatment at variation temperatures: 500 ° C, 600 ° C, 700 ° C and held for 2 hours.

IOP Conf. Series: Materials Science and Engineering **237** (2017) 012043 doi:10.1088/1757-899X/237/1/012043

- Polished them with sandpaper in roughness consecutive sequence : 120, 400, 800, 1200, to 2000.
- Topped them at velvet pabric samples on diamond paste 0.25 lm.
- Etched their surface with 3% Nital solution in 15 seconds until the surface shiny like mirror. Nital solution was made from a mixture between 3 ml HNO_3 with 97 ml methanol technical.
- Dried them with blow dryer at room temperature until completely dry.
- Tested them for microstructure by optical microscope, for crystal structure and grain size by X-ray diffractometer and for hardness by Vickers test method.

The x-ray diffractometer with a Cu-K wavelength ( ) at approximately 1,5405Å belongs to central laboratory UIN Ciputat, Tangerang. optic microscopy tool and hardnes tester Vickers method at engineering Laboratory FT, UKI, Jakarta.

4. Results and discussion

4.1a. Analysis crystal structure and grain size of the alloy material Zr $_{96,2}$ Sn $_{2,3}$Nb$_{1.1}$ Fe$_{0,4}$
Ingot alloy materials visually formed a solid alloy with a little oxidation at the surface and homogeneous melted. Smelting alloy synthesis decreased Gibbs free energy of each element for made a particular phase at alloy. In this condition, both elements Nb and Fe has melted at 1800^{0}C and has diffused to zirconium matrix formed phase / two new compounds Zr $(Nb,Fe)_2$ and $(Zr, Nb)_3Fe$ [9].

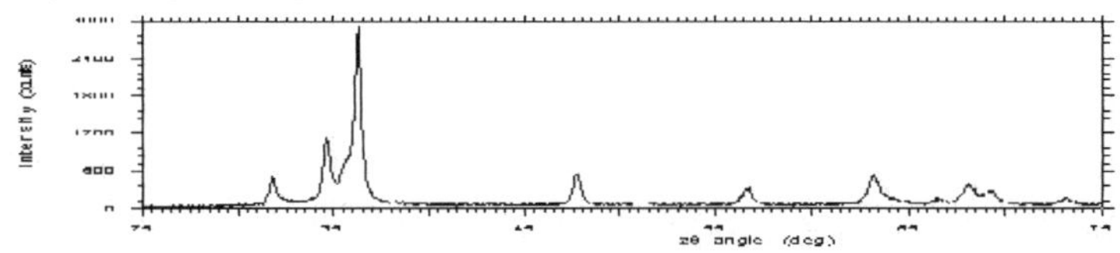

(a)

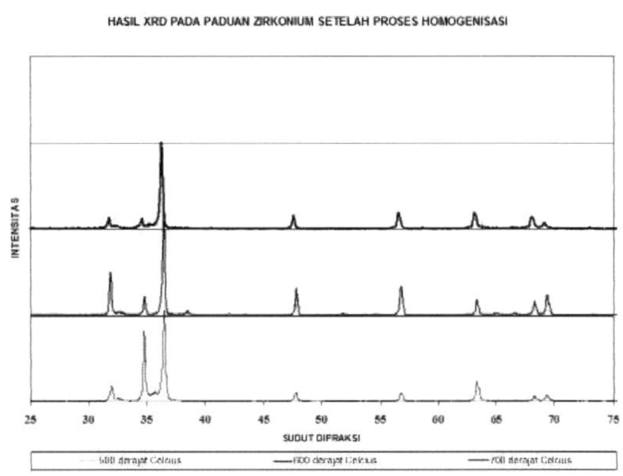

(b) (c) (d)

Figure 4.1 diffractograms of alloy materials Zr $_{96,2}$ Sn $_{2,3}$Nb$_{1.1}$ Fe$_{0,4}$ a) Without heat treatment and after heat treatment at temperature (b). 700°C, (c) 600°C, (d). 500°C.

Figure 4.1 showed a diffractogram pattern of alloy materials Zr $_{96,2}$ Sn $_{2,3}$Nb$_{1.1}$ Fe$_{0,4}$; the measurement results used x-ray diffractometer with a Cu-K wavelength ( ) around 1,5405Å. Adequate X-ray beam penetration power can displayed a best profile peak to background ratio. Diffraction peak intensity very clearly difference and sharply visible in Figure 4.1 (a) - 1(d), like needles. At the scattered start peak looked widen because of the

IOP Conf. Series: Materials Science and Engineering **237** (2017) 012043 doi:10.1088/1757-899X/237/1/012043

influence of fluorescence material. The whole pattern of diffraction peaks dominated by the phase zirconium cladding as the primary matrix material. Three main first peak in dominant phase were (10 0), (0002) and (10 1) looked at the angle 2 = 32.080^0 , 34.840^0 and 36.650^0. The highest intensity for 2010 counts / second (cps) was saw at field (10 1) belongs to zirconium phase. The diffraction pattern showed two predominant phase which formed before heat treatment were treated as follows: phase zirconium (Zr) in hexagonal structure, new phases or compounds Zr (Nb, Fe) $_2$ and (Zr, Nb)$_3$Fe in tetragonal structure. Some peak intensity shrinkaged, even there was only one dominant peak at 700 ° C. it was caused by atom diffusion of alloy material Zr $_{96,2}$ Sn $_{2,3}$Nb$_{1.1}$ Fe$_{0,4}$.

Table 4.1 Results matter of the grain size of the alloy material Zr $_{96,2}$ Sn $_{2,3}$Nb$_{1.1}$ Fe$_{0,4}$

Temperature (°C)	2 (deg)	d(Å)	FWHM (deg)	The Grain Size (Å)
Original (yet H.T)	36,251	2,4759	0,448	2.989
500	36,465	2,4620	0,200	7.299
600	36,455	2,4728	0,180	8.117
700	36,285	2,4627	0,160	9.354

Table 4.1 showed the heating process made the grain size becomes larger, the higher temperature the larger size grain. It can be happen because of atom diffusion at boundaries grain towards grain when it heated. There was a striking changeable depend on heat treatment temperatures: 500 ° C, 600 ° C and 700 ° C, the change of crystal structure direct to the single crystal formation in the plane *hkl* (101). There were some shrinkage peak intensity when the temperatures changed from 500 ° C to 600 ° C, and at 700 ° C there was a dominant peak with the angle θ 36.285^0. It indicated the material direct to homogeneous single crystal structure [6]. It become zirconium alloy would be better able as fuel cladding at high temperature. At high temperature the atoms will be retained their crystal structure, so there was no creep or fractures [5].

4.2 Analysis of the microstructure of alloy material Zr$_{96,2}$Sn$_{2,3}$Nb$_{1,1}$Fe$_{0,4}$

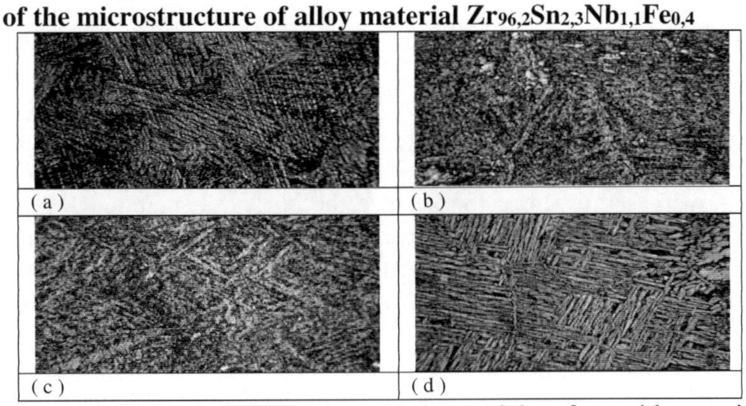

Figure 4.2 Photogram Zr $_{96,2}$ Sn $_{2,3}$Nb$_{1.1}$ Fe$_{0,4}$ alloy material surface with magnification of 500 X: (a) the original sample (not heat treatment), after the heat treatment (b) 500 ° C, (c) 600 ° C, and (d) 700 ° C.

Figure 4.2 appeared the microstructure of alloy material Zr $_{96,2}$ Sn $_{2,3}$Nb$_{1.1}$ Fe$_{0,4}$ before and after heat treatment at different shape of micro structure. Atom diffusion easily occurred at the material surface because it has lower stability than atom in crystal. This is due to the coordination of surface atoms was equal to the coordination of crystal atoms. Therefore the surface atoms have a higher free energy and a less robust bond [5]. There were black spots on grain and boundaries grain. It indicated the presence of material precipitation. Previously [9.14] precipitation of alloy material can be happen if there was intermetallic compounds in material such as Zr (Nb, $Fe)_2$ and $(Zr, Nb)_3Fe$. The precipitation may be made in the process to make alloy Zikonium or when heat treatment was given either at normalization or homogenization. When precipitation processed some element constituent material will secede and made intermetallic phases. Generally precipitation has a high energy and inhibit deformation rate of the pressure received from the material [11-13].

4.3 Analysis hardness of alloy material $Zr_{96,2}Sn_{2,3}Nb_{1,1}Fe_{0,4}$

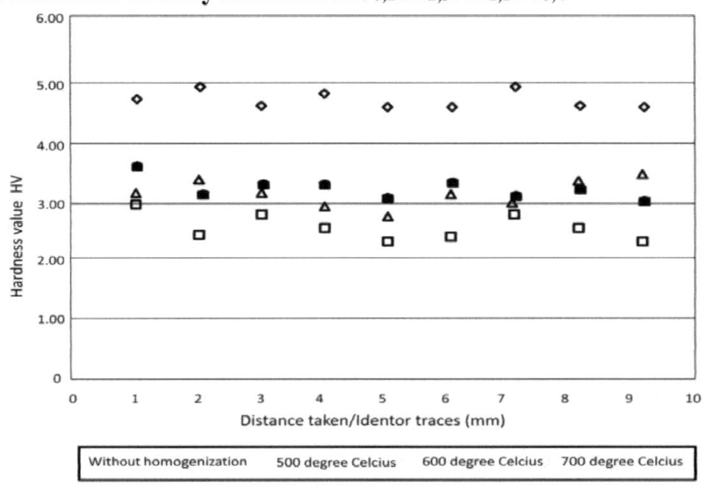

Figure 4.3 Hardness Test Results on alloy material $Zr_{96,2}Sn_{2,3}Nb_{1,1}Fe_{0,4}$

Figure 4.3 showed the higher temperature the more decreased hardness of the alloy. It was happen because of atom diffusion, there was migration from boundary to limit and make space; the consequences of violence will diminish [5]. The violence reduction as well as solid dislocation usually inhibit the creep propagation and pressure deformation was reduced. It become material hardness was reduced too. Figure 4.3 showed after homogenization, the material hardness increased, even in the surface the hardness scattered flat. If the hardness have flatten the material would be better, because the thermal expansion distribute quietly. At high temperatures dislocation of grain boundary made creep spread. Because the atoms get heat energy made the distance between atoms larger and propagated the creep [5]. Heat treatment multiplied scattering wheat field. This field made the crystals to their dominant orientation (preferred orientation); was field (10 1). Facts showed that the larger of grain size was 2989 (before heat treatment) for 9354 (after the heat treatment) caused a shift to the field (10 1). The hardness 329.6 HV after heat treatment the alloy material $Zr_{96,2}Sn_{2,3}Nb_{1,1}Fe_{0,4}$ smaller than before homogenization process 477,11HV. However it was feasible to be used as fuel cladding material, due to the hardness value 76.27 HV as fuel cladding material hardness minimum [8:13].

Conclusion

- Material alloys after heat treatment Zr96,2Sn2,3Nb1,1Fe0,4 made two dominant phases: zirconium phase hexagonal structure and intermetallic phases were Zr (Nb, Fe)$_2$ and (Zr,Nb)$_3$Fe the tetragonal structure. Also occured width grain from 2,989 (before heat treatment) to 9,354 , after heat treatment in the field (10 1).
- A violence decreased in $Zr_{96,2}Sn_{2,3}Nb_{1,1}Fe_{0,4}$ alloy material after heat treatment of 477.11 HV before heat treatment decreased to 261.44 HV at a temperature of 500^0C.

References

1. Slamet Pribadi, Joko Kisworo, Yatno. (2005) Synthesis of Zr-Sn alloy-Mo to get a New Material for Nuclear Fuel Element Cladding. Journal of Nuclear Materials Engineering. January; 1: 1-14.
2. Bandriyana B, Ismoyo A.H, Parikin. (2008) Rolling Process and Characterization of Alloy Zr-Nb-Mo-Ge for Nuclear Fuel Cladding Materials. Journal of Materials Science Indonesia; December Special Edition; p. 93-98.
3. Sungkono.2004. "The influence of homogenization process parameters on the microstructure and hardness Zr-0,4Mo-0,5Fe-0,5Cr" Solid .Konduktor Vol. 5. BATAN. Serpong.Tangerang
4. Dieter. 1996. "Mechanical Metallurgy". Volume 1.Erlangga: Jakarta
5. Vlack, Van L H.1994. "Science and Technology of Materials (Metals and Non Metals Science)." Erland Fifth Edition. Jakarta
6. Cullity, B. D., 1977, "Elements of X-Ray Diffraction", Addison-Wesley publishing company.California.
7. Lustman, B and Kerze, F. 1955. "The Metallurgy of Zirconium". Mc Graw Hill Book Company, INC.
8. Sigit and Elin Nuraini, 1998, "Characterization AIMgSi as Structural Materials Research Reactor" ISSN 1410-1998 Proceedings Presentations / lmiah Nuclear Fuel Cycle III, PEBN-BA TAN Jakarta.
9. Sugondo and Futichah. 2005. "Characterization Crystallite Size, Micro Strain and Yield Strength Sn6,1% Zr91% Nb2,2% and 0.7% Fe by X-ray Diffraction" Indonesia .Jurnal Materials Science, Vol. 6, No. 2. BATAN. Indonesia.
10. Sugondo and Futichah. 2004. "Effects of Annealing on Strain and Residual Stress Alloys Nb2,2 Zr91% Sn6.1% and 0.7% Fe" .Procceding of the 6th National and X-Ray Scattering, ISSN 1410-7686.
11. Martoyo, Nusin Naidoo, Usman Sudjadi. (2008) Hardness Test and Inspection . Slamet Pribadi, Joko Kisworo, Yatno. (2005) Synthesis of Zr-Sn alloy-Mo to get a New Material for Nuclear Fuel Element Cladding. Journal of Nuclear Materials Engineering. January; 1: 1-14.
12. Bandriyana B, Ismoyo A.H, Parikin. (2008) Rolling Process and Characterization of Alloy Zr-Nb-Mo-Ge for Nuclear Fuel Cladding Materials. Journal of Materials Science Indonesia; December Special Edition; p. 93-98.
13. Ismoyo A.H, Parikin, Bandriyana B. (2009) Synthesis alloys Zr-Nb-Mo-Ge with Ge Item Variation. Journal of Materials Science Indonesia; February; 10: 199-202.
14. Parikin, Andika Dawn, Ismoyo A.H, Bandriyana. Kuntoro I. (2010) Neutron Diffraction Technique on the Structural Identification of Zr-Nb-Mo-Ge alloy. International Conference on Materials Science and Technology; October 19-23; Serpong, Indonesia. Jakarta: PTBIN-BATAN & MRS-INA; Page. 91-97.

IOP Conf. Series: Materials Science and Engineering **237** (2017) 012044 doi:10.1088/1757-899X/237/1/012044

Behavior of Sn-$_{0.7}$Cu-xZn lead free solder on physical properties and micro structure

Erwin Siahaan*
Mechanical Engineering Department, School of Engineering , Tarumanagara University, Jakarta Indonesia

* mkgg_siahaan@yahoo.co.uk

Abstract. The issues to substitute Tin-Lead Solders is concerning the health and environmental hazards that is caused by lead, and also legislative actions around the world regarding lead toxicity, which has prompted the research community to attempt to replace solder alloys for the traditional Sn-Pb alloys lead which has been used by industrial worker throughout history because it is easily extracted and refined at a relatively low energy cost and also has a range of useful properties. Traditional industry lead has been used in soldering materials for electronic applications because it has low melting point and a soft, malleable nature, when combined with tin at the eutectic composition which causes the alloy to flow easily in the liquid state and solidifies over a very small range of temperature. One of the potential candidate to replace tin-lead solder is Sn-Cu-Zn eutectic alloy as it has a lower melting temperature. Consequently, it is of interest to determine what reactions can occur in ternary systems derived from the Sn-Cu-Zn eutectic. One such system is Sn-$_{0.7}$Cu-xZn .The specimen was elaborated on physical properties. The chemical content was analyzed by using Shimadzu XRD and melting point was analyzed by using Differential Scanning Calorimeter (DSC). The results has shown that the highest addition of Zinc content (15%Zn) will decrease the melting temperatur to 189°C compared to Sn-Pb at 183°C Increasing the amount of Zn on Sn$_{0.7}$Cu-xZn alloys will decrease Cu_3Sn intermetallic coumpound.

I. INTRODUCTION

Lead Free Solder has been developed in recent years to substitute lead-tin soldering alloys widely used in Microelectronics. The substitution of Lead-Tin solder has been triggered rapidly by European Union's (EU) proposal to reduce waste from electrical and electronic equipment (WEEE) and RoHS regualation[1,2,3,4]. The research was elaborated to substitute Lead as a toxic material[5,6,7]. Several candidates has been identified as a possible substitute for Lead-Tin as it requires a close melting point value as Lead-Tin which is at 183°C [8,9,10].These candidates of Lead Free Solder have a typical melting temperature between 198°C and 227°C [11,12,13,14]. These candidates had been selected to obtain Lead Free Solder which has similar behavior to Lead-Tin Solder [6,7,12,14] A System containing Zn, Cu, Ag, Bi, In and other elements have been the subjects of experimental investigation, but many of them has not yet been explored[15,16,17] .

There are some characteristics which are considere for lead-free solders as substitute of tin lead solders used in electronic soldering, which are a lower melting temperature of solder, good wetting properties, adequate strength, good thermal conductivity, low cost and also the most important characteristic is its effects on the environment such as its toxicity [8,10,12].

The environmental effect is a major consideration to design Lead Free Solders for interconnects in electronic applications to replace Lead-Tin solders. Basically all the research which have been focused in the development of Lead Free Solders to replace eutectic or near eutectic Lead-Tin solders that melts at 183°C, have not resulted in a development of a satisfactory solder substitute[18,19,20] Currently, Lead Free Solders such as Sn-Ag, Sn-Ag-Cu eutectic solders with the melting temperature of 221, 227 and 217 °C respectively are being used[6,8,12]. One candidate which can very likely be used in the long term is Sn-Zn based solders which are found extensively in the applications of Lead Free Solders electronic components[8,10] Sn-Zn base solders are much less costly than SnAg-Cu solders and they melt at 189°C. Furthermore it is possible to design Zn –containing lead-free solders with melting or solidification characteristics similar to near eutectic Sn-Pb solders. This can help the existing manufacturing infrastructure in device fabrication to find suitable interconnect material for the chip to chip carrier application. One of the important mechanical behaviour of Sn-Zn is that it has a superior isothermal fatigue compared to Sn-Pb eutectic solder[10,11] . This is another reason why Sn-Zn eutectic solder is of interest to the electronics industry.These Sn-Zn also has a lower melting point and improves wetting behaviours [12,13,14]. In this research Lead Free Solder Sn-Cu-Zn was designed by melting Sn-$_{0.7}$Cu with Zn. The present paper focuses on the structure and melting temperature of Sn-$_{0.7}$Cu-xZn cold rolling alloy. This alloy is one of the important alternative alloys recommended to substitute Tin-lead solders. This recommendation is due to its close lower melting point compared to tin lead solders at 183°C eutectic alloy. Moreover Sn-$_{0.7}$Cu-xZn alloy is advantageous from the economic point of view, because Zn is a low cost metal and is abundant in the world.

II. Method and Materials

The Sn-$_{0,7}$Cu alloy which has a purities value of 99.8% and Zn which has a variety percentage of weight were obtained from the industrial market and were weighted by a liquid metallurgy route. The alloys were casted then solidified in cast-iron in the form of 8 mm diameter x 2 mm thick. The casting was done in a graphite mould at a temperature of 600°C and thermally agitated to perform the homogenization. The resulting alloys were then cold rolling into long sheets of about 3 mm in width and 1 mm in thickness. The alloy samples are illustrated in Table 1. X-Ray diffraction analysis was performed on the flat surface of all alloys by using an X-ray diffractometer (Dx- 30, Shimadzu, Japan) of Cu-Kλ radiation and Ni-filter in the range from 20° to 100° of 2θ value (λ = 0.154056 nm, 4.5kV and 35mA).The samples were made to weigh 2.500 mg in the nitrogen atmosphere with 30[ml/min] flow rate.

Melting temperature is a critical solder characteristic because it determines the maximum operating temperature of the system and the minimum processing temperature in which the components can survive. DSC analysis was carried out in order to investigate the fundamental thermal reactions on heating of the solder alloy. Fig. 3.2 shows the thermographs relation between heat flow (V) and temperature (°C)) of the used alloys.

Sn$_{0.7}$CuxZn solders were mechanically polished with 1 m diamond paste. The etching solution contained 93% methanol, 5% nitric acid, and 2% hydrochloric acid. The microstructure of these solders were examined by Shimadzu optical microscope (OM). The Optical Microscope was utilized to analyse the spatial distribution of chemical species for identification of intermetallics SnZn and CuSn.

IOP Conf. Series: Materials Science and Engineering **237** (2017) 012044 doi:10.1088/1757-899X/237/1/012044

III. Results and Discussion

3.1. Structure of Sn-$_{0.7}$Cu-xZn Alloy Samples

Table 1. Chemical analysis of the Sn-Cu alloy Containing Zn

Sample	Chem.Form	Zn	Sn	Code
1	Sn-0,7Cu	11%	Balance	$SnCuZn_1$
2	Sn-0,7Cu	12%	Balance	$SnCuZn_2$
3	Sn-0,7Cu	14%	Balance	$SnCuZn_3$
4	Sn-0,7Cu	15%	Balance	$SnCuZn_4$

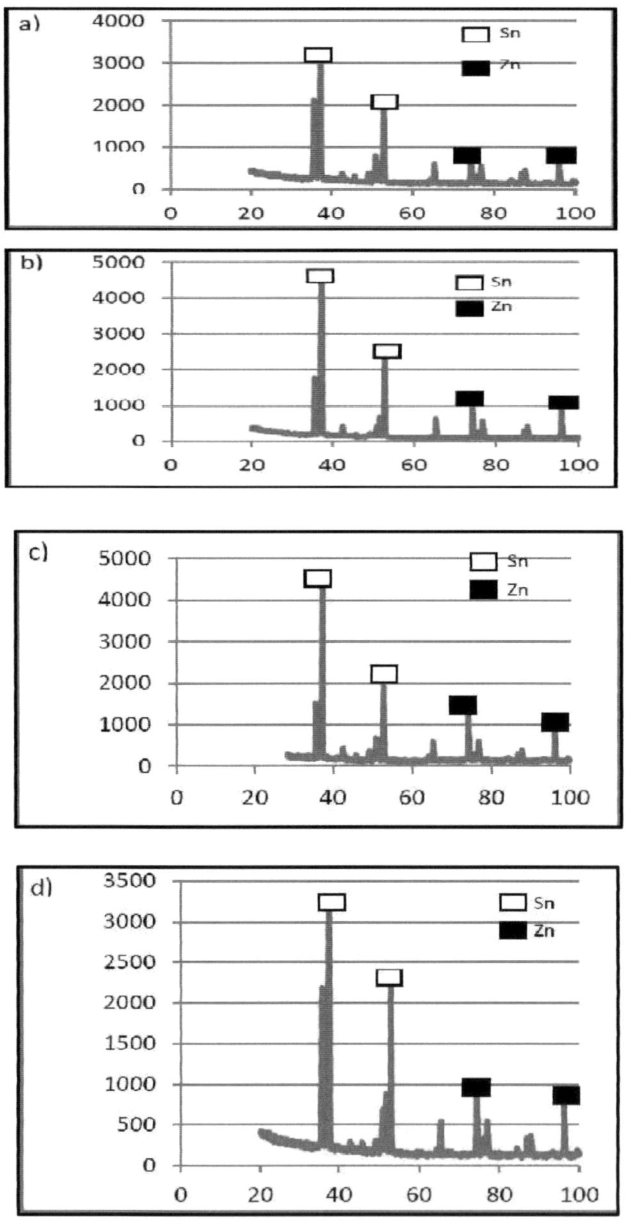

S

Figure 3.1 a. XRD patterns of (a) $SnCuZn_1$, (b) $SnCuZn_2$, (c) $CuZn_3$ and (d) $SnCuZn_4$ alloys.

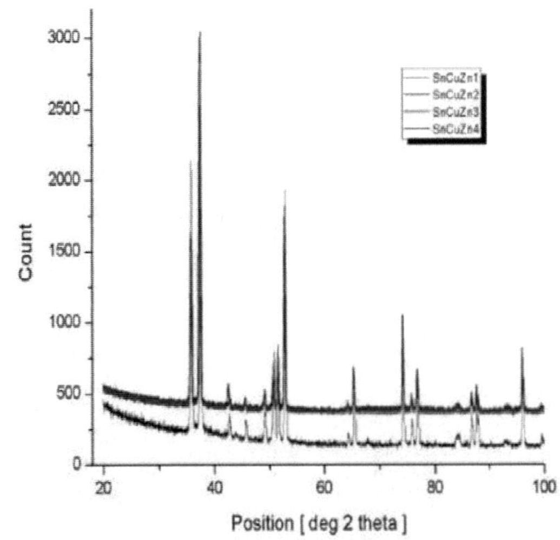

Figure 3 1.b. XRD patterns of all samples as combined curves

X-ray diffraction method was used to determine the phases of the four solder alloys by using Dx- 30, Shimadzu, Japan, see Fig.3.1 (a-d). The pattern in Fig.3.1(a-d) indicates the formation of both phases of Sn and Zn. However, Fig.3.1 (a-d) shows that Sn and Zn phases are the main constituents. From Fig. 1 (a-d), it is clear that Cu is absent which means a complete solubility of Cu in Sn matrix. The addition of wt. % Zn leads to the formation of the intermetallic compounds $SnZn_2$ and $SnZn_3$ respectively.

3.2 Melting Temperatures

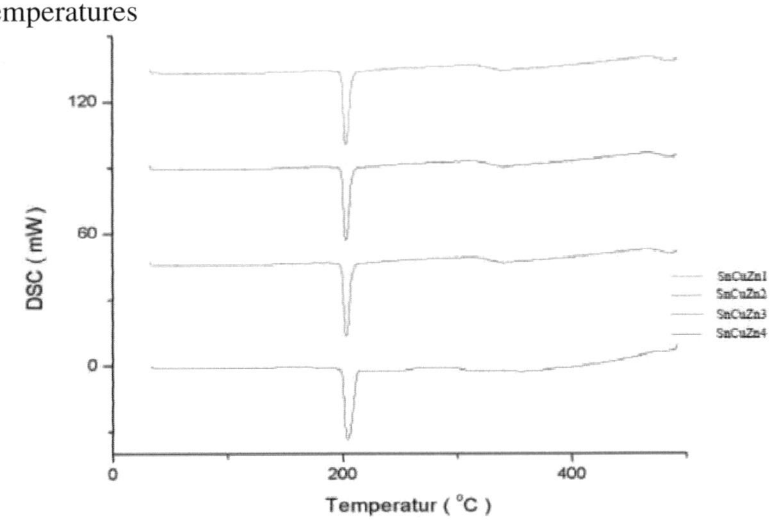

Figure 3.2 DSC pattern on variety Of Zn containing of lead-free solders.

Fig 3.2 shows the melting temperatures of the investigated $SnCuZn_1$, $SnCuZn_2$, $SnCuZn_3$ and $SnCuZn_4$ alloys. With the addition %wt of Zn element, the melting temperature slightly decreased to 189°C as compared to 217 °C of SnCu eutectic alloy which will improve the soldering properties in electronic packaging. The amount of wt.% of Zn addition lowers the melting temperature of the solder significantly. The melting temperature was decreased from 217°C of SnCu as compared to 15%Zn at 189°C for $SnCuZn_4$ alloys respectively.

3.3 Microstructures

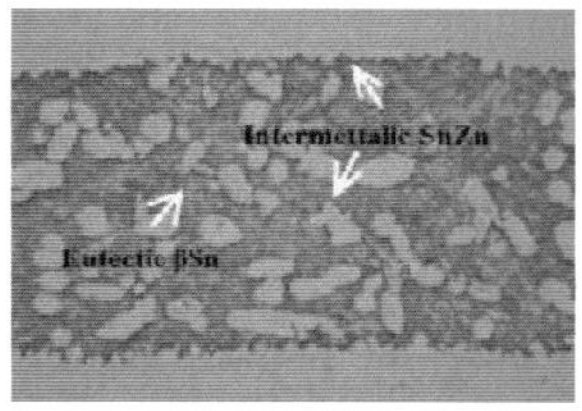

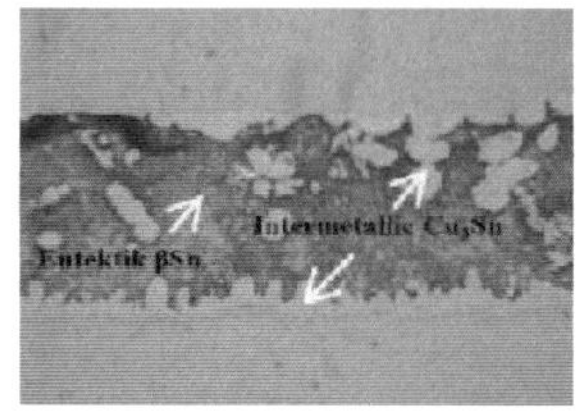

Figure 3.3. Sample of $SnCuZn_1$ **Figure 3.4.** Sample of $SnCuZn_2$

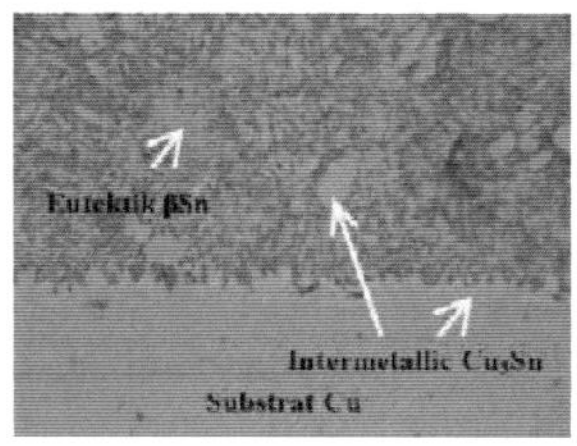

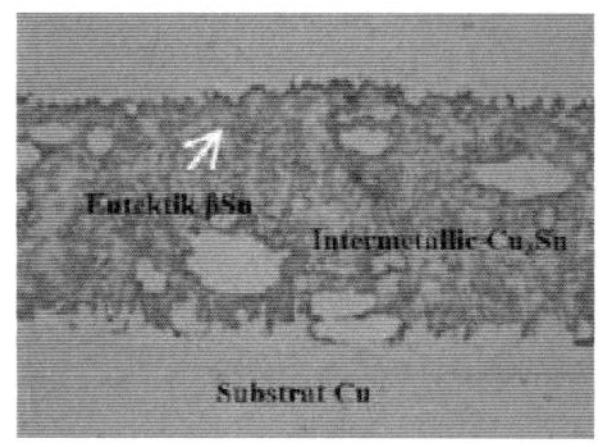

Figure 3.5. Sample of $SnCuZn_3$ **Figure 3.6.** Sample of $SnCuZn_4$

In this case, during the soldering operation, Cu is dissolved into liquid from the substrate, forming in the $Sn_{-0.7}CuxZn$ alloy system and thereby introducing the possibility of ternary phase reactions. The intermetallic Cu_6Sn_5 is important due to the large number of tin-lead and lead-free solder joints formed directly to copper. This intermetallic compound forms in an interfacial layer and can be found in the bulk microstructure of tin-lead solder joints where excessive time and temperature are involved during the soldering process. In addition, the Cu_6Sn_5 intermetallic is a primary feature in the microstructure of lead-free solder joints such as SAC 305 alloy (96.5Sn-3Ag-0.5Cu)[20,21].

The intermetallic Cu3Sn is important as it forms an interfacial layer between the copper and Cu_6Sn_5 IMC layer in tin-lead and lead-free solder joints formed directly to copper. The combination of the Cu_3Sn & Cu_6Sn_5 layers seems to result in a very strong bond between the solder and the copper [22,23,24,25].

Based on the microstructures of lead-free solders as above the apparent increase of Zn in $Sn_{-0.7}CuxZn$ has a decreasing effect on the intermetallic compound of Cu_3Sn. It will decrease its Hardness number. Also with a solubility of Zn in solid SnCu about 15 wt% Zn will result in a eutectic temperature, 189°C. The spalling phenomenon has been reported to occur also when Sn-Zn solders are used on a copper substrate.

IV. Conclusion

In this paper, the increase of Zinc %wt to $Sn_{0.7}CuxZn$ lead-free solder has proved to be an effective way to improve the microstructure and physical properties of $Sn_{0.7}CuxZn$. Melting temperature results shows an improvement with a higher amount of Zint in the solder $Sn_{0.7}CuxZn$, and optimal melting temperature was observed with the content of 15wt.% Zinc.The influence of Zinc content on Sn-Cu alloys lead-free solder at 15 wt % has showed a lowered melting temperature when compared to SAC led-free solder by approximately 28 °C .It appears that melting temperature of $Sn_{0.7}CuxZn$ lead free solder is closer to SnPb eutectic.

IOP Conf. Series: Materials Science and Engineering **237** (2017) 012044 doi:10.1088/1757-899X/237/1/012044

References

[1]. Vaynman, G. Ghosh* and M. E. Fine, 2004, Materials Transactions, Vol. 45,No. 3 pp. 630 to 636.
[2]. Ervina Efzan M. N.1, Aisyah Marini A, 2012, International Journal of Engineering and Applied Sciences. Vol. 1, No.1
[3]. Arun Vasantha Geethan, K1 and Jose, 2011, International Journal of Current Research Vol. 33, Issue, 3, pp.184-190.
[4]. Marek NOVOTNÝ, Ivan SZENDIUCH, 2006, Acta Electrotechnica et Informatica No. 4, Vol. 6.
[5]. Chung-Liang Cheng □, Jia-Syu Lin, Yang-Fang Chen, 2008, Science Direct, Materials Letters 62 pp.666–669.
[6]. H. Fallahi , M.S. Nurulakmal, A. Fallahi Arezodar, Jamaluddin Abdullah, 2012, Materials Science and Engineering A 553 pp.22– 31
[7]. Hiren R. Kotadia Philip D. Howes, Samjid H. Mannan, 2014, MicroelectronReliability 54 pp.1253–1273
[8]. Xiaowu Hu,⥣ , Ke Li ⥣ , Zhixian Min, 2013, Journal of Alloys and Compounds 566 pp.239–245.
[9]. R. Mahmudi, A.R. Geranmayeh, H. Khanbareh, N. Jahangiri, 2009, Mater. Des. 30 pp.574– 580.
[10]. A.R. Geranmayeh, G. Nayyeri, R. Mahmudi 2012, Mater. Sci. Eng. A 547 pp.110–119
[11]. Xiaowu Hu, Yulong Li, Zhixian Min, 2013, J. Mater. Sci.: Mater. Electron., http://dx.doi.org/10.1007/s10854-012-1052-.
[12]. Naoyuki Hamada, Tokuteru Useugi, Yorinobu Takigawa, Kenji Higashi, 2012, Journal Alloys Comp. 527 pp.226–232.
[13]. Islam, M.N., Chan, Y.C., Rizvi, M.J., Jillek W., 2005, Journal of Alloys and Compounds 400 p.136.
[14]. Liu, C.Y., Chen, Y.R., Li, W.L., Hon M.H., Wang, M.C. J. 2007, Elec. Mater ials, 36 p. 11
[15]. S.C. Yang, Y.W. Wang, C.C. Chang and C.R. Kao, 2008, Journal of Electronic Materials, Vol. 37, p. 1591.
[16]. Sengupta, S., Soda, H., McLean, A. *J.* 2002, *Mater. Sci.* 37 p.1747
[17]. Kim, Y.S., Kim, K.S., Hwang, C.H., Suganuma, K. 2003, Journal of Alloys and Compounds 352
[18]. Snugovsky, P. Snugovsky, D.D. Perovic and J.W.Rutter, 2009,Materials Science and Technology, Vol. 25, p.1296.
[19]. A H Habib, M G Ondeck, K J Miller, R Swaminathan and M E Mchenry, 2010, Magnetics, IEEE Transactions on, 46
[20]. A U Telang and T R Bieler, 2005,Journal of Materials, 57 p.44,
[21]. T.R. Bieler and Tae Kyu Lee, 2010,Encyclopedia of Materials Science and Technology .
[22]. N. Moelans*, K.C. Hari Kumar, P. Wollants. 2003, Journal of Alloys and Compounds 360 pp.98–106
[23]. W. Wendlandt, 1986, "Thermal analysis," Wiley, New York, pp. 271-274.
[24]. M. Kamal and E. S. Gouda, 2006, Materials and Manufacturing Processes,Vol. 21, pp. 736-740.
[25]. M. Kamal and E. S. Gouda, 2006,Crystal Re-search and Technology, Vol. 41, No. 12, pp. 1210-1213.

IOP Conf. Series: Materials Science and Engineering **237** (2017) 012045 doi:10.1088/1757-899X/237/1/012045

The influence of inhibitor Na_2CrO_4 and $NaNO_3$ on corrosion and micro structure on AISI 348 steel material

Rahmad Samosir[1,*], Budiarto[1], and Sutan L.M.H. Simanjuntak[2]
[1]Mechanical Engineering Study Programe, Faculty of Engineering, Universitas Kristen Indonesia, Jl. Major General Sutoyo, no.2, Cawang, Jakarta, Indonesia
[2]Mechinical Engineering Department, Nommensen HKBP University, Jl. Sutomo, no. 4a, Medan 20234, Indonesia.

* rahmadsamosir@yahoo.com

Abstract. The influence of inhibitor Na_2CrO_4 and $NaNO_3$ on corrotion and microstructure on AISI 348 steel material have been carried out. A series of corrosion tests were conducted with polarization and potentiodynamic and microstructure resistance methods on AISI steel 348 for secondary cooling pipes at steam power plants. This research was carried out in demineralized water at room temperature (25°C) using Na_2CrO_4 and $NaNO_3$ inhibitors whose concentration varied between 0-150 ppm for corrosion rate inhibitors. Observations of microstructures were performed using optical microscopy and SEM. The corrosion test showed that in the $NaNO_3$ inhibitor solution, the highest corrosion rate was 0.05 mpy at 15 ppm and the lowest corrosion rate was 0.0058 mpy at 30 ppm, and the highest tail was 174.89 mV at 75 ppm and the lowest tail 103 , 04 mV at concentration 0 ppm. While in Na_2CrO_4 inhibitor solution obtained the highest corrosion rate 0,0606 mpy at concentration 75 ppm and lowest corrosion rate 0,0197 mpy at concentration 15 ppm, and highest tail 264 mV at concentration 30 ppm and lowest tail -128 mV at concentration 75 ppm from Inhibitor Na_2CrO_4. The optimum concentration of $NaNO_3$ was 30 ppm with an efficiency of 72.1%, and the optimum concentration of Na_2CrO_4 was 150 ppm with an efficiency of 1.9%. The result of microstructure observation showed that on the surface of the sample was formed a protective film of oxide, and on corrosion of AISI 348 steel corrosion of well form.

1. Introduction

The secondary cooling tube located in the interior of the reactor building is austenic steel, while located on the outside of the reactor building is carbon steel [1,2]. The problems encountered in secondary cooling systems are corrosion, microorganisms and deposits. Corrosion should be watched and controlled, as it may result in decreased equipment efficiency, leakage in piping systems, explosions in high pressure process equipment that could cause loss of life or environmental pollution or even equipment malfunction. To counteract the above, it has been investigated by the addition of Nalco inhibitor products with urea and thiourea media to the behavior of carbon steel DIN ST 35.8 [3]. A corrosion inhibitor is a chemical which, when added in a small concentration to a corrosive environment, will be very effective in lowering corrosion rate, even inhibitors can slow down corrosion rates by increasing anodic or cathodic polarization, reducing displacement or diffusion of ions to metal surfaces , As well as enlarge the electrical resistance at the metal surface [4]. Corrosion control with inhibitors is a controlling process by slowing corrosion speed, since inhibitors can chemically react with metallic surfaces or environments around the metal to provide a certain level of protection against the metal. Use of inhibitors is said to be effective when it can protect metals from corrosion attack at low concentrations and does not cause deposits (crust) on metal surfaces.

The AISI 348 steel which is steel with a ferritic phase has a bcc crystal structure, containing 14-18% chromium with a carbon content of 0.1%. Ferritic steels are not resistant to reducing acids such as HCl, have very strong and hard strength properties, but their resistance to corrosion is reduced, and resistant to oxidation processes at high temperatures, since they have a melting point between 1427^{0}C and 1510^{0}C [5]. In this study used AISI 348 steel with the addition of inorganic inhibitor Na_2CrO_4 and $NaNO_3$ which concentration varied between 0-150 ppm, through polarization and potentiodynamic resistance method to know the rate of corrosion [9,10], as well as observation of microstructure before and after corrosion by using optical microscope and SEM . This research was conducted in DM water (demineralization) at room temperature (25 ^{0}C).

2. Research Methodology

2.1. Materials and Equipment

Material : the materials used in this experiment are steel plate AISI 348, which is made in the form of a circle of 15.0 mm in diameter, with the composition as written in Table 1.

Table 1. AISI-348 steel composition with AA Spectrometer tool

Name of element	C	Mn	Ni	Cr	P	S	Fe
Percentage (%)	0.039	0.84	8.0	18.0	0.07	0.002	73.12

The corrosion medium used was demineralized water with pH = 7.13 at room temperature (25 ^{0}C), whereas the inhibitor used in this study was Sodium Chromate (Na_2CrO_4) and Merck's Sodium Nitrate ($NaNO_3$).

Equipment: The corrosion analyzers used in this study are as follows:
- M342, commonly referred to as "Softcorr Corrosion Measurement, Software" equipped with
-M273 Potensiostat / Galvanostat made by EG & G Princeton , Applied Research Corporation Buehler Samplmet2 cutting machine, Struers Dap-V sanding machine,
-Analytical scales of Ohaus brand, and Corrosion cells equipped with saturated columnelled electrodes (EKJ), auxiliary electrodes and cup holders, SEM and EDXS.

2.2. Working procedure

– Sample preparation

Preparation of the sample is done by forming AISI-348 Steel according to ASTM 262 with diameter 15,0 mm as many as 60 pieces mμ. After that the specimens are polished by using a SiC sandpaper with fineness no.120, 600, 800, 1000, and 1500, and followed by washing according to ASTM G1-90 using hot alcohol to sample fat free. Ready-use sample storage is done in the desiccator.

– Preparation of the solution

For the corrosion test solution is divided into two types, namely the solution containing Sodium Chromate inhibitors (Na_2CrO_4) and Sodium Nitrate Inhibitors ($NaNO_3$) with concentrations varying from 0-150 ppm.

– Corrosion Testing

First, the preparation of the potentiostat instrument and the preparation of the solution, after which the sample to be dissolved is placed on the trailer holder. Then the sample was immersed into corrosion cells containing a solution or a corrosive medium of 600 ml. Inside corrosion cells, this sample will act as a working electrode. Thereafter, the auxiliary electrode and the comparative electrode (EKJ) are fed into the corrosion cell, for which the three electrodes are then connected to the potentiostat instrument. The tail measurements were then measured against the current density with a scan speed of 0.1 mV / s, to obtain the Tafel curve and the pacific curve to determine the characteristics of the sample passivation. From this measurement will also get the density of corrosion current (I_{kor}), corrosion potential (Tail), polarization resistance (Rp), and corrosion rate.

– Observation of microstructure

After corrosion testing, the surface of the sample was analyzed by using Scanning Electron Microscope (SEM-EDXS) to see the corrosion forms on the surface of each sample.

3. Results and discussion

3.1. Analysis of polarization resistance

Corrosion testing by using polarization resistance technique (Polarisation Resistance) is intended to see the resistance of the sample to oxidation when given the outside potential. Polarization resistance is a fast method for determining the corrosion rate without damaging the metal and the measurement results more accurately. The experimental results using the polarization resistance method are as shown in Table 2 and Table 3.

Table 2. Corrosion test results in Na_2CrO_4 inhibitor solution

Cocentration (*ppm*)	Na_2CrO_4			
	E_{kor} (*mV*)	i_{kor} ($\mu A/cm^2$)	Rp ($K.ohm/cm^2$)	Corrosion rate (*mpy*)
0	103.04	0,08	273,78	0,0208
15	40.45	0,08	289,04	0,0197
30	264	0,1	207,11	0,0275
75	-128.66	0,23	94	0,0606
150	-89.12	0,08	279,36	0,0204

Tabel 3. Corrosion test results in $NaNO_3$ inhibitor solution

Cocentration (*ppm*)	$NaNO_3$			
	E_{kor} (*mV*)	i_{kor} ($\mu A/cm^2$)	Rp ($K.ohm/cm^2$)	Corrosion rate (*mpy*)
0	103,04	0,08	273,78	0,0208
15	105,78	0,19	113,95	0,05
30	147,79	0,02	982,86	0,0058
75	174,89	0,03	1087,56	0,0075
150	108,52	0,03	789,25	0,0072

The results in Tables 2 and 3 show that the value of free corrosion potential (Tails), the density of the corrosion current (kink), the polarization resistance (Rp) and the corrosion rate of each sample metal vary in each type and the concentration of inhibitor used . When the metal is introduced into the solution there will be electrochemical reactions at the interface between the metal and the solution. This reaction produces an electrochemical potential called corrosion potential (Tails). This potential is determined by the amount of negative charge formed when the metal is introduced into the solution. At the tail, the oxidation rate is equal to the reduction rate so that the system is called equilibrium. The ired reduction stream, occurs in a reduction process. Aqueous corrosion current occurs by an oxidation process. On the Tail, $I_{red} = I_x$ and $I_{total} - I\ red + I_{oks} = 0$.

The measured current on the instrument is the total current. If a potential that is not equal to the tail is given to a system there will be polarization resulting in a reduction reaction and oxidation, thus the I_{red} and I_{ox} on the tail can be determined. This potential can not be measured directly, but can only be compared with the potential of a known reference electrode.

3.2. Analysis of Corrosion rate

In accordance with the results of corrosion tests showed that the different in concentration of water-soluble inhibitors affect the size of the metal dissolution rate. At certain concentrations there is an indication of the addition of corrosion rate from the initial corrosion rate price before being given the inhibitor as shown in Fig. 1. Basically AISI-348 is a type of metal that is resistant to corrosion in the water environment because the chromium content present in AISI-348 can react in corrosive media to form chromium oxide membrane, but that does not mean it cannot be corroded. This can be seen from the very small corrosion rate when corrosion test on pure water media is 0.0208 mpy. The corrosion rate of corrosion test result with potentiostat as shown in Figure 1 obtained the largest and smallest corrosion rate for the sample which was dissolved in two different solutions ie $NaNO_3$ and Na_2CrO_4. For the sample which was dissolved in $NaNO_3$ solution obtained the highest corrosion rate of 0.05 mpy at a concentration of 15 ppm (2.4 times greater than corrosion without inhibitor) and the lowest corrosion rate of 0.0058 mpy at a concentration of 30 ppm. In Figure 2, the curve rises up to 15 ppm and falls after 15 ppm, this is because $NaNO_3$ is one of the inorganic inorganic metals inhibitors. In theory, if the addition of passive element is added in an insufficient quantity into a corrosive medium there will be an increase in corrosion rate because the process of pacification or decreasing corrosion rate can only be reduced if the added inhibitor has reached the minimum concentration to pacify the metal. What if the minimum concentration for passage is not reached then the protective layer formed cannot protect the entire surface of the sample so that the part having the oxide protective layer will be cathodic and the part that is not covered by the oxide protective film will be anodic, thus increasing the corrosion process in the sample. The decline in corrosion rate at concentrations above 30 ppm $NaNO_3$ because at that concentration has perfectly formed a protective film that is evenly distributed across the surface of the metal sample so as to cover the entire metal surface. While at concentrations below 30 ppm there is an increase in corrosion rate which is the result of not reaching a minimum concentration for inhibitors to inhibit the rate of corrosion, resulting in membranes formed only in certain parts of the metal surface to form anodic and cathodic parts in Metal surfaces as a larger corrosion trigger. The presence of sulfate elements that enter with the addition of $NaNO_3$ inhibitors also accelerate the destruction of the oxide protective membrane because the sulfate element is an active element in attacking the metal at concentrations below the minimum.

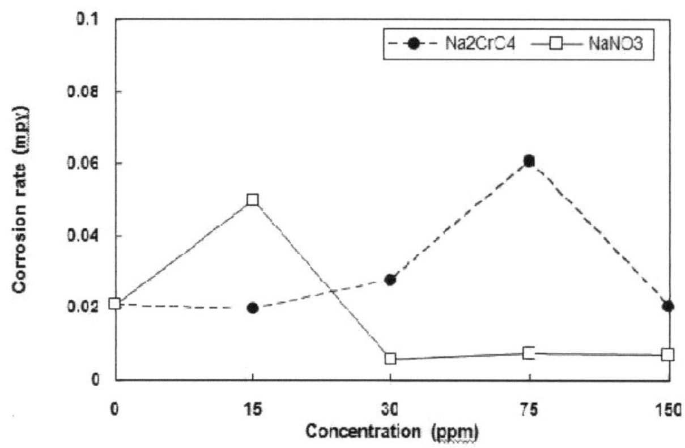

Figure 1. Graph of the relationship between corrosion rate and $NaNO_3$ concentration And Na_2CrO_4

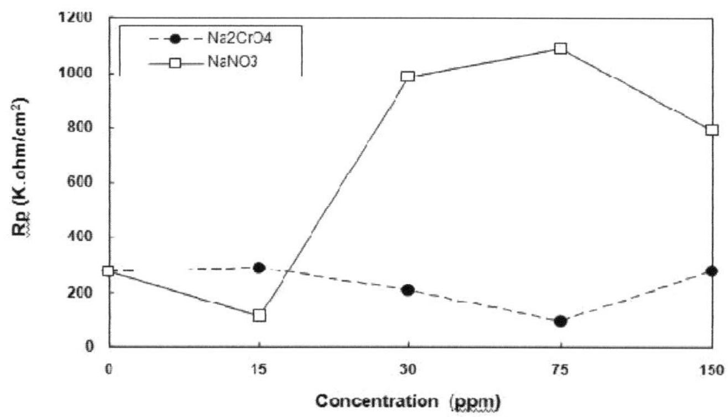

Figure 2. Graph of the relationship between polarization resistance and $NaNO_3$ and Na_2CrO_4 concentrations

In accordance with its function Na_2CrO_4 is one inorganic inhibitor that is also a passive on metal, so that when added in an amount that has not met the minimum limit for the passage it will increase the rate of corrosion. Based on the theory can be explained that at the concentration of 15 ppm Na_2CrO_4 deviation from the price that should be. The occurrence of deviation at 15 ppm Na_2CrO_4 is possible due to the presence of active corrosion-causing elements that follow along with Na_2CrO_4 such as chloride (Cl) because the inhibitor of Na_2CrO_4 used contains a number of chloride elements which are inevitable so that at higher concentrations the chloride content will also Higher and will have a great effect on sample corrosion. The magnitude of the corrosion rate is determined by the magnitude of the sample polarization resistance and the corrosion current density, as shown in Fig.2. According to the current-generated corrosion mechanism, when the resistance per unit area is large, the small unity current is small. Increasing the price of the polarization resistance on the sample surface causes the diffusion of the ions and electrons apart from the metal surface to decrease so that the resulting current is small and the corrosion rate will decrease vice versa in the sample having the price of the small polarization resistance having a large corrosion rate.

The decline in corrosion rate at concentration of 150 ppm Na_2CrO_4 because at this concentration formed a thin protective film that is evenly distributed across the surface of the metal sample so that the oxidation process becomes reduced. If the entire metal surface has been covered by a protective layer then the metal is more cathodic and the corrosion rate will decrease.

3.3 Analysis of potentiodynamic

To determine the active-passive behavior of the sample was conducted corrosion test using potentiodynamic techniques. The results of the measurements are shown in the form of a cassette-like curve shown in Figures 3, and 4.

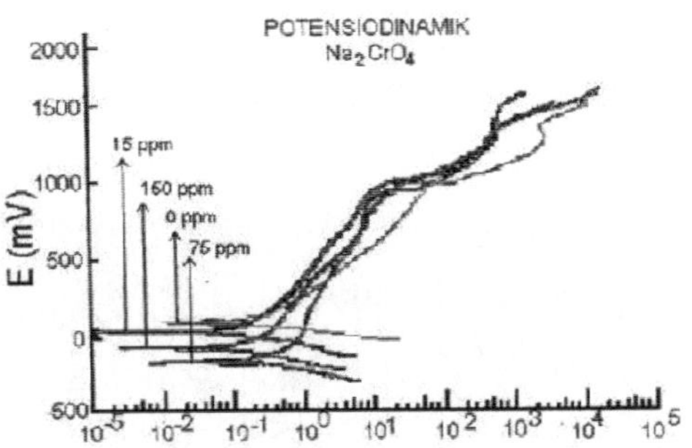

Figure 3 Active-passive curves at concentrations of 0 ppm, 15 ppm, 30 ppm, and 150 ppm Na_2CrO_4 inhibitors

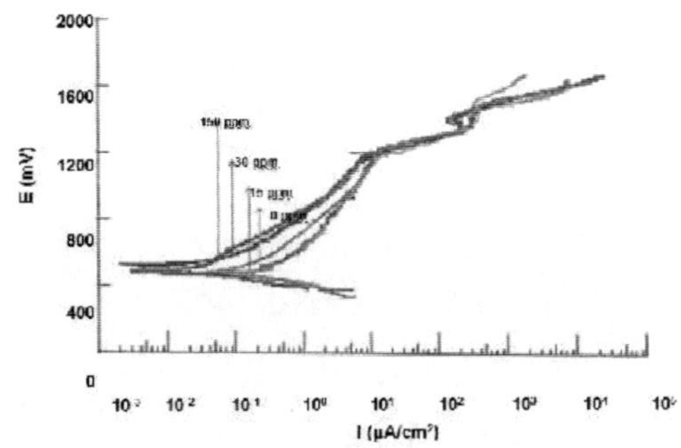

Figure 4. Active-passive curves at concentrations of 0 ppm, 15 ppm, 75 ppm, and 150 ppm of Inhibitor $NaNO_3$

The potentialododynamic curves for both Na_2CrO_4 and $NaNO_3$ inhibitors show different curve shapes. The difference in the curve's slope and the corrosion potential of the tail depends on the protective layer formed on the sample surface by both types of inhibitors. In a sample measurement of 0 ppm (without inhibitor), 2 passivations occur by the protective membrane formed on the metal surface where the second passive region has a greater potential range than the first passive region. The first passive potential occurs at a potential of +100 mV with a very small passive potential range so that it is only seen as a straight line on the curve. This instantaneous passage is characterized by a rapidly decreasing current density. This decrease in current density occurs due to the cessation of the corrosion process in the metal sample. The second passive potential of Epp occurs at + 1250 mV and is then followed by the formation of the transverse region. This second passage also does not last long because the formed membranes are also easily damaged by the potential changes that are given toward the positive potential. Overall it is seen that the sample has a high rate of passivity on the water medium. In the samples tested in a 0 ppm sodium chromate inhibitor solution, 15ppm, 30ppm, 75ppm and 150ppm of Na_2CrO_4 inhibitors as shown in Figures 3 was no passive region. This occurs because there is no perfect layer of protective oxide and evenly distributed on the metal surface, thus never reaching the passive potential of Epp, whereas at 15 ppm there is a passive region at +1230 mV potential. The 15 ppm passing on consentration indicates that at that concentration an oxide protective film covering the entire surface of the sample is capable of halting corrosion in the potential range from +1230 to +1350 mV, whereas at potentials above +1350 mV the protective film will break and the sample will re-corrodate.

In the sample tested in a solution of 0ppm sodium nitrate inhibitor, 15 ppm, 30 ppm 75 ppm and 150 ppm of $NaNO_3$ inhibitor as in Figures 4. The occurrence of passivity at these sodium nitrate concentrations resulting from the protective layer of oxide formed throughout the metal surface is able to isolate or protect the sample from its corrosive solution completely so that at +1250 mV potential the sample is cathodic and the corrosion process stops. The passive region formed at these sodium nitrate concentrations has the same passive potential value of + 1250 mV and has the same potential passive range

3.4. Analysis microstructure

For AISI 348 steel samples of Na_2CrO_4 inhibitor and AISI 348 solutions $NaNO_3$ inhibitor solution by Scanning Electron Microscope 500X and 1500X magnification is seen in Fig. 5a-5b, and composition unsure with EDXS.(Fig.5c)

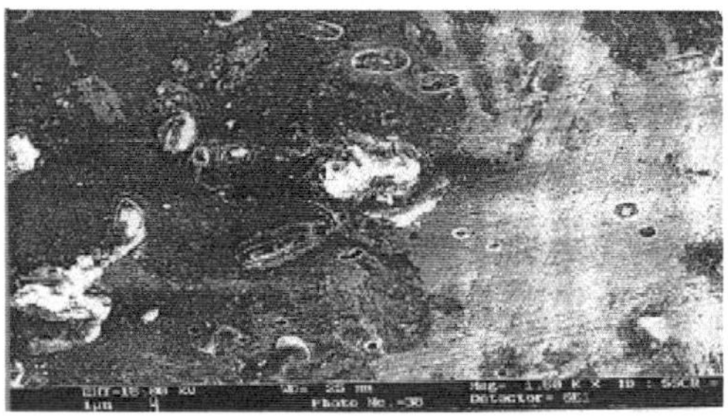

(a)

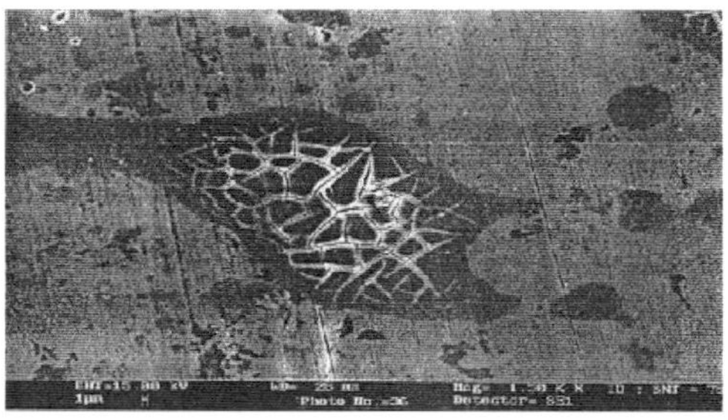

(b)

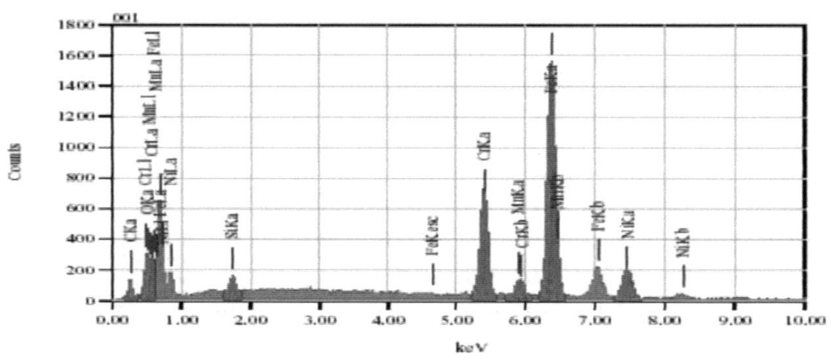

(c)

Figure 5. Surface Morphology with SEM-EDXS, a). AISI 348 steel in Na_2CrO_4 solution (Magn.1500X), b). AISI348 steel in $NaNO_3$ solution (Magn.1500X) c). EDS from AISI 348 steel in Na_2CrO_4 result of sample bulk of grey area dominated by Cr and Fe element beside oxygen, showing the possibility of chrome-oxide forming there is a possibility that chrome oxide or iron oxide are present on the spot.

The cracked region is an area of the oxide membrane that has begun to deteriorate due to a higher potential supply (see Figure 5a). Cracking can also occur by attack of chloride and sulfate ions. Around the cracked surface there has been corrosion characterized by a blackish color which means that corrosion processes have occurred when the protective membrane begins to crack or break. Areas with irregularly blackish colors are areas that have been corroded because their protective oxidant membranes have been damaged (see Figure 5b). In this area the corrosion process is fast so that it will form a basin / overdraft. EDS pattern (Fig. 5c) on a selected bulk spot in this very small surface area part reveals the possibility of iron oxide (Fe_2O_3) and/or chrome oxide (Cr_2O_3) being present in the sample. Based on the microstructure investigation supplemented by EDS data it could be claimed that in AISI 348 steel, pitting/erosion corrosion have certainly occur.

4. Conclusion

The conclusions can be drawn from the results of the discussion and analysis conducted on the research data are as follows:

- The highest corrosion rate in $NaNO_3$ solution inhibitor was 0.05 mpy at 15 ppm concentration and the lowest corrosion rate was 0.0058 mpy at 30 ppm, whereas in the Na_2CrO_4 solution the highest corrosion rate was 0.0606 mpy at 75 ppm and The lowest of 0.0197 mpy at 15 ppm concentration.
- The optimum concentration of $NaNO_3$ was concentrated 30 ppm with the inhibition efficiency of 72,1% and the optimum concentration of Na_2CrO_4 at concentration 150 ppm with the efficiency of inhibition equal to 1,9%.
- From the active-passive curve of the potentiodynamic measurement, there is a passive region at +100 mV and + 1250 mV in a sample that is corroded in pure water media, +1230 mV in addition of 15 ppm Na_2CrO_4 and + 1250mV in addition of 15 ppm to 150 ppm $NaNO_3$. In general the rate of sample passing on pure water media with $NaNO_3$ addition is higher than that of Na_2CrO Na_2CrO_4 and without inhibitor.
- From the results of micro structural observations using SEM it is seen that all samples of corrosion test result have a corrosion attack well.

References

[1]. BATAN, Safety Analysis Report, Vol.2, Copy No.11, Multipurpose Research Reactor G.A. Siwabessy, Jakarta, September 1989.
[2]. Munarka, A.C. And Pustandyo, W., Presentation Meeting of Scientific Research on Basic Science of Nuclear Technology, PPNy, Batan, Yogyakarta, 1992.
[3]. Grace T.S., Effect of urea and thiourea inhibitors on corrosion rate of carbon steel DIN ST 35-8, Thesis S-2, Graduate Program, UI, 1992, p.2-20.
[4]. Pierre, R.R., Handbook of Corrosion Engineering, New York, 1999, page. 69-121
[5]. ASM International Handbook Committe, Vol.1, 10th ed., 1990, Properties and Selection: Irons, Steels and High Performance Alloys, p.38-46
[6]. Trethwey, K.R., Corrosion, for Science and Engineering Students, Alex T.K.W,Grame dia Pustaka Utama, Jakarta, 1991.
[7]. Boffardi, B.G, Corrosion Control of Industrial Cooling Water Systems. MaterialPerformance, Vol.23, No.11, November, 1984, p.126-129.
[8]. Robitaille, D.R., Sodium Molybdate as Corrosion Inhibitor in Cooling Tower Water, Material Performance, No.11, November, 1976, p.25-30.
[9]. ASTM Standard G 59-78, Standard Practice for Conducting Potentiodynamic Polarization Resistance Measurements, American Society for Testing and Materials, Philadelphia, 1991, p.15-47
[10]. FONTANA G., Corrosion Engineering (1987), McGraw Hill Book Co., third ed.Singapore, p. 12-38.

Shielding gas effect to diffusion activities of magnesium and copper on aluminum clad

Charles SP Manurung*, Richard AM Napitupulu
Department of Mechanical Engineering, Nommensen HKBP University, Medan 20234, Indonesia

* charles.manurung1972@gmail.com

Abstract. Aluminum is the second most metal used in many application, because of its corrosion resistance. The Aluminum will be damaged in over time if it's not maintained in good condition. That is important to give protection to the Aluminums surface. Cladding process is one of surface protection methodes, especially for metals. Aluminum clad copper (Al/Cu) or copper clad aluminum (Cu/Al) composite metals have been widely used for many years. These mature protection method and well tested clad metal systems are used industrially in a variety application. The inherent properties and behavior of both copper and aluminum combine to provide unique performance advantages. In this paper Aluminum 2024 series will be covered with Aluminum 1100 series by hot rolling process. Observations will focus on diffusion activities of Mg and Cu that not present on Aluminum 1100 series. The differences of clad material samples is the use of shielding gas during heating before hot rolling process. The metallurgical characteristics will be examined by using optical microscopy. Transition zone from the interface cannot be observed but from Energy Dispersive Spectrometry it's found that Mg and Cu are diffused from base metal (Al 2024) to the clad metal (Al 1100). Hardness test proved that base metals hardness to interface was decrease.

1. Introduction

Aluminum is a material that is widely used in various fields. Starting from household appliances, construction, electricity, aviation to reflector equipment on military vehicles. Aluminum has a Face Center Cubic (FCC) crystal structure. Aluminum is known as a light metal with 2.7 g/cm^3 of density compared to steel with 7.9 g/cm^3 of density[2]. Aluminum has good electrical and heat conductivity as well as corrosion resistance in the general environment with melting point is 660°C (1220 °F). Most of the Aluminum alloys can easily be formed because they have high ductility. Aluminum remains ductile even at low temperature. Therefore Aluminum is also used as a air condition elements on vehicles. Although Aluminum has good corrosion resistance, but basically Aluminum can't be detached from the oxidation process, especially when Aluminum is in an environment containing a lot of oxygen, water or other oxides. At room temperature, the thickness of oxide layer can reach 2.5 ~ 3 nm (25 ~ 30Å). But the oxide layer that is formed is passive and does not result in great damage to the surface so it can be a protector of Aluminum from environment.

Aluminum alloy is very widely used because of its superior mechanical properties is Al-Cu alloy. The strength of this alloy is obtained through a solution heat treatment process at eutectic temperature. With that solution heat treatment, the θ phase formed and distributed

IOP Conf. Series: Materials Science and Engineering **237** (2017) 012046 doi:10.1088/1757-899X/237/1/012046

throughout the Al-Cu. The θ phase is a CuAl2 compound that formed as precipitate which can serve as a barrier to the dislocation movement.

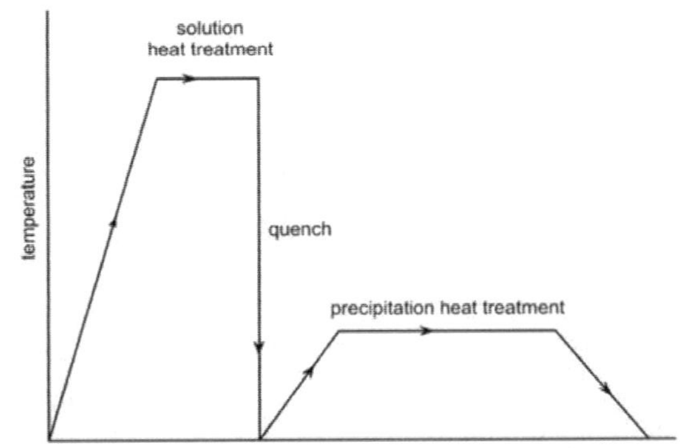

Figure 1 Solution heat treatment process fo Al-Cu alloys

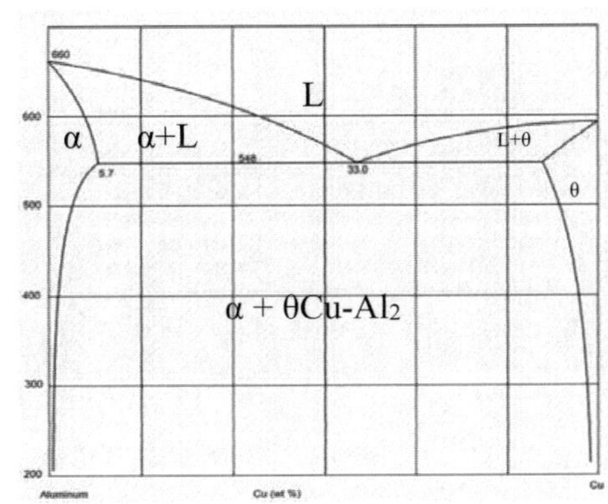

Figure 2 Part of Al-Cu phase diagram

Al-Cu alloys have good strength but low corrosion resistance due to galvanic phenomenon on Aluminum and on the Al-Cu due to potential difference of two types of metal. Overcome that problem Aluminum alloy need to be coating with a better metal such as pure Aluminum. Metal coating is performed to provide corrosion resistance to metals, obtaining heat and electric conductivity properties and to improve the appearance of a material.

The problem are often encountered in the coating process is the adhesion between coating materials with base metal. In metallic materials, the quality of the coating can be assessed by the bonding quality between two metals, the base metal and the coating material. The type of bond that is formed also depends on the coating technique. In this research, the coating is done by cladding method using hot rolling process. This process is most widely used given the high production amount especially when using continuous rolling.

At high temperatures, aluminum is highly vulnerable to oxidation, so to produce a good coating quality the surface should be avoided from the influence of oxygen around it. This is

IOP Conf. Series: Materials Science and Engineering **237** (2017) 012046 doi:10.1088/1757-899X/237/1/012046

to prevent the formation of oxide layer on the surface which will inhibit the diffusion process between atoms from metal coatings with base metal.

2. Materials and Methode

In this research, we use Aluminum alloy from 2024 series (Al-Cu) as base metal, and Aluminum 1100 series as clad metal. Cladding material will be form by hot rolling process. Emission Spectrometry is used to get the information of alloying element on both materials as shown on table 1.

Table 1 Elements composition of the material

Comp.(%wt) / Material	Al	Cu	Fe	Si	Mn	Mg	Cr	Zn	Ti	others
Al-Cu 2024	93.59	4.2	0.08	0.05	0.53	1.4	0.05	0.03	-	0.07
Al 1100	99.17	-	0.4	-	0.00019	-	0.0002	0.0166	0.0403	-

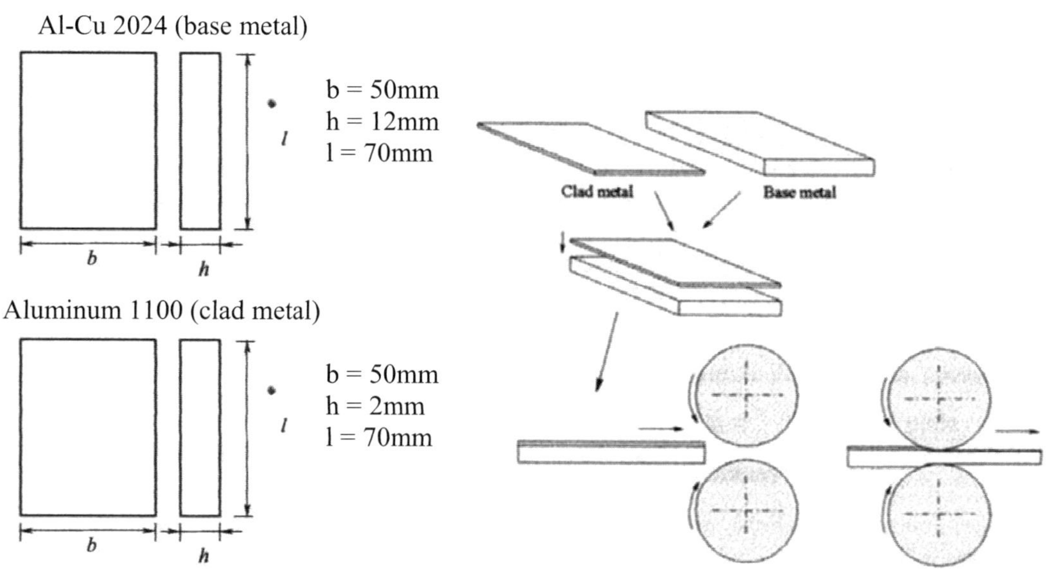

Figure 3 Dimension and rolling scheme

Surface of both materials used should be prepared to obtain good bonding qualities in the interface. This research was conduct from sample preparation to element examined distributing.

1. Sample Preparation.
   The sample preparation begins from the material cutting and then the surface preparation to produce a good contact area between the two materials to be cladded. We use abrasive paper with. We use abrasive paper from the level of roughness 800, 1000, 1500 up to 2000, followed by a polishing process using an abrasive material with a roughness of 0.3μm. Sample surface cleaned by pickling process by using HNO_3 to remove oxide layer on the

IOP Conf. Series: Materials Science and Engineering **237** (2017) 012046 doi:10.1088/1757-899X/237/1/012046

both of samples contact area. After the pikling process, the samples are stored in a special packaging to prevent the formation of oxide layer before the rolling process is performed.

2. Hot Rolling Process
   Hot rolling was done by a reduction of 5% by 9 pass. The heating process is divided into 2 variables by using protective gas and without protective gas. The protective gas used is a pure Argon gas with a flow rate of 2 liters / minute. The sample heating temperature is 450°C and 500°C and after each rolling each sample fitting is reheated for 10 minutes at predetermined temperature.
3. Metallographic test
   Metallographic testing is performed to observe the interface area of both materials. The etching solution used was the modification of Keller solution with a composition of 10ml HNO3 + 1.5 ml HCL + 1.0 ml HF + 87.5 ml Aquadest.
4. Emission Dispersive Spectrometry Test (EDS Test)
   The EDS test is performed to find out where the presence of atoms diffuses during the heating and rolling process. This test is aimed at the Mg and Cu atoms that are clearly not found in the clad material of Al 1100. The method used is line analysis method ranging from 600μm from the base metal interface (Al-Cu 2024) to 300μm from the interface to the clad metal (Al 1100). The test is continued by the spot area method to calibrate the elements obtained from the line analysis method.
5. Hardness test
   Hardness testing is done from base metal area until clad metal passes through interface area. This test is performed to determine the change in the distribution of violence due to the displacement of atoms due to diffusion.

3. Result and Discussion

Based on the results of metallographic testing on the interface area can be seen differences in the thickness of each sample as shown in the following table:

Table 2. Interface area thickness of each sample

Sample				
T (°C)	450°C		500°C	
Shielding gas	Yes	No	Yes	No
Speciment Number	1	2	3	4
Thickness (μm)	1.06 ± 0.34	2.06 ± 0.61	0.95 ± 0.21	1.59 ± 0.56

IOP Conf. Series: Materials Science and Engineering **237** (2017) 012046 doi:10.1088/1757-899X/237/1/012046

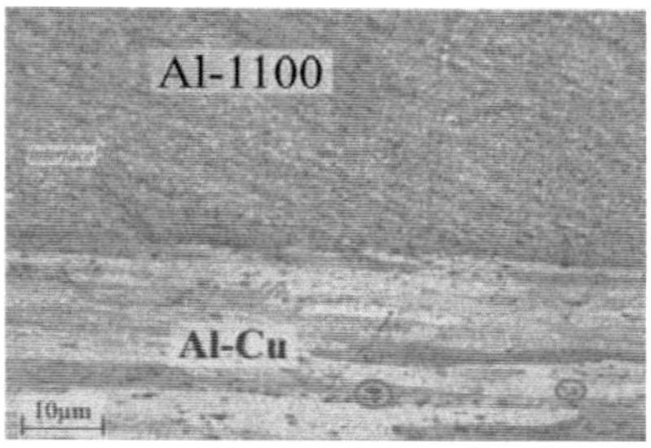

a. Temp. 450°C with shielding gas

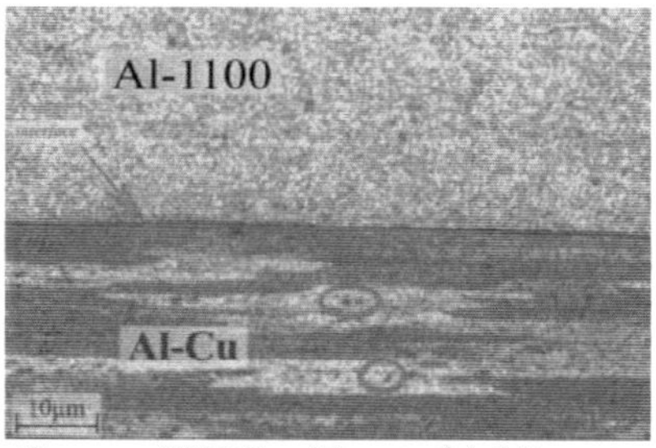

b. Temp. 450°C without shielding gas

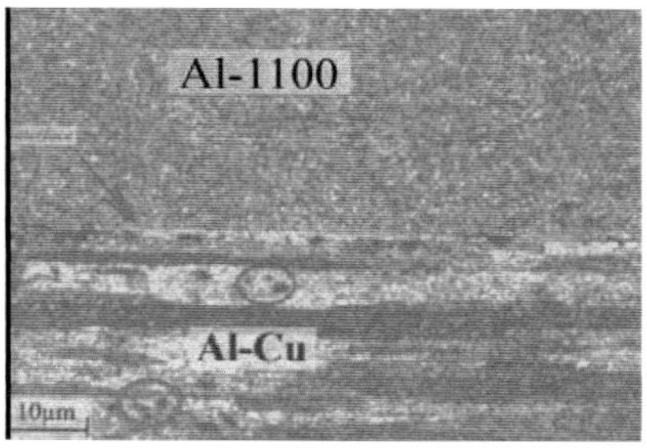

c. Temp. 500°C with shielding gas

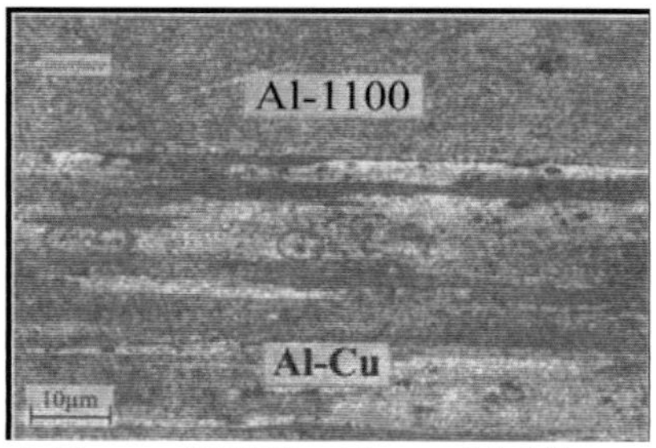

d. Temp. 500°C without shielding gas

Figure 4. Microstructure of cladding material sample

From the observations it can be concluded that the heating conditions without the use of protective gas resulted in the presence of oxide layers that block the unification process of both types of materials. This can be seen both at the heating temperature of 450°C and 500°C. In the heated sample using protective gas, it appears that the interface area has a better transition region.

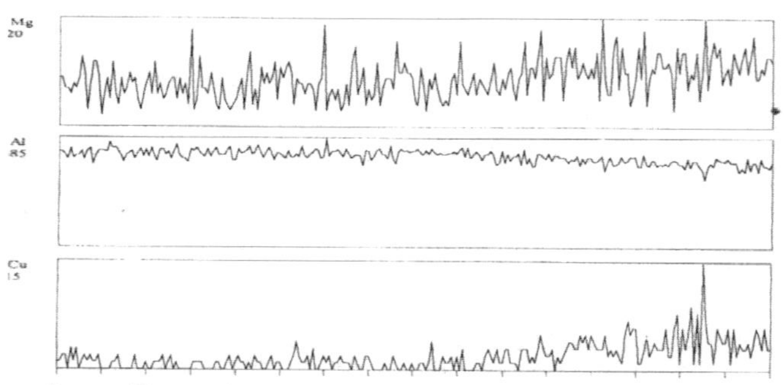

(a)

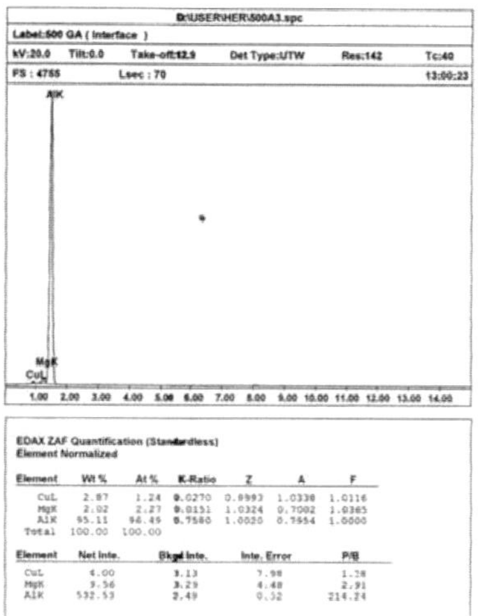

(b)

Figure 5. Part of EDS & EDAX examination
a. line analysis method
b. selected area to calibrate the elements

EDS tests shown the distribution of Cu and MG elements that diffuse as shown in the following table.

Table 3. The distribution of Cu and Mg elements in the observation area

Temperature (°C)	Shielding gas	Element (%)	Distance from Interface					
			To base metal				To clad metal	
			600 µm	300 µm	150 µm	interface	150 µm	300 µm
450	Yes	Cu	6.64	5.92	4.49	2.61	2.30	2.27
		Mg	3.51	3.38	2.76	2.10	2.05	1.96
450	No	Cu	5.60	5.13	4.79	3.80	1.91	1.78
		Mg	3.33	3.20	3.15	2.67	1.95	1.93
500	Yes	Cu	6.62	5.99	5.34	2.87	2.27	0.26
		Mg	3.53	3.30	3.02	2.02	1.98	1.88
500	No	Cu	6.68	5.53	3.76	3.63	2.16	0.00
		Mg	3.47	3.26	2.56	2.41	1.97	1.45

Table 3 shown that use of protective gas at the time of heating can reduce the formation of oxide layer on the surface. Thus the diffusion process is not obstructed, seen from the amount of Cu and Mg elements diffusing towards Al 1100 which previously did not contain elements Cu and Mg.

In samples that do not use protective gas during heating there is an increase in Cu and Mg elements in the interface area. It can be concluded that there is an obstruction of oxide to the diffusion process. Testing Hardness in the transverse direction from base metal to clad metal can be seen in the following table.

Table 4. Distribution of hardness throughout the observation area

Temperature (°C)	Shielding gas	Vickers Hardness Number (VHN)				
		Distance from Interface				
		Base Metal				Clad Metal
		2 mm	300µm	200 µm	100 µm	100 µm
450	Yes	100 ± 2.9	97 ± 6.2	93 ± 4.8	85 ± 10.4	28 ± 2.3
450	No	139 ± 15.4	135 ± 117	122 ± 15	114 ± 6.0	39 ± 1.4
500	Yes	96 ± 1.7	93 ± 3.6	92 ± 3.1	86 ± 4.6	25 ± 1.9
500	No	97 ± 1.8	92 ± 2.2	84 ± 2.5	84 ± 2.4	31 ± 2.7

Hardness test result show a decrease in hardness in the base metal area leading to the interface. This is due to the decrease in metal properties due to reduced alloying elements in the area because it has diffused into clad metal.

4. Conclusion

1. The use of protective gas when heating the material sample can suppress the formation of the oxide layer in the interface area. This is evidenced by the diffusion activity of Cu and Mg atoms. In Table 3 for heated samples 450°C and 500°C without the use of protective

IOP Conf. Series: Materials Science and Engineering **237** (2017) 012046 doi:10.1088/1757-899X/237/1/012046

gases it appears that the activation of Cu and Mg atoms is retained in the interface area. While the sample using protective gas seen diffusion of Cu and Mg atoms more to the clad metal area (Al 1100).

2. The thickness of the interface area is affected by the use of protective gases . This is due to the number of oxide layers on the surface of the interfaces being formed more frequently during heating. The heated sample without the use of a protective gas has a thicker interface area.
3. Base metal hardness in the interface direction decreases due to atom diffusion from base metal to clad metal.

References

1. Gabe, D.R., 1978, Principles of Metal Surface Treatment and Protection, 2^{nd}ed, Pergamon Press.
2. Callister,W.D., 1996, Material Science and Engineering, An Introduction, 4th ed., Jhon Willey & Sons.
3. ASM Handbook, Vol. 9 Metallography and Microstructures, 9th ed., 1992.
4. ASM Handbook, Vol. 10 Material Characterization, 9th ed., 1992.
5. Hatch, J.E., 1983, Aluminum Properties and Physical Metallurgy, Aluminum Associaation, Inc. & ASM.

IOP Conf. Series: Materials Science and Engineering **237** (2017) 012047 doi:10.1088/1757-899X/237/1/012047

The influence of concentration and pH on corrotion rate in stainless steels – 316 solution HNO_3 medium

Rahmad Samosir[1,*] **and Sutan L.M.H. Simanjuntak**[2]
[1]Departement of Mechanical Engineering, Faculty of Engineering, UKI Jakarta
[2]Departement of Mechanical Engineering, Faculty of Engineering, Nommensen HKBP University, Medan Indonesia.

* rachmadsamosir@yahoo.com

Abstract. The Influence of concentration and pH on corrosion rate on commercial stainless steels-316 materials have been carried out. The experiments were carried out in HNO_3 medium with a concentration variation of 0.1 mol, 0.2 mol, 0.3 mol and 0.4 mol corresponding to pH values of 1.0, 0.7, 0.523, and 0.4 respectively. The experiments were carried out using a type M-273 EG&G galvanometer test instrument. The post-corrosion samples' microstructure has analyzed with the aid of energy dispersive spectroscopy (EDS) equipped SEM instrument to detect the presence of any viable corrosion by products. Experimental results confirm that stainless steels-316 commercial alloys immersed in a nitric acid corrosion medium with a variation of concentration experience very little or almost no corrosion and that according to the so-called Fontana's criteria these test-materials turn out to have an excellent resistance toward nitric acid corrosion. This is also evidenced by the very low corrosion rate value measured in this study. SEM-EDS study, Its indicate that the possible ensuing corrosion by products are iron oxides, and chrome oxides.

1. Introduction

One of the motivate behind the fabrication of stainless steels are their excellent resistance to corrosion facilitated by the presence of chromium. One of the commercially available stainless steel is the stainless steels-316 which is an austenite-type low-carbon steel. Stainless steels-316 contains chromium as the main alloying element in iron and steel for inhibiting corrosion combines with other alloying elements such as nickel, silicon, titanium, and carbon with a composition listed [1]. Kyung Seon Min *et al.* have discussed in details all aspects covering the influence of carbide grain boundary density on SS-321 alloy [2], and carbide grain boundary feature correlation with respect to its creep and fatigue properties [3]. Corrosion measurements on SS-321 samples coated with ion implantation and ceramic film have been carried out by F. Noli *et al.* [4], where H_2SO_4 solution was used as the corrosion medium. High temperature corrosion measurements on SS-321 in atmospheric environment have been carried out by Nurdin Effendi *et al.* [5,6].

This experiment would be supplemental to the similar corrosion experiments carried out earlier in an atmospheric medium. In stainless steels-316 alloys nickel acts both as an austenite phase forming agent as well as an anti-corrosion agent [10,11]. Chrome is also known as an excellent alloying element because of its anti-corrosion property. Silicon is used as a nucleation agent because of its high melting and freezing points, as well as its anti-corrosion property [10]. Manganese has an excellent anti-corrosion property and acts as a phase-stabilizer for the austenite phase[10,11], carbon is an excellent hardness-increasing element, and finally titanium functions as a sensitizing element in the alloy [10,11]. The

IOP Conf. Series: Materials Science and Engineering **237** (2017) 012047 doi:10.1088/1757-899X/237/1/012047

purpose of this experiment is to investigate corrosion behavior of stainless steels-316 alloys in nitric-acid solution medium, and the damage of the sample due to the penetration of nitric element.

2. Theory

Corrosion reaction could be viewed upon a thermodynamic system surrounded by an environment forming an imaginary wall [7]. Corrosive reaction in a certain type of environment occurs rather spontaneously by releasing free-energy from the system into the environment. Normally, a system is endowed with a certain amount of internal-energy, and most of this energy could be transformed into another form of energy, called the free energy [7, 8]. Corrosive reaction is a transitional condition, and is often labeled as a transition reaction. According to a theory first formulated by Arrhenius, corrosion rate is determined by two main parameters, free activation energy and temperature. Both parameters form a formula expressing the reaction constant, k_{corr}, otherwise known as Arrhenius constant [7]. Activation free-energy is a barrier-energy between the alloy and its corrosion product. For this reason it becomes clear that the condition where the alloy is at a higher energy level compared to its surrounding element is undesirable. In this system, corrosion occurs whenever free-energy is available as the potential to generate the flow of electric current (electrons) between anode and cathode in the sample. In this case the test sample should have more anodicity than the reference potential, to enable the cathode to collect the electron flow generated by the test sample [7, 8]. This electronic current or flow is the main cause of corrosion in the test sample. The resistance to corrosion is measured using the resistance polarization to produce the polarization resistance parameter (Rp). Polarization resistance is a measure of the specimen resistance toward corrosion of the specimen while connected to an external potential source. The main function of Rp is to calculate the corrosion current i_{corr}, which is to be determined prior to calculating the corrosion rate. The following expression elucidate the relation between R_P, Tafel constants, and corrosion current i_{corr} ,

$$\Delta E/\Delta I_{\text{spt}} = R_P = \beta a\ \beta c/2.3\ (i_{corr})\ (\beta a + \beta c) \quad \text{......... (1)}$$

with : $\Delta E/\Delta I_{\text{spt}} = R_p$ = polarization curve slope

βa = polarization tafel constant

βc = cathodic tafel constant

i_{corr} = corrosion current

After i_{corr} is calculated. the corrosion rate could be determined as follows :

Corrosion rate (mpy) = $0.13 \text{ x } i_{corr}.\text{x } E_W/A.d$

.... (2)

with mpy = mili-inch per year

E_W = weight equivalent (g/equivalent)

A = area (cm^2)

d = density (g/cm^3)

0.13 = conversion factor

Therefore according to equation 2, corrosion current determines the corrosion rate. In this work the *polarization resistance / polarization linear* technique is carried out first to obtain the polarization resistance (Rp). The polarization technique is superior with respect to other techniques because it is faster and it uses a relatively smaller potential scanning which causes insignificant changes or damages to the samples surface.

The potentio dynamic technique is utilized to determine the active and passive region of the measured alloy system. The overall form of the curve (Figure 2) is an indication of the corrosion behavior of the samples immersed in the test-solution. This technique is capable of producing reliable data if carried out at a relatively low scanning rate. Passivation is defined as the loss of chemical reactivity in materials while subjected to certain environment or conditions. Passivation generally occurs because of the formation of a protective layer in the alloy-electrolyte interface therefore preventing a direct contact between the alloy s surface and the electrolyte.

The measurement of an alloy tendency toward passivation when it is immersed inside a certain type of environment is very important in order to obtain information on the corrosion behavior or characteristics of that particular alloy. The measurement is accomplished by drawing the characteristic anodic polarization curve of the sample. The halogenic ions, especially the chloride ion could destroy the passivation in the spots or local points on the sample which is passive with respect to the air environment. (e.g. Cr, stainless steels). The local spot of the active alloy acts as an anode whereas the cathode forms the wider region of the passive alloy.

In a potentio dynamic-test measurement, the assigned potential values cover a far broader range compared to the potential values used in a polarization resistance measurement. with the consequent that samples usually experience some damages after a potentio dynamic experiment. The corroding medium concentration also influences the corrosion rate because the concentration will modify the pH value.

3. Experimental Method

3.1 Materials

The test samples used in this work are plates of Stainless Steels-316 alloys with a thickness of 5 mm. The applied solution used as the corrosion-medium is chloric-acid solution with a concentration-variation of 0.1 moles to 0.4 moles with a steady increment in step of 0.1 moles. The choice of this variation is strongly delimited by the instrumental precision limit, i.e. above 0.4 moles the corrosion rate increases steeply outside the range of the detection limits of the instrument.

3.2 Equipment

This experimental work uses a Buehler cutting machine with a Buehler-type disc blade, a double polishing machine fabricated by Karl Kolb (Denmark); sand papers of various grades for polishing. The corrosion-testing machine used is the potensiostat M273 assembled by EG & G Princeton Applied Research Corporation. Equipment with test-beakers, vials and pipe test tube. The supporting analysis instruments for the corrosion-samples are SEM-EDXS (for surface microstructure observations) equipped.

3.3. Methods

Sample preparation is carried out in compliance with the standard procedure [9]. Samples are cut into equal sized and shaped into round-forms by polishing, to a final size of 15 - 16 mm diameters. Afterwards the surfaces of the samples were polished in a polishing machine until a relatively uniform and surface's smoothness and evenness is achieved. The main objective is to get rid of the oxide-layer in the surface. The first experimental-test performed is the polarization technique experiment. The experimental procedure is outlined as follows: First, the sample is immersed into the three-electrode corrosion-test instrument, which has been filled with a 600 ml of corrosion-fluid, and standard saturated calomel is used as the reference electrode, two carbon-bars are used as supporting electrodes. The sample functions as the working electrode. To start the experimental measurement, the corrosion cell is connected to

IOP Conf. Series: Materials Science and Engineering **237** (2017) 012047 doi:10.1088/1757-899X/237/1/012047

the potentiostat by applying a potential 20mV below the corrosion potential and 20 mV above the corrosion potential, and the scanning rate is set at 0.1 mVs-1; from this measurement both *Rp* values and anodic-cathodic Tafel constants will automatically be obtained. Potentio dynamic-technique is carried out next. The primary aim of this measurement is to map the active as well as passive characteristic areas of the specimen-solution system by applying a working potential in the range of 500mVbelowand 1600mV above the corrosion potential with a scanning rate of 0.5 mVs-1. Finally these experimental curves are analyzed. To analyze the sample s surface and to determine its corrosion products, the corrosion-test samples are observed under an EDS equipped Scanning Electronic Microscope (SEM). In order to complete the classification of the resulting types of corrosion-product.

4. Result and Disccution

4.1.Analysis of polarization and potentiodynamic.

The curves obtained from the corrosion-test measurements on stainless steels-316 samples using the polarization resistance technique in an HNO_3 medium for each pH or concentration value should generate the curves presented in figures 1-a through 1-d and the experimentally measured parameters available in Table 1. The experimental results clearly show that for a stainless steels-316 sample immersed in an HNO_3 solution with 0.1- 0.4molar concentration (equivalent to a 1.0 – 4.0 pH value) the lowest corrosion potential of -344.47 mV has been recorded for the pH value of 0.4. The corrosion rate in the sample tends to peak at this pH value, which is around 2.5 mpy (Table-3). According to the rule established by Fontana [7, 8], this particular corrosion rate value should rank stainless steels-316 in a class of materials with excellent corrosion-resistance with respect to nitric acid solution corroding medium.

Figures 2-a to 2-d show several potentiodynamic curves of a stainless steels-316 sample in a nitric acid medium with a variation of pH, the concurrently generated experimental parameters are presented in Table 2. In a measurement procedure of potentiodynamic technique, the assigned potential range is far broader than the range used in a polarization-technique experiment, and this would ultimately lead to the sample(s) being damaged or ruined in the process.

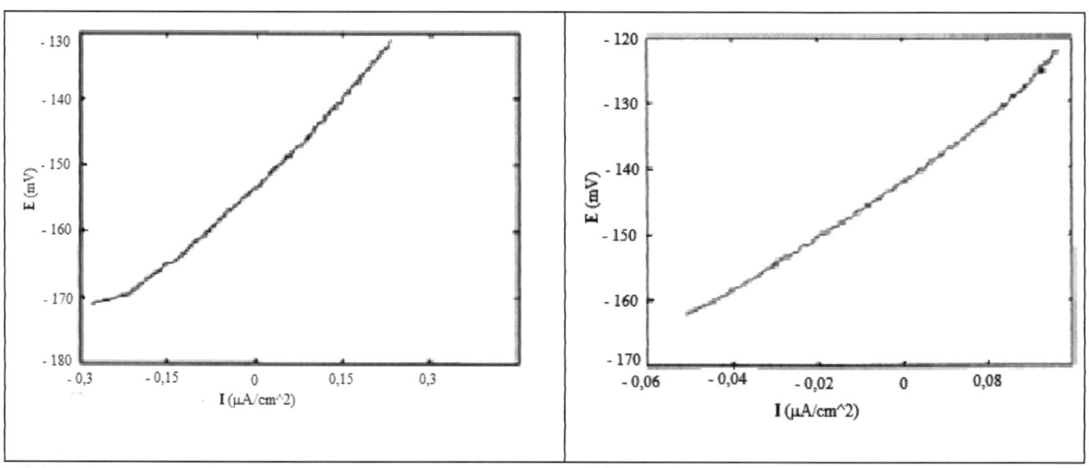

(a) (b)

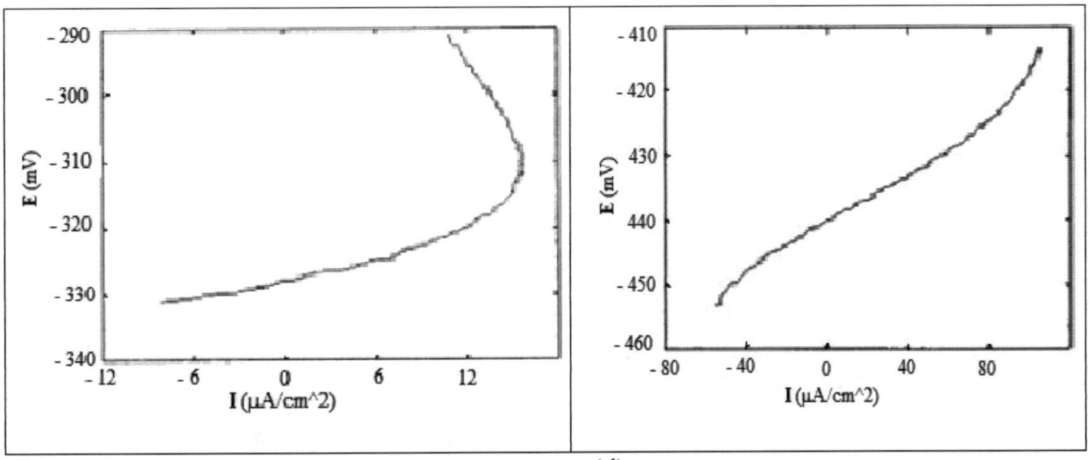

(c) (d)

Figure 1. Polarization-resistance potential versus current curve pattern obtained from stainless steels-316 sample immersed in a nitric-acid medium : (a). pH = 0.4 (b). pH = 0.522 (c). pH = 0.7 (d). pH = 1.0

The potentiodynamic curves would display the following general pattern: initially the samples are cathodic, because of cathodic protection process initiated by the anticorrosion elements, and in this potential range and up to its free corrosion potential (E_{corr}), the current will concurrently decrease. The free corrosion potential is the scale which eventually determines the virtual equilibrium condition; now for this particular condition the alloy would experience deliberate corrosion without any external interference in the form of additional input of potential. After the free-corrosion potential has been reached, then to generate the corrosion process in the sample, the overall potential must be steadily increased, which in turn causes the corrosion current to increase to generate corrosion in the sample. The free corrosion-potential is an indicator whether a sample immersed in a corrosion-medium is easy or difficult to corrode; the higher the free corrosion potential, the more difficult it is for a sample to experience corrosion.

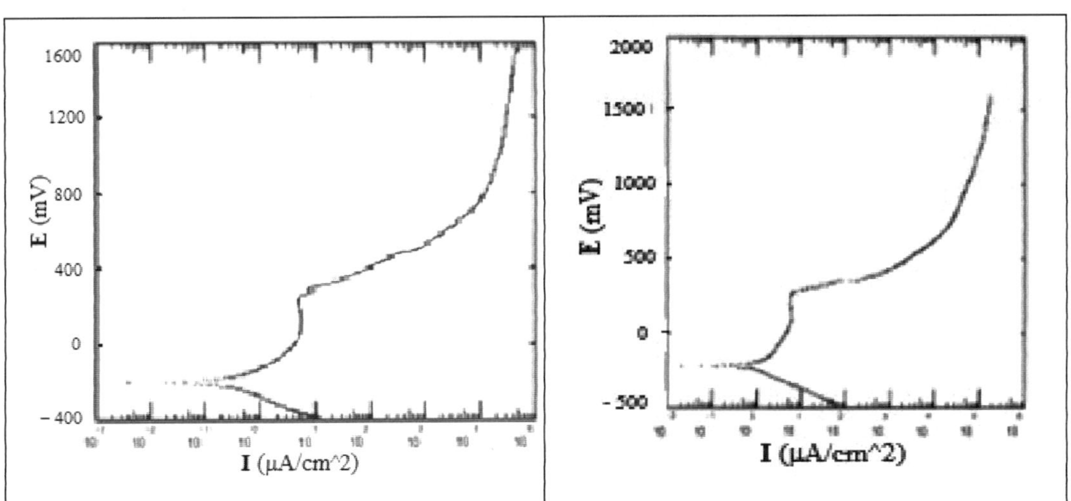

(a) (b)

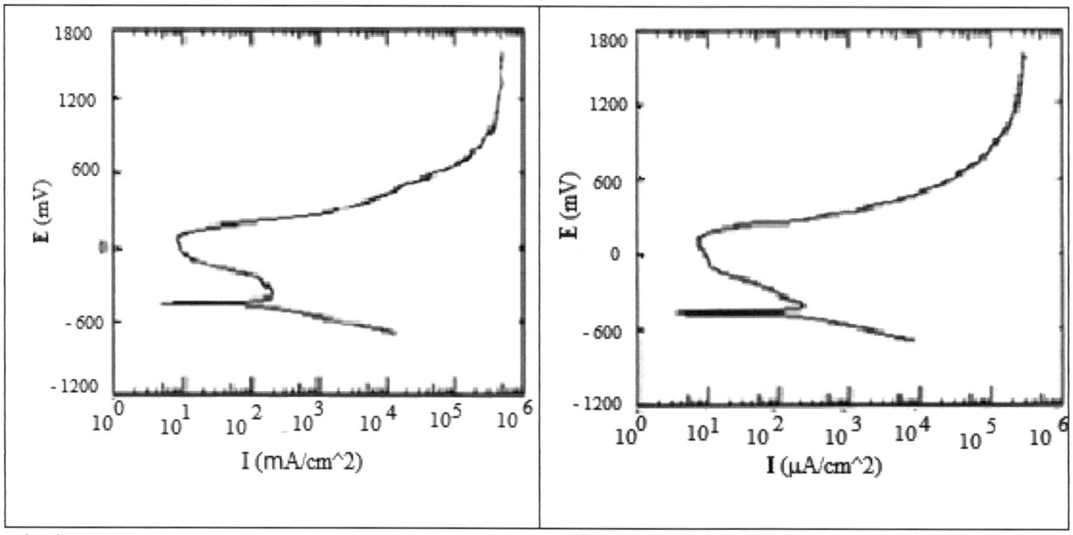

(c) (d)

Figure 2. Potentiodynamic potential versus current curve pattern obtained from stainless steels-316 sample immersed in a chloride-acid medium: (a). pH = 0.4. (b) pH = 0.522, (c) pH = 0.7, (d) pH = 1.0.

Experimental results of potentiodynamic technique measurements show that the lowest corrosion potential at pH equals to 0.522, which is equivalent to a 0.3 mol concentration chloride acid solution has been recorded to be -465.69 mV. It is clear that for pH values equal to 0.522 and 0.4 the passivation potential is of the same magnitude which is about -400 mV; on the other hand for pH values equal to 0.522 and 0.4 the transpassive potential is measured to be +175 mV and +125 mV respectively. At pH equals to 0.4 the free corrosion-potential increases from -465.69 mV at pH 0.522 to a value of around -445mV; the reason behind this is an increase of cathodic protection and also because of a stronger acid medium, the exchange current generation becomes more significant forming more abundant hydrogen bubbles which tend to block the corrosion current increases [8].

Table 1. Experimental results of a stainless steels-316 sample immersed in an HNO_3 solution medium Environment with a variation of molar concentration or pH using inside the brackets are pH values.

Experimental Parameters	Molar Concentration (pH)			
	0.1 (1.0)	0.2 (0.7)	0.3 (0.523)	0.4 (0.4)
E(I=0) (mV)	-153.06	-141.9	-327.96	-344.47
Rp (k cm^{-2})	83.8	44.12	0.4549	0.2158
I_{corr} (Acm^{-2})	0.12.	0.35	7.16	9.68
Corr.Rate (mpy)	0.0312	0.0907	1.85	2.5
Correlation	1.0	1.0	1.0	1.0
Range of Exp. Pot. E(I=0) mV	10	10	10	10

IOP Conf. Series: Materials Science and Engineering **237** (2017) 012047 doi:10.1088/1757-899X/237/1/012047

Table 2. Experime ntal results of a stanless steels-316 sample immersed in an HNO_3 solution medium environment with a variation of molar concentration or pH using the potentiodynamic method.

Molar Concentration	pH	Free corrotion potensial (mV)	Passivation Potensial(mV)	Transpassive Potential(mV)
0.1	1.0	=-198.26	+70	+300
0.2	0.7	-211.74	+100	+310
0.3	0.522	-465.469	-400	+175
0.4	0.4	-455.15	-400	+125

The corrosion rate tends to decrease with respect to pH; this means that corrosion rate is proportional to the solution's acid concentration. The higher the concentration of nitric acid applied in the corrosion solution, the higher the corrosion rate is obtained.

4.2.Analysis of microstructure

Results of microstructure measurements on a corrosion sample in chloride acid mediumwith pH equal to 0.4 is shown in Figure 5:micrograms taken with SEM.

IOP Conf. Series: Materials Science and Engineering **237** (2017) 012047 doi:10.1088/1757-899X/237/1/012047

A.2.100 SISI BAWAH

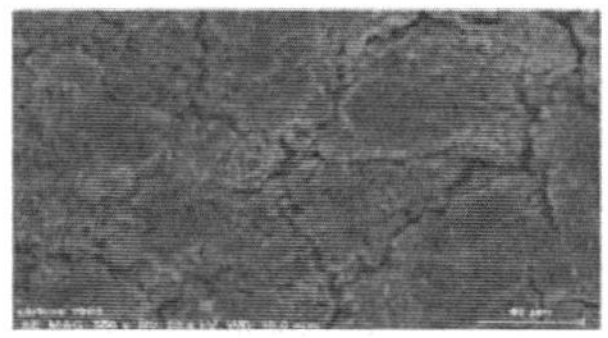

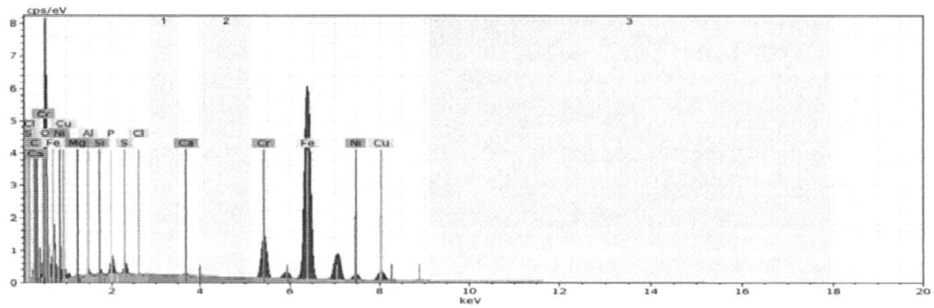

Spectrum: Acquisition

Element	Series	unn. C [wt.%]	norm. C [wt.%]	Atom. C [at.%]	Error (1 Sigma) [wt.%]
Carbon	K-series	6.76	7.39	16.73	1.24
Oxygen	K-series	28.00	30.62	52.01	3.46
Magnesium	K-series	0.18	0.20	0.22	0.04
Aluminium	K-series	0.33	0.36	0.36	0.05
Silicon	K-series	0.18	0.20	0.19	0.04
Phosphorus	K-series	1.15	1.26	1.10	0.08
Sulfur	K-series	0.63	0.69	0.58	0.05
Chlorine	K-series	0.09	0.09	0.07	0.03
Calcium	K-series	0.19	0.21	0.14	0.04
Chromium	K-series	5.31	5.81	3.03	0.18
Iron	K-series	42.45	46.42	22.59	1.16
Nickel	K-series	1.95	2.14	0.99	0.10
Copper	K-series	4.23	4.62	1.98	0.16
	Total:	91.45	100.00	100.00	

26-Apr-16 Page 1 / 3

Figure 3. SEM-EDX Micrograms of a corrosion sample immersed in a pH = 0.4 HNO_3 medium. The microgram gives a clear indication of surface erosion caused by the corrosion medium.

Here it is obvious that the sample's surface has been eroded by the corrosion medium consequently showing corrosion erosion. EDS pattern (Figure 3) on a selected spot in the surface reveals The possibility of chrome oxide (Cr_2O_3) or iron oxide (Fe_2O_3) being present in the sample. Finally the same could be concluded for yet another selected spot, namely that chrome oxide, iron oxide, and iron chloride possibly coexist in the sample. Based on both microstructural investigation supplemented by EDS data it could be claimed that in a nitric acid medium, pitting/erosion corrosion have certainly occur.

5. Conclusion

Concluding from the above discussion and analysis, it could be argued that for commercial stainless steels-316 alloys undergoing corrosion in an nitric-acid medium with a pH variation of 1.0 , 0.7, 0.523 , and 0.4 the corrosion effect with measured corrosion values of only 0.0312, 0.0907, 1.85, and 2.5.

The ensuing type of corrosion is erosion corrosion preceded by pitting corrosion. Corrosion rate will continue to decrease with respect to corrosion medium pH values. On the other hand corrosion potential and passivation potential tend to increase with respect to the corrosion medium H values.

Potentiodynamic measurements yield the following potentials for pH value of 1.0, 0.7, 0.522, and 0.4 are -198, -214, -465, and -445 mV respectively. The passivation potential for pH value of 1.0, 0.7, 0.522, and 0.4 are +70, +100, -400, and -400 mV respectively. EDS results reveal that the corrosion byproducts are possibly chrome oxide, iron oxide, and nickel hydride oxide.

References

[1]. Advanced Material and Processes, Composition of Standard Stainless Steels. (2000)

[2]. KYUNG SEON MIN, KI JAE KIM, SOO WOO NAM, Investigation of the effect of the types and densities of grain boundary carbides on grain boundary cavitation resistance of AISI-321 stainless steel under creep-fatigue interaction, *Journal of Alloys and Compounds*, Elsevier, (2003)

[3]. KYUNGSEONMIN. SOOWOONAM.Corelation between characteristics of grain boundary carbides and creep-fatigue properties inAISI-321 stainless steel. *Journal of Nuclear Materials* **322** (2003)91-97

[4]. F. NOLI. P. MISAELIDES. H. BAUMANN. And A. HATZIDIMITRIOU. The preparation. characterization and corrosion behavior of ion implanted and ceramic-coated AISI-321 steel samples. *Corrsion Science*, **38** (12) 2235-2246,(1996)

[5]. NURDINEFFENDI.WIHATMOKOW. ,WAGIYO, M. DANI, and A.K. JAHJA, Corrosion Experiment Of AISI-321Material at Temperature of 640°C and 740°C For 60 h in Artificial Air Medium, *Jurna Mesin Univ. Trisakti*. **7** (3) (2005)

[6]. NURDIN EFFENDI and MOH. DANI. Corrosion Experiment ofAISI-321Material at Temperature of 840°C and 940°C for 60 h in Artificial Air Medium. *Indocor. Majalah Korosi dan Material,* **6** (1)(2006)

[7]. TRETHEWEY. R.K. and CHAMBERLIN. *Korosi Untuk Mahasiswa Sains Dan Rekayasawan*, Alih bahasaWidodo, Gramedia Pustaka Utama. Jakarta. (1988)

[8]. FONTANA. G.. *Corrosion Engineering*..McGraw Hill BookCo.. third ed.. Singapore. (1987)

[9]. J.GOODHEW, *Specimen Preparation InMaterials Science*, North Holland.American Elsevier, (1973)

[10]. TATASURDIA, SHINROKU SAITO, *Pengetahuan Bahan Teknik*, Pradnya Paramita, Pustaka Teknologi dan Informasi, (1992)

[11]. ASM vol..2, Properties and Selection: Nonferroes Alloys, and Special Purpose Materials, Tenth edition, (1992)

IOP Conf. Series: Materials Science and Engineering **237** (2017) 012048 doi:10.1088/1757-899X/237/1/012048

Study of the near band-edge excitonic transition energies of the mixed layered compounds of $Mo_{1-x}W_x Se_2$ single crystals

Mula Sigiro
Department of Physics Education, Faculty of Teacher Training and Education, Universitas HKBP Nommensen, Medan 20234, Indonesia

mulasigiro@gmail.com

Abstract
The temperature dependence of the spectral features in the vicinity of the direct band edge of $Mo_{1-x}W_xSe_2$ single crystals was measured in the temperature range of 25–300 K using piezoreflectance (PzR). The near band-edge excitonic transition energies of the mixed layered compounds of $Mo_{1-x}W_xSe_2$ single crystals determined accurately from a detailed line-shape fit of the PzR spectra. From a detailed line-shape fit of the PzR spectra, the temperature dependence of the energies of the band-edge excitons are determined accurately.

1. Introduction

Layered-structure transition metal dichalcogenides have the formula MX_2, where M = Mo, W and X=S, Se that exhibit many physical properties with a pronounced two-dimensional character.[1-3] The peculiar properties of these materials result from their layered structure, consisting of mostly covalently bonded X–M–X sheets linked by weak Van der Waals (VdW) forces. The layered transition metal dichalcogenides MX_2 semiconductors have been extensively investigated due to the potential application of the materials as solid lubricants, catalysts and photovoltaic solar cell materials.[4-7] Agarwal *et al.* have successfully described the growth and some of the key properties of $Mo_{1-x}W_xSe_2$, $Mo_{0.5}W_{0.5}Se_2$ and WS_xSe_{2-x} single crystals.[8-11] Similar mixed-layered crystals of compounds such as $Mo_xW_{1-x}Se_2$, $Mo_{0.5}W_{0.5}Se_2$ and $Mo_{0.02}W_{0.98}Se_2$ by using various methods have been published.[12-14] Our previously work, $Mo_{1-x}W_xS_2$ single crystals have been experimentally realized to vary their bandgap values and temperature dependence of energies and broadening parameters of the band-edge excitons.[15,16]

In this paper we report a detailed study of the temperature dependence of the piezoreflectance (PzR) measurements in the spectral range near the direct band gap of $Mo_{1-x}W_xSe_2$ single crystals in the temperature range 25–300 K. The PzR spectra are fitted with a form of the Aspnes equation of the first-derivative Lorentzian line shape.[17] From a detailed lineshape fit we are able to determine accurately the energies and broadening parameters of the excitonic transitions.

2. Experiment

Single crystals of $Mo_{1-x}W_xSe_2$ solid solutions were grown by chemical-vapour transport method. The chemical transport was achieved with Br_2 as transport agent in the amount of about 10 mg/cm^3. Prior to the crystal growth the powdered compounds of the series were prepared from the elements (Mo: 99.99%; W: 99.99%; and Se:99.99%) by reaction at 1000 ^{0}C for 10 days in evacuated quartz ampoules. To improve the stoichiometry, sulphur 2 mol% in excess was added with respect to the stoichiometry mixture of the constituent elements. The mixture was slowly heated to 1000 ^{0}C. This slow heating is necessary to avoid any explosions due to the strongly exothermic reaction between the elements. The growth temperature was about 1050 °C in a temperature gradient of about 3 °C/cm and the growth time was about 20 days. The crystals had the shape of thin layer plates with a thickness of 20–100 μm and a surface area of 5–200 mm^2. X-ray analysis confirmed that the samples were single-phase materials of two hexagonal structure over the entire range of the W composition x. The lattice constants determined for the crystals were practically identical to those of the stoichiometric polycrystalline starting materials.

The PzR measurements were achieved by gluing the thin single crystal specimens on a 0.15 cm thick lead-zirconate-titanate (PZT) piezoelectric transducer driven by a 200 V_{rms} sinusoidal wave at 200 Hz. The alternating expansion and contraction of the transducer subject the sample to an alternating strain with a typical rms $\Delta l/l$ value of ~ 10^{-5}. A 150 W tungsten-halogen lamp filtered by a 0.25 m monochromator provided the monochromatic light. The reflected light was detected by a silicon detector. The DC output of the silicon photodiode was maintained constant by a servo mechanism of a variable neutral density filter. A dual-phase lock-in amplifier was used to measure the detected signal. Modulated spectra were normalized to the reflectance to obtain $\Delta R/R$. A close-cycle cryogenic refrigerator equipped with a digital thermometer controller was used to control the measurement temperature between 25 and 300 K with a temperature stability of 0.5 K or better.

3. Results and discussion

Displayed by the dashed curves in Figs. 1(a) and 1(b) are the PzR spectra in the vicinity of direct band edge for several mixed crystals of $Mo_{1-x}W_xSe_2$ at 25 and 300 K. The nature of the line shape indicates the presence of two oscillators on the high-energy side of the spectra. The oscillations on the lower-energy side are caused by the interference effects. The experimental curves have been fitted to a functional form appropriate for excitonic transitions that can be expressed as a Lorentzian line-shape function of the form.[17]

$$\frac{\Delta R}{R} = \mathrm{Re}\left[\sum_{i=1}^{n} A_i^{ex} e^{j\varphi_i^{ex}} \left(E - E_i^{ex} + j\Gamma_i^{ex}\right)^{-2}\right] \quad , \qquad (1)$$

where A_i^{ex} and φ_i^{ex} are the amplitude and phase of the line shape, E_i^{ex} and Γ_i^{ex} are the energy and broadening parameters of the interband excitonic transitions, respectively. Shown by the solid curves in Fig. 1 are the least-squares fits using Eq. (1). The fits yield the parameters A_i^{ex} , E_i^{ex} and Γ_i^{ex}. The obtained values of E_i^{ex} are indicated by arrows and denoted as A_1, A_2 and B in the figures. However, at 300 K (see Fig. 1(b)), only A_1 and B features are clearly visible over most of the x values except that of $x = 0$, $x = 0.2$, $x = 0.3$ and $x = 0.4$ where a weak shoulder on the higher-energy side of A_1 excitonic transitions is visible. The weak shoulder corresponds to the A_2 peak. The energies and their splitting of the A, B excitons of $Mo_{1-x}W_xSe_2$ are listed in Table II. For comparison purposes, we have also listed numbers for MoS_2,[18,19] $MoSe_2$,[19] $MoTe_2$,[19] WS_2,[20,21] WSe_2[21] and our previous work of $Mo_{1-x}W_xS_2$.[22]

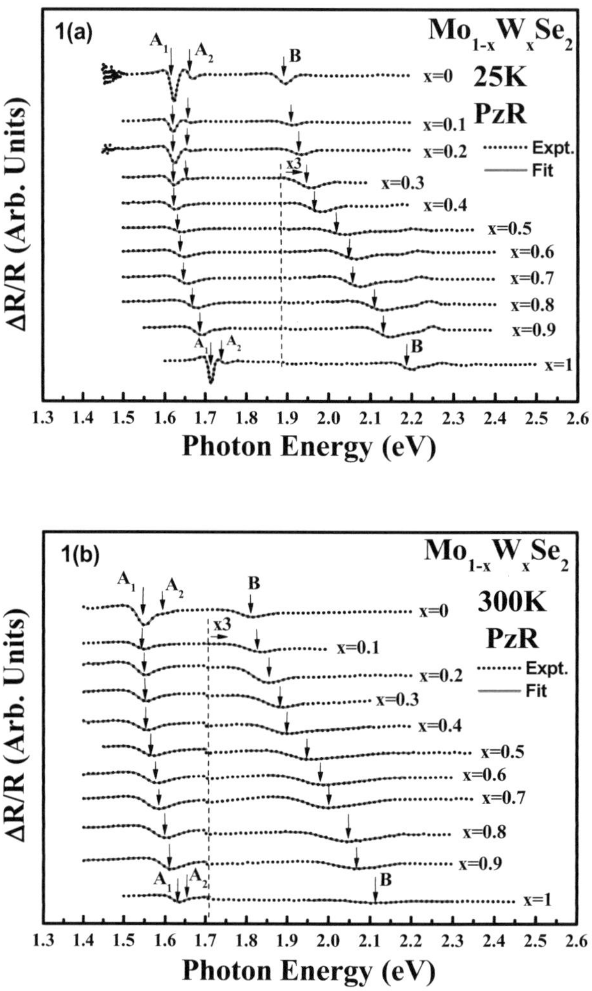

Figure. 1 The experimental PzR spectra (dashed curves) of $Mo_{1-x}W_xSe_2$ at (a) 25 K and (b) 300 K. The solid curves are least-squares fits to Eq. (1) which yields the excitonic transition energies indicated by the arrows.

Table.1 Energy of A, B excitons and their splittings in $Mo_{1-x}W_xSe_2$, $Mo_{1-x}W_xS_2$, $MoSe_2$, $MoTe_2$ and WSe_2.

Materials	E_A (eV)	E_B (eV)	E_{A-B} (eV)	Temperature (K)
$MoSe_2$[a]	1.622 ± 0.005	1.892 ± 0.008	0.270 ± 0.013	25
	1.546 ± 0.008	1.799 ± 0.01	0.253 ± 0.018	300
$MoSe_2$[b]	1.57	1.82	0.25	300
$Mo_{0.9}W_{0.1}Se_2$[a]	1.619 ± 0.005	1.909 ± 0.008	0.290 ± 0.013	25
	1.544 ± 0.008	1.816 ± 0.01	0.272 ± 0.018	300

$Mo_{0.8}W_{0.2}Se_2$[a]	1.620 ± 0.005	1.929 ± 0.008	0.309 ± 0.013	25
	1.550 ± 0.008	1.834 ± 0.01	0.284 ± 0.018	300
$Mo_{0.7}W_{0.3}Se_2$[a]	1.621 ± 0.005	1.947 ± 0.008	0.326 ± 0.013	25
	1.552 ± 0.008	1.871 ± 0.01	0.319 ± 0.018	300
$Mo_{0.6}W_{0.4}Se_2$[a]	1.622 ± 0.005	1.966 ± 0.008	0.344 ± 0.013	25
	1.553 ± 0.008	1.903 ± 0.01	0.350 ± 0.018	300
$Mo_{0.5}W_{0.5}Se_2$[a]	1.631 ± 0.005	2.018 ± 0.008	0.387 ± 0.013	25
	1.555 ± 0.008	1.942 ± 0.01	0.387 ± 0.018	300
$Mo_{0.4}W_{0.6}Se_2$[a]	1.638 ± 0.005	2.044 ± 0.008	0.406 ± 0.013	25
	1.561 ± 0.008	1.960 ± 0.01	0.399 ± 0.018	300
$Mo_{0.3}W_{0.7}Se_2$[a]	1.645 ± 0.005	2.057 ± 0.008	0.421 ± 0.013	25
	1.569 ± 0.008	1.980 ± 0.01	0.411 ± 0.018	300
$Mo_{0.2}W_{0.8}Se_2$[a]	1.668 ± 0.005	2.110 ± 0.008	0.442 ± 0.013	25
	1.583 ± 0.008	2.001 ± 0.01	0.418 ± 0.018	300
$Mo_{0.1}W_{0.9}Se_2$[a]	1.687 ± 0.005	2.130 ± 0.008	0.443 ± 0.013	25
	1.600 ± 0.008	2.043 ± 0.01	0.443 ± 0.018	300
WSe_2[a]	1.714 ± 0.005	2.187 ± 0.008	0.473 ± 0.013	25
	1.629 ± 0.008	2.113 ± 0.01	0.484 ± 0.018	300
2H-WSe_2[c]	1.71	2.30	0.59	77
$Mo_{0.7}W_{0.3}S_2$[d]	1.932 ± 0.005	2.223 ± 0.008	0.291 ± 0.013	25
	1.858 ± 0.008	2.168 ± 0.01	0.290 ± 0.018	300
$Mo_{0.5}W_{0.5}S_2$[d]	1.943 ± 0.005	2.252 ± 0.008	0.309 ± 0.013	25
	1.860 ± 0.008	2.170 ± 0.01	0.310 ± 0.018	300
$Mo_{0.3}W_{0.7}S_2$[d]	1.976 ± 0.005	2.333 ± 0.008	0.357 ± 0.013	25
	1.902 ± 0.008	2.253 ± 0.01	0.351 ± 0.018	300
MoS_2[d]	1.929 ± 0.005	2.136 ± 0.008	0.207 ± 0.013	25
	1.845 ± 0.008	2.053 ± 0.01	0.208 ± 0.018	300
WS_2[d]	2.064 ± 0.005	2.489 ± 0.008	0.425 ± 0.013	25
	1.981 ± 0.008	2.401 ± 0.01	0.420 ± 0.018	300
WS_2[e]	2.06	2.47	0.41	77
	2.0	2.386	0.386	300
3R-WS_2[f]	2.06	2.50	0.44	77
$MoTe_2$[g]	1.71	2.30	0.59	77
MoS_2[h]	1.9255	2.137	0.2115	4.2
MoS_2[i]	1.92	2.124	0.204	4.2

[a]this work [d]Reference 22. [g]Reference 19.
[b]Reference 19. [e]Reference 21. [h]Reference 18.
[c]Reference 20. [f]Reference 20. [i]Reference 18.

The A and B exciton peaks near the optical absorption edge are characteristic features of the optical spectra of all layered molybdenum and tungsten dichalcogenides.[22] The nature of the direct gaps has been investigated by studying these exciton pairs. Following Wilson and Yoffe[22] many authors have attributed the A–B exciton pair to transitions at the Γ point, split by spin–orbit splitting, as suggested by the dependence of the splittings on the masses of the constituent elements.

4. Conclusions

In conclusions we have measured the temperature dependence of the energies and broadening parameters of the direct band-edge excitonic transitions of $Mo_{1-x}W_xSe_2$ using PzR in the temperature range between 25 and 300 K. The transition energies of A, B excitons and their separation at various temperatures change smoothly with W composition x, indicating that the nature of the direct band edges of $Mo_{1-x}W_xSe_2$ are similar, even for $Mo_{1-x}W_xS_2$. The splitting of excitons A and B for $Mo_{1-x}W_xSe_2$ is larger than $Mo_{1-x}W_xS_2$ at various temperatures.

Acknowledgements

The author would like to acknowledge Professor Ying-Sheng Huang and Professor Ching-Hwa Ho for advising and providing materials and experimental equipment at Semiconductor Characterization Laboratory, National Taiwan University of Science and Technology (NTUST).

References

[1] A. R. Beal and J. C. Knights, J. Phys. C: Solid State Phys. **5**, 3540 (1972).
[2] W. Y. Liang, J. Phys. C: Solid State Phys. **6**, 551 (1973).
[3] W. J. Schutte, J. L. De Boer, and F. Jellinek, J. Solid State Chem. **70**, 207 (1987).
[4] P. D. Fleischauer, Thin Solid Films **154**, 309 (1987).
[5] J. M. Martin, C. Donnet, J. Le Mogne, and T. Epicier, Phys. Rev. B **48**, 10583 (1993).
[6] K. K. Kam and B. A. Parkinson, J. Phys. Chem. **86**, 463 (1982).
[7] A. Jager-Waldau, M. Lux-Steiner, R. Jager-Waldau, R. Burkhardt, and E. Bucher, Thin Solid Films **189**, 339 (1990).
[8] M. K. Agarwal and P. A. Wani, Mater. Res. Bull. **14**, 82513 (1979).
[9] M. K. Agarwal, P. A. Wani, and P. D. Patel, J. Cryst. Growth **49**, 693 (1980).
[10] M. K. Agarwal and L. T. Talell, Solid State Commun. **76**, 549 (1986).
[11] M. K. Agarwal, M. N. Vashi, and A. R. Jani, J. Cryst. Growth **71**, 414 (1985).
[12] W.K. Hofmann, H.J. Lewerern, and C. Pettenkofer, Solar Energy Mater. **17**, 165 (1988).
[13] A. A. Patil, P. P. Hankare, A. B. Gaikawad, and K. M. Garadkar, J. Mater. Sci.: Mater. Electron. **23**, 909 (2012).
[14] G. Kline, K. K. Kam, R. Ziegler, and B. A. Parkinson, Solar Energy Mater. **6**, 337 (1982).
[15] Y. Chen, C. Xi, D. O. Dumcenco, Z. Liu, K. Suenaga, D. Wang, Z. Shuai, Y. S. Huang, and L. Xie, ACS Nano **7**, 4610 (2013).
[16] M. Tanaka, H. Fukutani, and G. Kuwabara, Solid State Commun. **26**, 911 (1978).
[17] D. E. Aspnes, in Handbook on Semiconductors, edited by M. Balkanski (North-Holland, Amsterdam), 1980, p. 109
[18] E. Fortin and F. Raga, Phys. Rev. B **11**, 905 (1975).
[19] A. R. Beal, W.Y. Liang, and H.P. Hughes, J. Phys. C: Solid State Phys. **9**, 2449 (1976).
[20] A. R. Beal and H. P. Hughes, J. Phys. C: Solid State Phys. **12**, 881 (1979).
[21] S. J. Li, J. C. Bernede, J. Pouzet, and M. Jamali, J. Phys.: Condens. Matter **8**, 2291 (1996).
[22] J. A. Wilson and A. D. Yoffe, Adv. Phys. **18**, 193 (1969).

Institute of Physics
Dirac House, Temple Back
Bristol BS1 6BE UK

ISSN: 1757-8981
ISBN 978-1-5108-4869-6

CURRAN ASSOCIATES INC.
proceedings.com